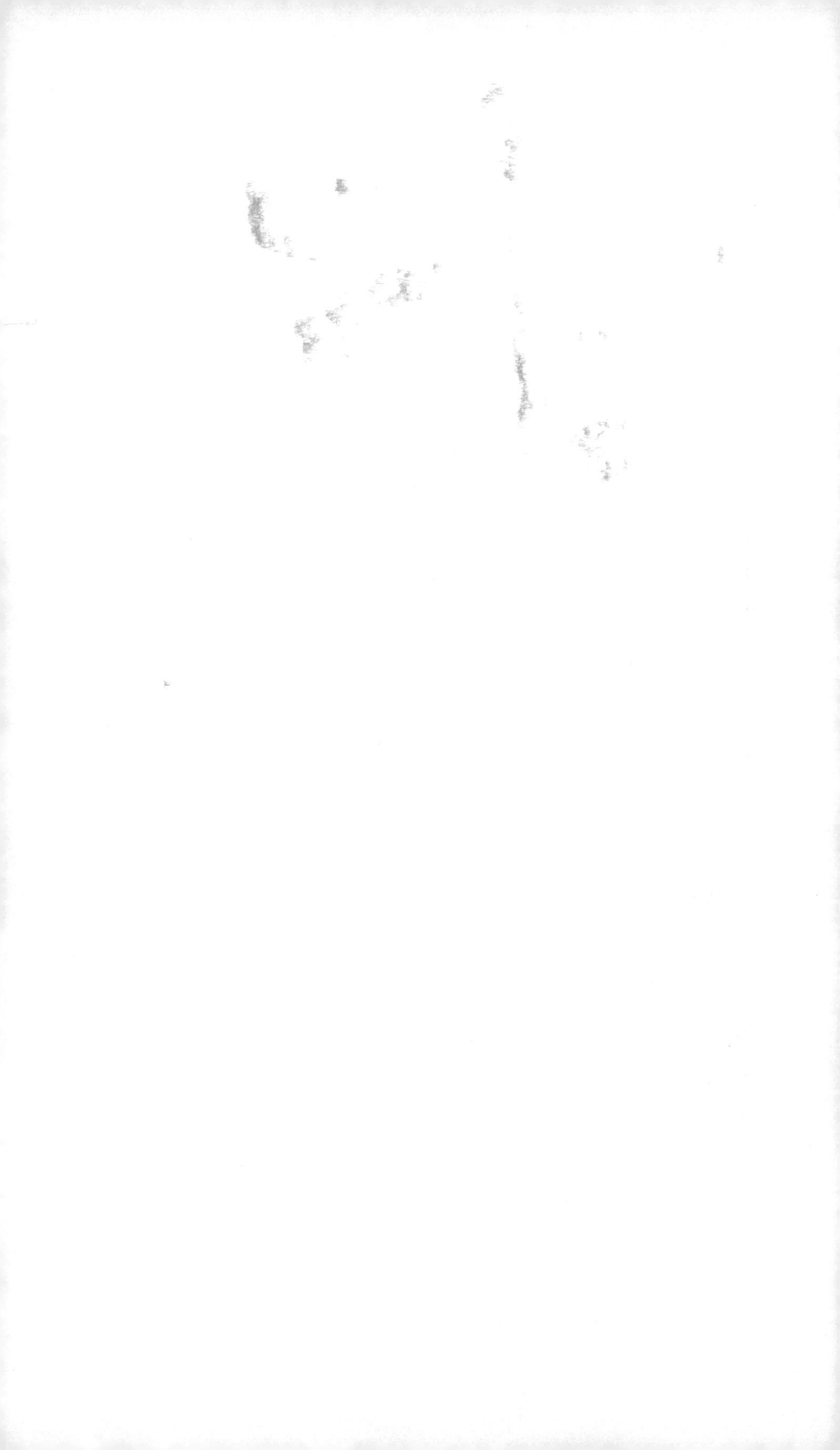

Advance Praise for

SCIENCE UNDER SIEGE

"Science is indeed under siege, and that's not good for any of us. Here, Peter Hotez and Michael Mann name names. They describe the role of Russia; the Supreme Court Justices, who've lost sight of the forest for the disappearing trees; and the like-minded officials who promote nutty conspiracy theories. It's not too late to do something; it's time to get things done. Read on."   —Bill Nye, science educator

"In standing up for evidence and truth, Drs. Mann and Hotez exemplify the citizen scientists who have been integral to making our country healthier, more sustainable, and—yes—great. *Science Under Siege* informs us of the peril of not combatting attacks on science and scientists, and urges all of us who care about our shared future to join the fight."

—Chelsea Clinton, vice chair, Clinton Foundation and
Clinton Health Access Initiative

"Once upon a time, scientists could assume that people would take their hard work, believe it, and put it to use. No longer—as Michael Mann and Peter Hotez make painfully clear, a concentrated assault by special interests has made the work of science painful and perilous. *Science Under Siege* is a well-researched guide and offers some powerful ideas about how to fight back."

—Bill McKibben, author *Here Comes the Sun*

"Climate change and cyberattacks threaten our very survival, but an even more insidious force is eating democracy from within: disinformation. Propagandists, lobbyists, and political operatives have hijacked science, turning fact into fiction until millions of Americans reject the reality of climate change—even as wildfires rage, oceans rise, and the planet becomes increasingly uninhabitable. In *Science Under Siege*, Michael Mann and Peter Hotez expose the machinery behind these deadly lies and call us to fight back—before truth itself becomes collateral damage."

—Nicole Perlroth, *New York Times*–bestselling author of *This Is How They Tell Me the World Ends*

"In our dark times, as we face a frontal assault on science, two revered scientists—Drs. Mann and Hotez—exemplify the warriors courageously speaking truth to power despite serious attacks."

—Eric Topol, author of *Deep Medicine*

"The vicious and unprincipled war on science is escalating on all fronts. Now two of our most distinguished scientists, who have both been subjected to abuse and threats for their diligence and work, have documented in devastating detail the forces undermining our efforts against climate change and deadly diseases. A must-read if you care about the planet and humanity."

—Norman J. Ornstein, coauthor of *It's Even Worse Than It Looks*

SCIENCE
UNDER SIEGE

SCIENCE UNDER SIEGE

HOW TO FIGHT THE FIVE
MOST POWERFUL FORCES
THAT THREATEN OUR WORLD

MICHAEL E. MANN AND **PETER J. HOTEZ**

PA
PUBLICAFFAIRS
New York

PublicAffairs
Hachette Book Group
1290 Avenue of the Americas, New York, NY 10104
www.publicaffairsbooks.com
@Public_Affairs

Printed in the United States of America

First Edition: September 2025

Published by PublicAffairs, an imprint of Hachette Book Group, Inc. The PublicAffairs name and logo is a trademark of the Hachette Book Group.

The Hachette Speakers Bureau provides a wide range of authors for speaking events. To find out more, go to www.hachettespeakersbureau.com or email HachetteSpeakers@hbgusa.com.

PublicAffairs books may be purchased in bulk for business, educational, or promotional use. For more information, please contact your local bookseller or the Hachette Book Group Special Markets Department at special.markets@hbgusa.com.

The publisher is not responsible for websites (or their content) that are not owned by the publisher.

Print book interior design by Bart Dawson.

Library of Congress Cataloging-in-Publication Data has been applied for.

ISBNs: 9781541705494 (hardcover), 9781541705517 (ebook)

LSC-C

Printing 2, 2025

Michael Mann dedicates this book to his wife, Lorraine Santy, and daughter, Megan Mann. He also dedicates the book to his father, Larry Mann; his brother Jay Mann; and to the memories of his brother Jonathan Mann and mother, Paula Finesod Mann.

Peter Hotez dedicates this book to his wife, Ann Hotez, and his sons and daughters (and their spouses), Matthew and Brooke Hotez, Emily Hotez and Yan Slavinskiy, Daniel and Alexandra Hotez, and Rachel Hotez, and to the memories of his brother, Richard Hotes, and his parents, Edward and Jean Hotez.

CONTENTS

PREFACE

O ver the past decade, the world has suffered catastrophic losses in human life from pandemics and irreparable damage to us and our planet from the effects of climate change. Jointly, these two crises pose an unprecedented threat to our future. To make matters worse, their impacts are mutually reinforcing and are gaining in both strength and potency. It is now clear that in the absence of concerted interventions, the situation will grow far worse in the years ahead.

These dual drivers of societal risk alone would constitute humankind's most daunting test. But the situation is even worse, for our efforts to respond to the existential challenges of pandemics and the climate crisis are hampered by a common underlying threat—politically and ideologically motivated opposition to science from powerful special interests. There is a highly organized, well-funded campaign of antiscience, and it has now emerged as its own deadly force that not only makes it nearly impossible to combat future pandemics or the climate crisis, but also touches on many other aspects of our daily lives. We must now contend with orchestrated disinformation on a diverse range of topics, ranging from genetically modified foods or needless nutritional supplements to fluoridation of our drinking water. The list grows day by day.

For more than two decades we have each waged our own private battle against the mounting forces of antiscience in America. We are both scientists and university faculty members who contribute to the understanding of human-caused climate change and projections of its impact (Mike) and the development of new vaccines to counter pandemics and neglected diseases (Peter). Over the years, we watched a vacuum develop in the scientific response to the growing antiscience attacks, a vacuum that was all too readily filled with misinformation and disinformation. As a result, we have each made a personal commitment to fighting the rising antiscience aggression that threatens our respective fields of research.

The attacks against us have been highly personal at times. When Peter wrote about his youngest daughter, who has autism and intellectual disabilities, in the context of explaining why vaccines do not cause autism, it meant his family came under attack too. This translated into death threats and in-person confrontations at his lectures and even at his home. The attacks on Mike have also included death threats and threats to his family, as well as demands for him to be fired from his job, all because his findings threaten the profits of the fossil-fuel industry. We are attacked online through email or social media and sometimes physically menaced. Fox News and other conservative news media outlets portray us as cartoon villains or public enemies. Extremist members of the US Senate and House denounce us for their political gain. When these high-profile attacks are waged against us, they serve as dog whistles for a pile-on. We frequently require the protection of law enforcement. If the forces of antiscience activists are willing to go this far, what comes next? What else are they willing to do? We have no choice but to do all we can to stop it now. That means working tirelessly to fight disinformation, ensure the safety of scientists, and, ultimately, prevent the dismantling of our nation's entire scientific infrastructure while communicating science to the public and policymakers in the most effective manner possible.

Despite the monumental threat, we see a silver lining. Through this journey both of us became experts in recognizing and defending against the weapons used to wage a disinformation war against science. We have had a front-row seat to how those forces have evolved over the past twenty years and the major actors pushing an antiscience agenda. For the sake of this book, we've identified the five principal forces of antiscience that threaten us today: the plutocrats, the authoritarian petrostates (assisted by polluters and the politicians who advocate for them), the "pros" who use their professional (or in some cases scholarly) credentials to deceive or promote unsupported contrarian views, the propagandists who amplify them on social media and other venues, and, increasingly, the press, including even the *mainstream* press. Together, these forces constitute a vast and complex web of hate speech, lies, and public deception. Even worse, this antiscience superstorm is now causing loss of human life on a massive scale, and it threatens our entire planet. We'll break down each of these forces in this book and identify ways we can fight back when we come across them.

We are two scientists who never dreamed during our training that we would come under these sorts of attacks and find ourselves reluctantly dragged into a war against those denigrating our fields of science. But we have embraced the platform this has given us and consider it a great privilege to be in a position to inform the public discourse over the greatest challenges we face as a civilization. Although the attacks against us are frightening at times, we also find this struggle meaningful and feel we have a responsibility to the public.

Luckily, we are both realists and optimists. Neither of us believes it is too late to avert disaster. So here we are to propose a path forward. We are determined to neutralize antiscience as much as humanly possible—and for a good and simple reason: unless we confront the forces of antiscience and make smart decisions about how to defuse them, we risk failing in our ability to prevent the next catastrophic pandemic or avert an era of catastrophic climate impacts. So, we've chosen to fight back. And this book is a tangible extension of that effort. We hope you'll join us.

In *Science Under Siege* we seek to provide a succinct yet detailed delineation of the five forces behind the modern-day antiscience movement (the five *p*'s, as we'll call them—the *plutocrats*, the *petro-states*, the *pros*, the *propagandists*, and our *press*). We draw upon our respective experiences on two different fronts of the war on science to identify and delineate the drivers and their financial backers. We provide a road map for dismantling the antiscience machine, through stories that at times are quite personal but speak to challenges and threats that are broad and sweeping. This book is a warning. But it is also a call to arms. While there is *urgency*—unlike any we've ever known—there is still *agency*. We can still avert disaster if we can understand the nature of the mounting antiscience threat and formulate a strategy to counter it.

1
—

THE 1-2-3 PUNCH

Let us fight for a world of reason, a world where science and
progress will lead to all men's happiness

—Sir Charles Chaplin,
The Great Dictator, August 1940

Our story is one of an unfolding, dark new chapter in modern
history. In it, we find a dual threat to humankind that is now
as grave as the prospect of nuclear annihilation at the height of the
Cold War. Together, the mutually reinforcing impacts of the climate
crisis and pandemics threaten massive loss in human life. Millions
have died from these two drivers in recent years,[1] and in the absence
of concerted mitigation efforts we can expect far worse in the coming
years and decades. While the United States has become an epicenter
of climate change and disastrous emerging infections, other regions
such as the Middle East, Central Africa, India and South Asia, and
central Latin America have been similarly impacted.[2]

Beyond the direct deaths and damage, however, there are even
broader collateral effects on political stability. Both climate and
pandemic responses have emerged as significant social wedge issues
and are exploited by bad actors to expand the already deep political
divide that exists in the United States and other nations. The Russian

government under President Vladimir Putin as well as other foreign (typically authoritarian) governments such as Saudi Arabia target science (and scientists) in their disinformation campaigns, wielding bots and troll armies to sow doubt and division about vaccines or global warming, for example, on social media.[3] America's foreign enemies have learned that disinformation and propaganda are convenient shortcuts to destabilizing our democracy and democratic republics across the world. The antiscience-embracing candidate Donald Trump—who promoted hydroxychloroquine during the pandemic and withdrew from the Paris climate agreement—prevailed in the 2024 US election, in substantial part because his voters had been misled into provably false beliefs about the state of the economy, crime, and immigration[4] that were actively promoted online by Russia. Chock up another win for disinformation. Which brings us to the third and perhaps the most daunting challenge of all. It is currently impossible for global leaders to take the urgent actions necessary to respond to the climate crisis and pandemic threats because they are thwarted by a common enemy—*antiscience*—that is politically and ideologically motivated opposition to any science that threatens powerful special interests and their political agendas.

The great scientist and communicator Carl Sagan, decades ago, warned of the threat posed by a rising tide of pseudoscience and antiscience, foretelling a future when "no one representing the public interest can even grasp the issues." He feared that the populace would increasingly be unable to differentiate between "what feels good and what's true," dependent on "media sound bites, predigested science, and pseudoscience." That premonition has unfortunately now been realized.[5] Wealthy, increasingly powerful plutocrats are today engaged in a full frontal assault on our academic and scientific institutions,[6] leaving citizens ever more vulnerable to misinformation and disinformation.

So, we find ourselves facing not just a one-two punch of pandemics and the climate crisis, but a one-two-three punch, with that third punch, antiscience, obstructing the needed response from governments and civil society. The future of humankind and the health of our planet now depend on surmounting the dark forces of antiscience.

Antiscience is such a huge obstacle because any solutions to the great challenges we face are intrinsically tied to acceptance of the underlying science—whether it's the science linking fossil-fuel burning and carbon pollution to the warming of the planet or the science of vaccine development and design. So far, the consequences of it have already proven deadly. In the United States alone, antiscientific rhetoric, policies, and actions caused hundreds of thousands of Americans to lose their lives during the pandemic by cajoling them into refusing safe and effective COVID-19 vaccines during the virus waves in 2021 and 2022 and resisting social distancing and mask wearing.[7] The deaths occurred mostly along a political partisan divide, with those living in Republican-majority ("red") states disproportionately suffering most of the deaths and disabilities as a consequence of being targeted by propaganda and disinformation from elected leaders, extremist media, and the modern political Right.[8] The antivaccine lobby is now so politically powerful and well financed and organized that it is hindering the adoption of COVID boosters while facilitating the return of measles and other vaccine-preventable childhood illnesses.[9] The GOP polling firm Rasmussen has turned into a "purveyor of anti-vaccine propaganda," according to Michael Hiltzik of the *Los Angeles Times*.[10] In fact, it's gotten so bad that Republicans in New Hampshire have moved to remove the requirement that children be vaccinated for measles or polio.[11] Reelected Republican president Donald Trump has threatened to take away funding from schools that have a vaccine mandate,[12] and members of his transition team support revocation of the polio vaccine.[13] Such escalating vaccine hesitancy may halt our ability to vaccinate human populations in the next pandemic. Antiscience activism has made it impossible for Americans to accept masking and other nonpharmaceutical interventions; it has killed, and will continue to kill, Americans on an unprecedented scale by halting substantive public health interventions.[14]

At the same time, efforts to address the climate crisis have been stymied because antiscience forces have blocked both national and international efforts to transition away from fossil fuels. As long as we continue to generate carbon pollution, the planet will continue to warm, and we'll see worse heat waves, droughts, wildfires, floods, and superstorms.[15] We have been given only a taste of what is in store if we fail to ramp down carbon emissions in the years ahead.

These crises are inextricably interwoven. Climate change—along with urbanization, poverty, human migrations, and other social determinants—is driving pandemics with increasing frequency. Whether it's the health of our people or the health of our planet, we are on an unsustainable path. To make matters worse, the rejection of science is being used to target the physical safety and careers of scientists committed to preventing the next pandemic and halting the progression of climate change.

When we first embarked on our doctoral training decades ago, we never thought we were signing up to be combatants in a war on science and scientists. That was not the America we knew. While scientists in certain areas such as the health impacts of pesticides or tobacco products had attracted dubious scrutiny from special interests looking to discredit their research,[16] antiscience—we thought—was something largely consigned to the past and mostly a legacy of past fascist regimes. We knew about Stalin and his persecution of scientists in the 1930s and '40s and how he promoted the pseudoscientific theories of Trofim Lysenko,[17] leading to massive failures in Soviet wheat crops killing thousands or millions. We knew how Albert Einstein's theory of relativity was dismissed by the Nazis as "Jewish science," as were Freud's psychoanalytic theories in Vienna.[18]

However, these matters were in the distant past. Our grandparents had thankfully found their way to America, leaving fascism and its antiscience leanings behind. In its place, America had built a strong nation that both thrived from and depended on science and technology. Our country excelled and achieved greatness because of its research universities, institutes, and corporations—a vast US academic-industrial complex—to produce the doses of penicillin required to heal war wounds or give us the radar technology needed to achieve military victories in World War II. Science gave us the air and space supremacy to launch NASA in the 1950s and '60s and send humans to the moon; it launched satellites to explore the outer reaches of the solar system; it assembled the James Webb Space Telescope and discovered the Higgs boson to reveal the time and space origins of the universe; it showed us how chemical bonds form between atoms to give us molecules or how molecules become macromolecules to give us life. Science and technology gave us transistors, computer chips and Silicon Valley, and the novel therapeutics behind a moon shot for cancer. Training to be a scientist was an expression of patriotism and love of country, and for our parents and families it was a source of pride and one of the highest expressions of human intellect for the common good.[19] For the most part, the American people in the last half of the twentieth century admired and even revered scientists.

Now, every week, sometimes daily, we each receive online threats or harassing phone calls to our offices. On multiple occasions, we have faced actual physical confrontations and stalking. We are not alone, as many scientists working in the areas of biomedicine and climate science, especially in the United States, now encounter similar threats.[20] You might assume that the aggressors are lone actors—cranks and conspiracists in their basements—but in many cases the attacks are government sanctioned, organized at the highest levels of the US Congress, with senators and members of the House of Representatives boasting of their efforts to intimidate prominent scientists.[21] During the 2024 US presidential election, two candidates, Robert F. Kennedy

Jr. and Ron DeSantis, publicly sought to discredit one of us (Peter). Politicians have in the past attacked us both on social media, made wanton or frivolous Freedom of Information Act (FOIA) requests for our emails or correspondence, and sent us veiled threats and unjustified subpoenas to appear before House or Senate committees.[22] They have told easily debunked lies about us in their speeches and in their self-published books.

Many of these attacks come from the Far Right and extremist elements of the Republican Party.[23] They do this for their political gain and, especially, sound bites on Fox News or other conservative news outlets and podcasts. Not since the House Un-American Activities Committee (HUAC) hearings during the 1950s when J. Robert Oppenheimer and his colleagues came under attack have productive and hardworking American scientists had to face such unfair and unrelenting assaults by elected officials, including humiliating congressional hearings televised live on C-SPAN. The public assault by the GOP has diminished the public trust of science among conservatives. As science historians Naomi Oreskes and Erik Conway note in describing the current political atmosphere, "Reaction to scientific findings is highly polarized, with Republican voters and self-identified conservatives far more likely than Democrats and self-identified liberals to reject consensus scientific findings, particularly in the areas of climate change and COVID-19 response."[24] It's no coincidence.

Attacks on established scientists like us are used to serve as a warning to other current and future scientists who might consider facing the antiscience juggernaut. Professional scientists are being put on notice; increasingly, they are being told they deserve public derision and should be treated as enemies of the state. We risk, as a result, losing a whole generation of American scientists because young people are frightened and disgusted by what they see online or in the media, or what their parents now tell them. Why would they choose a career that subjects them to such vitriol and venom? Mistrust in science is now escalating in certain demographics because of a targeted campaign against us—antiscience predation for someone else's financial or political gain.

Before we dive into the five forces of antiscience and how we might counter them, let's take a brief look at how the climate crisis and pandemics have emerged as the twenty-first century's dominant threats and why it is that we come to arrive in such a fragile time when facts are so often ignored or distorted. Any effort to counter this threat requires that we arm ourselves with evidence and stand firm in our knowledge of the truth.

THE CLIMATE CRISIS

The current warming of our planet is caused by carbon pollution, primarily from the continued burning of fossil fuels. Warming, in turn, is altering global wind patterns and ocean currents and shifting rainfall zones, while drying out many regions of the continents and melting sea ice in the Arctic and Southern Oceans. It is also causing the retreat of glaciers around the world and is initiating the collapse of the great continental ice sheets in Greenland and Antarctica, contributing to global sea-level rise. Collectively, these changes are termed *climate change*.

Multiple lines of evidence tell us that the observed warming is human caused.[25] First and foremost, climate models show that the planet should have cooled very slightly over the past century, rather than warmed substantially, in the absence of human-generated climate pollution. Only the human-generated rise in greenhouse gases explains the warming. We also know that the overall pattern of temperature changes, with warming of the lower atmosphere and cooling of the upper atmosphere, bears the fingerprint of greenhouse warming and is inconsistent with natural causes.

Furthermore, we know that this warming is unprecedented over a very long time frame and closely coincides with the Industrial Revolution. Back in the late 1990s, while still a postdoctoral researcher, Mike coauthored a now famous study demonstrating that the recent warmth is unmatched over at least the past millennium. The resulting graph, termed the "hockey stick" for its shape, became an icon of the climate-change debate (Fig. 1). In the decades since, the hockey stick has been replicated by other scientists many times over and extended further back in time.[26] Yet it—and Mike—continues to be a target of attack by climate-change deniers owing to the enduring potency of this symbol of the profound impact human activity is having on our climate.

We are now witnessing the profound consequences of unchecked human-caused warming, in the form of coastal inundation, deadly floods, hurricanes, wildfires, and of course extreme heat. These

The latest version of the "hockey stick" chart shows unprecedented warming in recent years.

Change in global surface temperature relative to 1850-1900 average

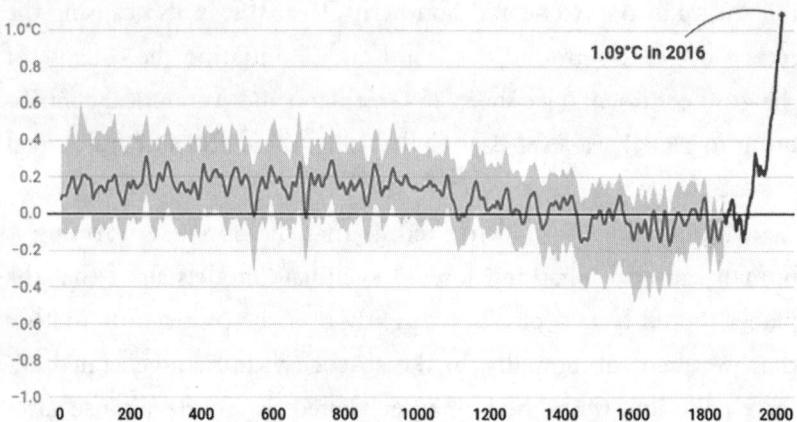

Numbers are observed from 1850–2016; for prior years, they are reconstructed using proxy records like tree rings, corals, and ice cores.

Chart: Elijah Wolfson for TIME • Source: IPCC, 2021: Summary for Policymakers   TIME

Figure 1. The latest incarnation of the "hockey stick" curve. Credit: From TIME. © 2021. TIME USA LLC. All rights reserved. Used under license. time.com.

impacts have reached such crisis proportions that the term *climate crisis* is frequently used to describe our predicament.

To avert ever more catastrophic impacts, we must rapidly decarbonize the global economy.

As we said in the Preface, there is *urgency*, but we have *agency* too. Take it from two experts. And recognizing that fact is increasingly critical for all of us. Today there is nearly as much danger posed by climate despair as there is by climate denial. And bad actors have weaponized that despair, actively fanning the flames of doomism in the hope that we'll throw up our hands in defeat. They don't want you on the front lines. They want you on the sidelines. And both denial and despair can potentially lead you there. The science, however, simply doesn't support a narrative of doom and inevitability.

There is still time to limit warming below truly catastrophic levels, but it requires dramatic action. To keep planetary warming below 1.5°C/3°F, where we'll see far worse consequences, we must reduce carbon emissions rapidly in the years ahead. The good news is that we're making progress as the world slowly but steadily transitions away from fossil fuels toward renewable energy. Carbon emissions have plateaued. The bad news is that carbon emissions have to come down substantially. They must drop by more than 50 percent this decade and reach zero by midcentury if we are to avert 1.5°C warming. This will be challenging, *but it is doable*. The obstacles aren't physical or technological. They are entirely political at this point.[27] If we miss the 1.5°C target, we don't drop off a cliff. It's more like a dangerous highway, and if we miss the 1.5°C exit, 2°C is still a whole lot better than 3°C. Every bit of progress makes us better off.

One of the arguments invoked by critics of climate action is that there is supposedly too much uncertainty to take action. First of all, the fact that the warming of the planet is human caused, and that it is associated with increasingly dangerous and damaging impacts, is about as certain as any scientific proposition there is.[28] Furthermore, uncertainty isn't our friend here. The impacts of climate change, in many respects, have already exceeded the climate-model predictions. The Greenland Ice Sheet and West Antarctic Ice Sheet are showing signs of disintegration far earlier than models predicted, contributing to sea-level rise ahead of schedule. We are also seeing far greater increases in extreme weather events than models have predicted. Some of Mike's own research demonstrates that climate models are not adequately capturing the factors behind the increase in persistent summer weather extremes we've seen in North America and Eurasia in recent years.[29] Uncertainty is a reason for taking more action, not less.

None of this comes as a surprise. Science warned us decades ago of impending calamity. James Hansen, the former director of the NASA Goddard Institute for Space Studies, famously testified to the US Senate on an unusually hot Washington, DC, day in June 1988 that "global warming is affecting our planet now."[30] In the ensuing decades the science became increasingly decisive. But fossil fuel–industry groups chose to engage in a millions-of-dollars disinformation campaign to undermine public faith in the science and the scientists. Meanwhile, the world's largest fossil-fuel companies already understood the climate threat. In an internal report from 1982 that was eventually leaked into the public domain, ExxonMobil's own scientists accurately predicted the increase in CO_2 concentrations and warming that would occur today in the absence of efforts to curtail fossil-fuel burning. They warned of the "catastrophic events" we would see. But rather than own up to the findings of their own scientists, ExxonMobil instead engaged in a hundreds of millions of dollars' campaign to attack and discredit independent climate scientists coming to the very same conclusions their own scientists had reached.[31] The opportunity cost of that disinformation campaign is almost incalculable. We are now experiencing those "catastrophic events." If we had begun decarbonizing our societal infrastructure decades ago, when we first knew there was a problem, the transition would have been smooth and easy. Now we must reduce emissions far more rapidly, with greater potential for disruption.

Industry groups like the George Marshall Institute (GMI) sought to attack and undermine mainstream climate scientists and their research findings in the 1990s. One example was Stephen Schneider, a physicist by training who turned to climate modeling in the 1980s and forged a path forward, developing methods of coupling climate models to models of societal and environmental impacts to inform policy decision-making. Critics love to claim that he predicted global cooling in the 1970s (he didn't—he simply pointed out that there was a possibility that reflective atmospheric pollutants from coal burning could outpace fossil fuel–generated carbon emissions) or that he advised exaggerating the climate threat to encourage climate action (he didn't—he advised that climate communicators could be both effective and faithful to the science at the same time). Ben Santer's work was critical to the historic finding of the 1995 Third Assessment Report of the United Nations Intergovernmental Panel on Climate Change (IPCC) that there was a "discernible human influence on the climate." Fred Singer, a fossil fuel–funded climate-science critic, led a public campaign—including an op-ed on the ultraconservative editorial pages of the *Wall Street Journal*—to impugn Santer's integrity, falsely accusing him of "scientific cleansing" in his role as a lead author of one of the key chapters of the report.[32]

The next major victim of the assault would be none other than the lead author behind the hockey stick—Mike. Quoting Santer himself, in a 2006 interview with the *New Scientist*: "There are people who believe that if they bring down Mike Mann, they can bring down the IPCC."[33] The modus operandi of the climate-denial machine, which is to pick one scientist, separate them from the rest of the community, and make an example of them for others who might consider speaking out, is what Mike has termed the *Serengeti Strategy*, akin to the way that lions isolate a vulnerable member of a pack of zebras at the edge of a herd. He was the latest target in that strategy, as he was subject to death threats, calls to be fired, congressional inquisitions, and more in the decades following the publication of the famous graph.[34]

It might be argued that the bad-faith assault on climate science in the 1990s was like a localized cancer that would later metastasize throughout our body politic writ large, infecting the entire public discourse over science. In 2020, as the COVID-19 pandemic played out, we witnessed a similar but remarkably accelerated version of a science-denial campaign play out not over decades but over months. We're talking, of course, about the ideologically motivated effort to deny the threat of the COVID-19 pandemic. Like climate, it began with dismissiveness of the science by prominent politicians with an agenda (Donald Trump, president at the time, was fearful that it might threaten his reelection hopes and downplayed the danger of a pandemic, insisting it would soon disappear all on its own).[35] Then it evolved to denial of the efficacy of proposed solutions (vaccines and masking), and of course then came the "blame-the-messenger" strategy of attacking and discrediting the scientists themselves, including Peter, and of course the nation's top infectious-disease expert and director of the nation's largest infectious-diseases research institute, the National Institute of Allergy and Infectious Diseases (NIAID), Anthony Fauci. Mike wrote in *Newsweek* at the time, "Climate scientists feel your pain, Dr. Fauci."[36]

But there is another connection between climate and pandemics, a direct and critical one. The climate crisis fundamentally impacts our lives in many ways—food, water, and land resources, as well as the stability of our infrastructure, homes, buildings, and transportation systems. It breeds conflict and threatens national security, as a growing global population competes for diminishing resources. And, like deadly pandemics, it threatens our lives. The extraction and burning of fossil fuels themselves lead to adverse health consequences and millions of lost lives per year from air and water pollution.[37] Deadly weather extremes exacerbated by human-caused warming—floods, storms, droughts, wildfires, and extreme heat—lead to many million more lost lives per year.[38]

Sometimes the impacts hit very close to home. Mike and his family were traveling back from a trip to the Grand Canyon in June 2017. Phoenix, Arizona, already known for its exceptionally hot summers, was experiencing a record heat wave. On June 20, 2017, temperatures at the Phoenix airport rose to 120°F, exceeding the maximum safe operating temperature (118°C) for aircraft takeoff. All flights were grounded.[39] Mike and his family spent the night at a hotel near the Phoenix airport before flying home the next morning. His then eleven-year-old daughter, Megan, and he went swimming in the outdoor pool that afternoon. They felt like they were diving into a bathtub. Treading barefoot on the pavement was like walking on hot coals. They quickly retreated to the oasis of their air-conditioned hotel room.

Mike and his wife were awakened by Megan around three in the morning. She was in distress, having difficulty breathing. They rushed her to the nearest hospital. The doctor put her on an inhaler, and that seemed to alleviate her symptoms for the time being. They took her to see the doctor once they were back home, and he diagnosed her as having suffered from "intermittent asthma," which was likely triggered by the dangerously high surface ozone levels resulting from the record heat. That was the moment when Mike and his family first came face-to-face with the deadly nature of the climate crisis.[40] There is nothing any of us cares more about than our health and the health of our loved ones. And the climate crisis poses a serious threat to both. But perhaps the most serious health consequence of climate change is the threat of more deadly pandemics.

PANDEMIC THREATS

As COVID-19 swept across the world in 2020 and began decimating human populations, many were surprised at the ferocity of this illness or how a new respiratory virus pathogen, known as a coronavirus, could catch so many global leaders unprepared. COVID-19 deaths rose precipitously, first in the Western Hemisphere and Europe in 2020 and then in India during the first half of 2021, when vaccines were not yet available and surges of the sick and dying overwhelmed limited numbers of trained personnel staffing hospital emergency rooms and intensive-care units (ICUs). Just one year after COVID-19 emerged in the United States, more than five hundred thousand Americans had died, while globally there were more than three million deaths. As awful as those numbers are, they may in fact be underestimated because of the inadequate reporting among depleted health systems in some of the world's low- and middle-income countries (LMICs). One estimate finds that more than two million people in India alone had lost their lives due to COVID by the second quarter of 2021.[41]

Among those professionals who anticipated the possibility of a devastating pandemic was a small community of virologists who were studying coronaviruses for more than a decade, and in some cases even longer than that. Peter and his research group based at two affiliated institutions in the Texas Medical Center in Houston— Texas Children's Hospital Center for Vaccine Development (CVD) and the National School of Tropical Medicine of Baylor College of Medicine—had already been developing new vaccines for coronavirus infections since 2012. Their scientists had already witnessed the devastation of two previous coronavirus outbreaks: severe acute respiratory syndrome (SARS) had arisen out of southern China in 2002 and caused hundreds of deaths before it traveled across the world to Toronto, Ontario, and later in 2012 another severe coronavirus infection known as Middle Eastern respiratory syndrome (MERS) emerged on the Arabian Peninsula, causing hundreds of deaths in Saudi Arabia and the Republic of Korea. They predicted it was just a matter of time before a third major epidemic or even coronavirus pandemic (global epidemic) was before us. And during the decade of the 2010s, the Texas Children's Hospital CVD research group discovered how to target the now well-known spike protein of the virus and deliver it to the immune system as a vaccine strategy.

Such studies were prescient because when they learned in January 2020 that a new lethal virus arising out of the city of Wuhan in central China was indeed a coronavirus that closely resembled the original SARS from almost twenty years before—in fact, it was appropriately named the SARS-2 coronavirus—that team of SARS and MERS vaccine scientists quickly pivoted to make a new version of the vaccine that was specific for COVID-19. Ultimately, the Texas Children's Hospital CVD scientists made two related COVID-19 vaccine technologies, leading to the production of a low-cost vaccine in India and Indonesia.[42] Almost one hundred million doses were administered to children and adults in these two countries. In so doing, they showed how it was possible to reach major milestones in the vaccine space without depending on a multinational vaccine producer, such as Pfizer or Moderna. They offered a new path to make safe, effective, and low-cost vaccines for the world's poor.

Almost as important as making cheap yet efficacious vaccines was the realization that COVID-19 or another serious coronavirus pandemic was in many respects both predicted and predictable. Coronavirus epidemics or pandemics appear regularly because of a confluence of twenty-first-century forces, with climate change among the multiple contributing factors. Many viruses capable of transmission to humans originate in bats before they eventually transmit to people. This includes coronaviruses. One possible scenario under investigation is how altered environmental conditions due to climate change can shift bat habitats, which together with deforestation (due to both climate change and other human activities) bring bats in closer proximity with humans or other mammals capable of transmitting coronaviruses. Thus, climate change now combines with socioeconomic factors including urbanization and shifts in human and animal migrations to promote the emergence of catastrophic epidemics.[43] The bottom line: pandemic threats or rising tropical infections in North America and Europe just became our new normal. These days when Peter appears on the cable news channels or podcasts, or delivers public lectures to general audiences, one of the most common questions he receives is something to the effect of "Hey, Doc, what the heck is going on?" Another way to phrase this question: What explains why now in this new century we are witnessing a regular cadence of catastrophic epidemics of viral and parasitic illnesses?

Some of the major illnesses we can anticipate might emerge or accelerate in the coming decades due to climate change and allied forces include mosquito-transmitted virus infections, parasitic infections, and Ebola viruses and coronaviruses, each of which are described below:

Mosquito-Transmitted Viruses (Arboviruses)

During the early history of the United States, one of the most feared diseases was yellow fever, a mosquito-transmitted virus infection that causes internal bleeding and often death. Throughout the nineteenth century, ships from the Caribbean carrying patients infected with yellow fever (or *Aedes aegypti* mosquitoes harboring the virus) entered southern American ports such as New Orleans, Mobile, and Galveston on the Gulf Coast or Charleston and Savannah along the Eastern Seaboard to cause horrific epidemics. Tropical medicine experts now worry that the US Gulf Coast region, especially in South Texas and South Florida, is increasingly vulnerable to a recurrence of yellow fever or other serious arboviral infections transmitted by the same *Ae.*

aegypti mosquitoes.[44] These include dengue fever, Zika virus infection, Mayaro virus infection, and chikungunya.[45]

The major drivers include climate change and urban poverty. With regards to the former, higher temperatures and relative humidity often favor mosquito development, reproduction, feeding, or pathogen replication, although this situation can vary for some species.[46] Climate-prediction maps indicate that an estimated one billion people globally will become at risk for *Aedes* mosquito-transmitted virus in the coming decades. This includes Texas on the US Gulf Coast. Another factor: mosquitoes thrive in urban environments. Houston, Texas (where Peter lives), is of concern because of its explosive population growth—it is one of the fastest-growing large metropolitan areas in the Western Hemisphere—and urbanization favoring *Ae. aegypti* combines with climate change to make arboviruses a particular threat.[47] Gulf Coast cities also host areas of extreme poverty. When visiting poor areas in Houston, one cannot help but notice all the discarded car or truck tires dumped into our low-income neighborhoods. Driving through these low-income neighborhoods, one is profoundly moved by their depth and breadth of poverty. The dilapidated housing without window screens, drainage ditches, and the discarded tires are particularly noteworthy. Tires with standing rainwater constitute ideal urban habitats for mosquitoes.[48] Peter's home city of Houston is a perfect storm of climate change, urbanization, poverty, and therefore mosquito-transmitted viruses.

Parasitic Infections

Climate change, urbanization, and political instability also promote parasitic infections caused by microorganisms that are more complex than viruses, such as the protozoa that cause human leishmaniasis or malaria, or even multicellular worm parasites such as hookworms and schistosomes. Parasites spread disease and are transmitted by insect vectors or through contaminated soil or water. One of the most common parasitic infections is human hookworm infection and anemia, which is caused by small worms (less than an inch long) in the human intestinal tract. The worms are adapted to feed on blood in the inner lining of the small intestine to produce blood loss that results in anemia.

Hookworm is now a leading cause of anemia among people who live in extreme poverty, especially in Africa, Southeast Asia, and the poorest parts of Latin America.[49] Individuals acquire hookworm infection from larvae that live in a state of biological dormancy in the warm and moist soils of tropical and subtropical regions of the world, especially in impoverished areas lacking sanitation or access to clean water. Another feature about hookworms is their ability to adapt to urban environments,[50] and hookworm infection occurs even in impoverished areas in the southern United States.[51] Peter began working to develop a hookworm-anemia vaccine forty years ago as a graduate student. Now, the scientists at the Texas Children's Hospital CVD are developing an updated version of this vaccine in the expectation that hookworm infection becomes the dominant parasitic disease of climate change, urbanization, and other twenty-first-century physical and social determinants.[52] The same holds true for some insect-transmitted parasitic infections. Malaria patterns are also shifting: local transmission was reported for both South Florida and South Texas,[53] and there are predictions that malaria will become more widespread in Europe,[54] by now a repeating theme we also saw for arbovirus infections, as well as hookworm infection in the southern United States.

Viruses Originating from Bats: Ebola Virus and Coronaviruses

We have seen two major Ebola epidemics, in 2014 in West Africa and in 2019 in Central Africa. Ebola viruses can be carried by bats. According to some conservation and ecology groups, shifting weather patterns, including droughts, altered temperatures, and storms, can promote significant bat migrations as they search for new food resources or as their habitats for reproduction and roosting close or expand. Through such climate shifts, Ebola-infected bats may have migrated into new areas of Africa. The next phase relates to other human activities, as economic deprivations caused individuals or groups to move into forested areas where they caught bats for food sources or sales. In this way, poverty couples with deforestation.[55] Upon becoming ill, those exposed to Ebola from bats seek help in depleted health centers where there is an inadequate supply of gloves or other personal protective equipment. Health-care workers become infected and return home or to more urbanized areas. Once Ebola virus enters cities, it can spread more easily. Therefore, forces of poverty, urbanization, and climate change accelerate Ebola virus infections, just as they do arbovirus infections, although through different mechanisms.

Just as human Ebola virus infections first originated from bats, the viruses that cause SARS, MERS, and COVID-19 also most likely originated in bats before spreading either directly to humans or to humans through other mammals. SARS coronavirus may have spread from bats to civet cats prior to human infections, MERS coronavirus from bats to dromedary camels, and SARS-2 coronavirus (the cause of COVID-19) from bats to raccoon dogs or pangolins. Viruses with high levels of genetic-sequence similarity to SARS-2 have been detected among bats in Cambodia, Laos, Malaysia, Thailand, and Yunnan Province in southwest China, including some coronaviruses with 98–99 percent sequence similarity (in their 3' UTR gene) to SARS-2.[56] The EcoHealth Alliance, based in New York City, finds that SARS-like coronaviruses are ubiquitous among bat populations in Asia and estimates that more than sixty thousand humans become infected with these viruses annually, either directly from bats or through other mammals.[57] In terms of twenty-first-century forces, the increased migrations of bats due to a variety of factors, including climate change, deforestation, and urbanization, may lead to more extensive bat and human interactions. Another important factor for both SARS and SARS-2 is the wholesale or wet markets in China that pack secondary animal hosts in cages and offer additional conduits for these viruses to spread to humans. One specific market, the Huanan Seafood Wholesale Market located in Wuhan, China, is generally accepted as the site where vendors sold infected live animals, and two strains of the virus emerged before humans first contracted COVID-19 in December 2019.[58] Following these events, the SARS-2 coronavirus spread rapidly to ignite a global pandemic. Perhaps the most disturbing pandemic driver was the disinformation campaign to convince Americans to resist COVID-19 vaccines after they became widely available in 2021, causing the deaths to climb even higher—a topic we will revisit throughout this book.

A CANDLE IN THE DARK

The great Carl Sagan, in his tome *The Demon-Haunted World*, explained why science is a "candle in the dark," a way to make informed decisions about the major challenges we collectively face. Only science can illuminate the way forward in tackling the climate crisis and combating novel pandemics. So we must both trust in it and push back against those seeking to sow distrust in it.

Some years ago, youth climate activist Greta Thunberg famously admonished critics to "listen to the scientists." While she was referring specifically to those failing to heed the warnings of climate scientists, it could have equally been said of public health scientists now under assault by antivaccine activists and COVID-19 conspiracy theorists. The notion that we should listen to and trust science doesn't mean exalting or deifying individual scientists. Rather, it means that we should place our trust in the collective judgment of the scientific community. Science is not unerring, but it offers our best understanding at any given point in time, subject of course to revision and modification as new observations or data become available, new hypotheses are tested, and an ever more robust and comprehensive understanding emerges.

There is a reason that scientists find themselves directly in the crosshairs of bad actors. Scientists rank among the most trusted messengers in modern society and are seen to possess both authority and integrity.[59] There are few bars higher than the standards of legitimate scientific pursuit. Ultimately, eroding truths in science or equating scientific mainstream findings with propaganda makes it easier to diminish the veracity of just about everything else. That makes both science and scientists prime targets.

A prophetic nineteenth-century illustration of how vested interests are led to portray scientists as enemies of the people is, well, *An Enemy of the People*—the Henrik Ibsen play (turned into a film starring Steve McQueen in 1978, just two years before his death). The fictional tale takes place in 1882 in a small Norwegian town whose economy is closely tied to tourism driven by their local hot springs. Thomas Stockmann is a medical officer living in the town. He is vilified by the townspeople for revealing how a tannery is polluting the baths of a recently opened spa. Ultimately, his family is widely ostracized and even considers leaving town, but they decide to stay in the hope that the townspeople will come to recognize and appreciate Stockmann's warnings. The ending suggests this is a false hope indeed. The story is a remarkably prescient cautionary tale for the toxic antiscientific sentiment we encounter today, where environmental and human health is threatened by special interests and the scientists speaking truth to power are portrayed as the enemy.

"An enemy of the people" is precisely the way bad actors want you to think about academics and scientists (and because the universe has a sense of irony, you can purchase a copy of *An Enemy of the People* with a foreword by none other than . . . Robert F. Kennedy Jr.).[60] Consider the case of J. D. Vance, Donald Trump's running mate in the 2024 election. Vance, an author turned politician, now serves as vice president of the United States. He is the product of a low-income upbringing who benefited from an elite education, including a law degree from Yale. As Katherine Knott details in *Inside Higher Ed*, "Vance has written and talked about how education is key to opportunity and how higher education helped to lift him out of poverty," as detailed in his memoir, *Hillbilly Elegy*, published in 2016.[61]

A vocal Trump critic back in 2016, Vance did an about-face by 2018, positioning himself as a loyal member of the MAGA (Make America Great Again) tribe,[62] running successfully for an Ohio Senate seat in 2022 as a Trump-loyalist Republican. Higher education, rather than being his personal savior as he'd avowed in the past, was now instead a villain to be maligned and undermined. In a 2021 speech at the National Conservatism Conference titled "The Universities Are the Enemy," he said that universities are the source of "deceit and lies, not . . . the truth."[63] And at the 2024 Republican National Convention in late July he went further, exclaiming that "professors are the enemy."[64] For the authoritarian Right, professors and universities are the enemy because they seek to inform and to widen perspectives. Authoritarian and populist regimes need their citizens to be aggrieved and ignorant, attributes that typically dissipate in the presence of a college education. And thus, we see figures such as J. D. Vance (Yale Law), Florida governor Ron DeSantis (Yale undergrad and Harvard Law), and Fox News anchor Laura Ingraham (Dartmouth College) flip their Ivy League educations on their sides to rail against the "educated elites." As we will see in Chapter 3, "The Petrostates," attacking science and scientists is an authoritarian propaganda practice that was first shaped in Stalinist Russia during the 1930s. Antiscience remains a signature of authoritarian regimes and aspirations.

In 2024, our mutual friend and colleague Tony Fauci published his memoir. In it, among other things, he recounts his experiences at the center of attacks by Trump supporters and Republicans for his role in the government response to COVID-19.[65] Mike happened to catch an interview he did about the book with Rachel Maddow on MSNBC. Almost as if to underscore the central thesis of this discussion, Maddow opined that it's the fact that Fauci "knows what he's talking about" that is so "toxic for their [Trump and Republicans] political project." Maddow added that "Dr. Fauci is in the bull's-eye of the Trump movement even today" because it "cannot abide public health expertise because it cannot abide *expertise.*" As she put it, anything that "challenges their leader" is answered "with menace." Look no further than MAGA firebrand Representative Marjorie Taylor Greene's (R-GA) insistence that Fauci "should be prosecuted for crimes against humanity" or Alex Jones's demand that he be imprisoned or even that COVID response leaders be executed.[66] "Enemy of the people, indeed. Or if you like, 'Public Enemy Number One.'" That's the way the *New Republic* referred to the Right's inflammatory caricature of Fauci.[67]

Something Tony described in the interview underscores the commonality of the threats faced by scientists today in the public health and climate arenas. He related an incident during the Trump years when he had opened a suspicious-looking envelope only to discover it contained white powder. After a tumultuous ordeal, which included quarantine and the worry he and his colleagues had been exposed to a deadly substance, it was found to be harmless. But that wasn't really the point. The intent was to scare him, to serve notice, to provide a warning. Just as they'd tried to do with Mike a decade earlier. As Mike recounts in *The Hockey Stick and the Climate Wars*:

> On August 18, 2010, I had to explain to colleagues in the Penn State University meteorology department, located in

the "happy valley" of Central Pennsylvania, why there was police tape over the door to my office. The immediate answer, is that the FBI [Federal Bureau of Investigation] had quarantined the room and sent a letter I'd received that afternoon off to their nearest testing facility to determine the nature of the white powder contained within it. At a more basic level, the answer was that this is what it means to be a prominent figure in the climate change debate in the U.S. today. The tests came back a few days later—luckily, the substance in the letter was corn starch. The sender had nonetheless committed a felony crime.

This confluence is not coincidental, for both are direct consequences of stochastic terrorism, a type of political violence inspired by bad actors fanning the flames of hatred and malice in a cynical effort to sow distrust in authority and anyone—especially scientists—who might stand in the way of their political and financial agenda. Physical threats and violence are not only an understood but in fact a *desired* consequence. Tony reports that Representative Taylor Greene's comments drove death threats against him.[68] Elon Musk has promoted the vilification of Fauci with tweets like "My pronouns are Prosecute/ Fauci," which has more than a million likes.[69] Six months later Musk attacked Peter for refusing to debate RFK Jr. about vaccines on the Joe Rogan podcast, insisting that Peter was "afraid of a public debate, because he knows he's wrong."[70] This egged on the darkest of antiscience forces, who saw Peter's defiance as a license not only to engage in a mass online social media attack on him, but also to stalk him at his home or at venues where he spoke, necessitating that he travel with security and protection from law enforcement. *Fortune* magazine reported that "after Elon Musk, Joe Rogan vaccine Twitter brawl, scientists say 'vile rhetoric and misinformation' is forcing them off the platform."[71] That's a double win for the forces of antiscience.

THE BATTLE WE FACE

The one-two punch of climate change and pandemics threatens humanity, ecosystems, and life on Earth as we know it. We are in a new age of metaphorical "global boiling" and recurrent deadly infectious-disease outbreaks. For future waves of global infections, we will need to employ an array of new technologies and enhance public health–control interventions. And to limit our exposure to the impacts of climate change, we will have to reduce carbon emissions dramatically, while funding and implementing adaptive measures to insulate us from the damaging consequences of coastal inundation, extreme weather events, dangerous heat, and of course more widespread infectious disease. Let us dig a bit deeper into that.

For the mosquito-transmitted viruses, we have the capacity to develop new vaccines, and some are already being either developed or deployed.[72] We also have new approaches to developing genetically modified mosquitoes or biological control methods using *Aedes* mosquitoes infected with *Wolbachia* bacteria.[73] It is unclear whether such innovations can even leave the starting gate given the widespread suspicion now generated by antiscience forces. Some of the same issues may apply to new antiparasitic-disease vaccines such as the human hookworm vaccine that Peter has developed, a lifelong quest that began when he was an MD/PhD student in the 1980s in New York City.[74] Even though that vaccine would likely be deployed only in LMICs, there are already indications that the antivaccine activism now pervasive in the United States is spilling over to slow malaria-vaccine uptake on the African continent.[75] There is excitement about the possibility of developing next-generation coronavirus vaccines, including a universal coronavirus vaccine.[76] However, it is unclear whether this also will be readily adopted. It is noteworthy that fewer than 20 percent of Americans eligible to receive a new annual immunization targeting the circulating variants at the end of 2023 and beginning of 2024 decided to take it, even though it was shown to be effective at reducing hospitalizations,[77] or even long-COVID signs and symptoms.[78] Will public health permanently suffer due to omnipresent antiscience and antivaccine activism?

In the chapters that follow, we detail how climate change and pandemics have acquired a new, self-sustaining fuel, namely, a veritable tsunami of antiscience disinformation that drives attacks on both science and scientists (Fig. 2). By themselves, climate change and pandemics pose a risk to the future of humankind and our planet. Now, antiscience has become an overriding third driver, which could make climate change and pandemics difficult if not impossible to fight. As Peter Katona, Kavita Patel, and Seth Freeman wrote in the *Hill* in 2023: "Rising anti-science attitudes in the U.S.—and anti-evidence generally—have become infused into the country's culture wars.

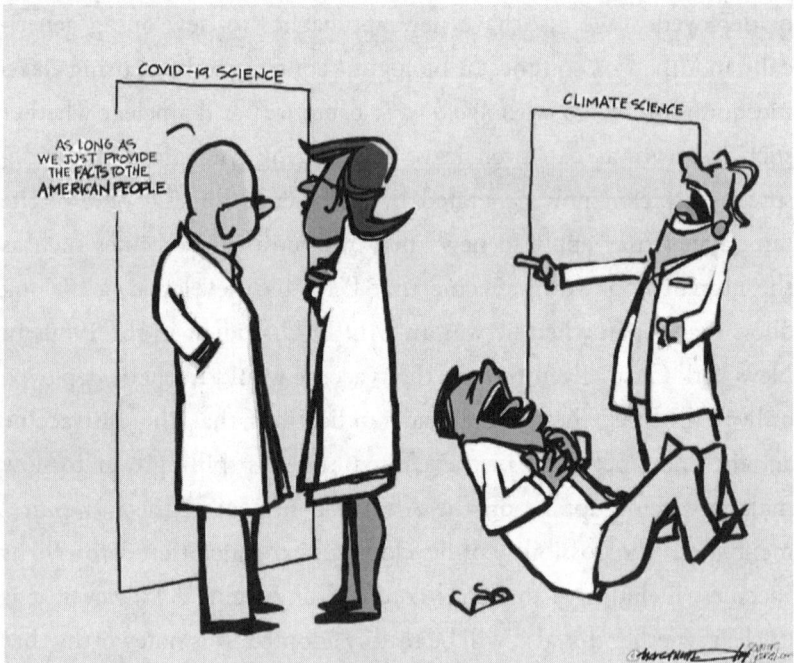

Figure 2. "As long as we just provide the facts to the American people." Marc Murphy for *The Courier Journal*

Anti-science sentiment impeded our pandemic response, and it prevents us from adequately addressing climate change and other societal problems. Anti-science is, quite literally, killing us."[79]

Unless we find a way to overcome antiscience, humankind will face its gravest threat yet—the collapse of civilization as we know it. The premise of this book, based on our own lived experience, is that antiscience has grown to constitute a vast ecosystem, supported by five major forces that favor and promote scientific disinformation, challenge the basic tenets of public and planetary health, and twist behaviors in a way that threatens human lives and livelihoods. It is a complex spiderweb of malevolence that thrives on dark money, bad actors, and the nebulous activities of state entities, and it is now a pervasive influence in America and beyond. The five major drivers or forces can be briefly summarized as follows:

- The *plutocrats*, that is, a group of billionaires who finance the antiscience empire, often through dark-money channels, while shaping an elaborate system of ultraconservative policy think tanks, political action committees (PACs), lobbying groups, and colleges.
- The *petrostates*, meaning state actors of nations typically propped up by extractive industries and especially fossil fuels, and the politicians who do their bidding by creating the legislation and using the power of the state to co-opt the media, push antiscience propaganda, create onerous antiscience policies, and intimidate scientists. They are assisted by polluters and their politician allies.
- The professionals, or "*pros*," who provide many of the key disinformative talking points or content. The pros include credentialed experts, oftentimes holding doctoral degrees, and in some cases (currently or formerly) major university or academic appointments. In climate

science they include industry shills who serve as talking heads on cable news or write op-eds and books promoting climate denial and inaction, while in biomedicine they include physicians and other health-care professionals serving the health and wellness empire, together with physician-associated organizations that tout or sell medications, cures, or nutritional supplements of unproven value. They also include a small cadre of contrarian medical school professors, with some now serving in the second Trump administration.

- The *propagandists*, who work hand in glove with the pros. In general, they comprise the ideologues on social media, blogs, and podcasts, including full-time "trolls" who hide behind the anonymity afforded them by social media; armies of "bots," often deployed by industry special interests or state actors who spread antiscience and smears; and individuals who present themselves as scientific experts, but lack the expertise relevant to the topic at hand and are often driven by ideology and axes to grind rather than any interest in the dissemination of objective science.

- The *press*, which includes not just the conservative media (for example, Fox News and the *Wall Street Journal* editorial page), but even elements of the mainstream media (MSM) who push dangerous false equivalencies to provide legitimacy to antiscience extremists (by no means do we mean that all journalists and all journalistic outlets are part of the problem—but far too many are today).

In some cases, there is a blurring of these categories, such as what it means to be a *pro* versus a *propagandist*, or a *plutocrat* versus a *propagandist*—think Elon Musk. So it should come as no surprise that the same names and groups might pop up in multiple chapters in this book. But there is no question that together, these five forces of antiscience have blazed a path of death and destruction across the United States and around the world. And we cannot confront and dismantle modern antiscience without first understanding it. Professor Marie Curie once said, "Nothing in life is to be feared, it is only to be understood. Now is the time to understand more, so that we may fear less."[80] While we're not sure we entirely agree with the first part of that statement (there is much to fear today when it comes to the pernicious impacts of antiscience), we nonetheless, in furtherance of understanding, provide a firsthand account of our experiences with the five forces of antiscience. We offer some practical steps to begin a long road to repair public understanding, to protect science and scientists from the five predatory forces of antiscience, and to ensure, for posterity, that young people continue to view science as a vital, noble, and worthy undertaking and profession.

THE PLUTOCRATS

[Rupert Murdoch] bears enormous responsibility for the world's lack of necessary action on climate change. His outlets have actively promoted skepticism about climate science that has undermined the need to act.

—Professor Lesley Hughes,
Guardian, September 2023

Public spaces must not be under the control of billionaires—public spaces belong under public control with public accountability.

—Professor Stephan Lewandowsky,
University of Bristol

Our first driver, leading the attack on modern science and scientists, is a group we call the plutocrats. The term is a portmanteau derived from ancient Greek—*ploutos*, or wealth, and *kratos*, or power. Therefore, a plutocracy is a society in which power is controlled by a small minority of high-net-worth individuals, and *plutocrats* are those individuals. Many civilizations throughout human history were partial or total plutocracies in some form. Wealthy patricians composed the Senate of Imperial Rome,[1] and the House of Medici ruled the Republic of Florence in the fifteenth century and eventually the Papal States on the Italian peninsula.[2] More recently, Mark Twain wrote and spoke about the United States at the end of the nineteenth century as a "Gilded Age," led by ultrawealthy industrialists, bankers, and railroad barons such as Andrew Carnegie, Henry Flagler, Henry Clay Frick, Jay Gould, Andrew Mellon, J. P. Morgan, Leland Stanford, and Cornelius Vanderbilt. While some of these Gilded Age plutocrats are remembered for their roles in violating workers' rights, busting unions, and other injustices, and were referred to as "robber barons," they also became important philanthropists.[3] They contributed to the creation of some of our nation's greatest research universities, including Carnegie Mellon, Stanford, and Vanderbilt Universities. The Gilded Age plutocrats also built great libraries, museums, and other institutions of cultural importance. In an 1889 essay on wealth, Andrew Carnegie famously wrote, "The man who dies thus rich dies disgraced."[4] American plutocrats, in short, have maintained a complicated status, using their wealth and power for both good and bad.

BENEVOLENT PLUTOCRATS?

Without question some of our most important and enduring American institutions have benefited from the largesse of plutocrats, who have contributed to our economic growth and success as a nation. So, while it might seem like an oxymoron, perhaps there really do exist *benevolent plutocrats*, that is, billionaires who have prioritized the common good and used their wealth and power to address some of society's ills and challenges. Plutocrats may give an initial illusion of benevolence, or they may genuinely start down a path of enlightenment, only to stray later, perhaps due to some triggering event. So we must be careful about entrusting too much wealth and power in the hands of a single individual or a group of oligarchs. That having been said, there does seem to be a subset of the plutocrat class that has contributed positively to the advancement of policy-relevant science.

One example today is former New York City mayor, media mogul, and businessman Michael Bloomberg. Through his philanthropic foundation,[5] Bloomberg has supported major health science and climate initiatives while using his media platform to promote proscience messaging on both human health and climate. He has also worked tirelessly to build his college alma mater, Johns Hopkins University, into one of our nation's top-ten educational and research institutions. Another is Eric Schmidt, former chief executive officer (CEO) of Google. He and his wife, Wendy Schmidt, established the Schmidt Family Foundation in 2006 to support environmental sustainability efforts.[6] They also fund educational programs at the University of Chicago and University of California, Berkeley (UC Berkeley) that support aspiring data scientists and guide them to the use of data science to serve the social good.

Sometimes it's a bit more complicated. Consider Mark Zuckerberg. He and his wife created the Chan-Zuckerberg Initiative for medical research and an important breakthrough prize for medical research. Peter frequently encourages his mentees to consider spending a part of their career at the Chan-Zuckerberg Biohub, a top-flight biomedical research institution based in San Francisco that conducts cutting-edge studies and taps into the intellectual strengths of Stanford University, UC San Francisco, and UC Berkeley. On the other hand, Zuckerberg has allowed his social media platform Facebook to be weaponized by bad actors promoting both climate denial[7] and vaccine misinformation.[8] And his mega company, Meta (which owns the social media outlets Facebook, Instagram, Threads, and WhatsApp), recently banned all links to a nonprofit news site that had the temerity to publish an article criticizing Meta's alleged suppression of climate-related content.[9] His recent embrace of Donald Trump has not helped matters.[10]

Then there's Bill Gates. The Bill & Melinda Gates Foundation has supported work for new parasitic-disease vaccines, including a malaria vaccine and funding for Peter's work to develop a hookworm-anemia vaccine. And they have supported innovations for mosquito control and funded a colleague[11] who collaborated with Mike on studies of the impact of climate change on the spread of malaria.[12] But Gates has downplayed the viability of a clean-energy transition and supported controversial and potentially dangerous "geoengineering" climate interventions.[13]

Finally, consider the famous industrialist John D. Rockefeller. Rockefeller made his fortune more than a century ago by forming the Standard Oil Company, the forerunner of today's ExxonMobil, along with Chevron, Marathon Oil, ConocoPhillips, and other oil and gas companies. In this sense, he contributed fundamentally to the climate crisis in which we now find ourselves enmeshed. But history has a sense of irony, and his legacy today has many positive attributes. Rockefeller also used his enormous wealth to create the University of Chicago and the Rockefeller Institute for Medical Research (now Rockefeller University),[14] where Peter did his graduate research on a vaccine for hookworm anemia, a major scourge of low- and middle-income countries. The Rockefeller Foundation, also founded by John D. Rockefeller, today funds critical human health research and promotes environmental initiatives, including action on climate.[15] Indeed, the Rockefeller Brothers Fund, created in 1940 by the sons of John D. Rockefeller and other Rockefeller family members, has provided philanthropic support for climate action. They have even played a leadership role in the global fossil fuel–divestment movement.[16]

Peter's work has in fact benefited directly and substantially from both Gates Foundation and Rockefeller Foundation support. One of Peter's proudest scientific accomplishments is the development and clinical testing of the first human hookworm-anemia vaccine. Because of climate change and extreme poverty on the African continent, especially in the Democratic Republic of the Congo and Nigeria, where eventually almost one-half of the world's population in extreme poverty will live by 2050,[17] human hookworm infection together with malaria could become the most widespread parasitic infections in that region.[18] Hookworm infection is even reemerging in the United States, possibly due to the combined effects of climate change and enduring poverty.[19]

As an MD/PhD doctoral student during the 1980s, some of Peter's hookworm-vaccine work received financial backing from a new Great Neglected Disease (GND) Program created by Dr. Kenneth Warren at the Rockefeller Foundation.[20] In the early twentieth century, John D. Rockefeller himself took a personal interest in hookworm anemia because of its debilitating effects, especially on childhood development, education, and worker productivity. Through the Rockefeller Sanitary Commission, he even launched efforts to control hookworm through improved sanitation and mass-treatment campaigns in the United States and by endowing a system of university-based schools of public health.[21] Later through the Rockefeller Foundation International Health Board, hookworm mass-treatment campaigns were extended globally and conducted in China, the Indian subcontinent, several South Pacific islands, and multiple Latin American and Caribbean nations.[22] While advances were ultimately insufficient to eradicate hookworm due to limited antiparasitic drug effectiveness or posttreatment reinfections, the work has saved countless lives, and it is Peter's aspiration to develop and distribute a new hookworm-anemia vaccine as a new breakthrough technology for global health.[23]

After completing his doctoral degrees, Peter relocated to Washington, DC, and George Washington University School of Medicine and Health Sciences. There, in the 1990s, he became president of the Sabin Vaccine Institute, which both affiliated with GWU and received large-scale support from the brand-new Bill & Melinda Gates Foundation. The Gates Foundation also receives important cofunding from Warren Buffett, yet another powerful and wealthy individual. Their critical and timely support over a period of almost fifteen years, beginning in 2000, allowed the discovery of additional promising hookworm-vaccine antigens and then transitioned them into vaccines for manufacture and clinical testing.[24] In those early days the Gates Foundation began supporting new vaccines for malaria and many other ancient global scourges, including Peter's *Na*-GST-1 human hookworm vaccine, currently accelerating through final development at the Texas Children's Hospital CVD.

The Gates Foundation also donated the extraordinary sum of $750 million in 1999 to launch Gavi, a global alliance for vaccines and immunization to expand childhood vaccination programs globally.[25] Today, millions of infants and children living in the world's LMICs receive their routine immunizations through Gavi-sponsored programs operating hand in glove with the World Health Organization (WHO), while a new malaria vaccine has been licensed for use and recently introduced in Africa.[26] The two takeaways: first, a project Peter began as a graduate and medical student forty years ago may soon materialize into a parasitic-disease vaccine used worldwide, and, second, wealthy people like Bill Gates have the capacity for doing enormous good in the world. There is no question that Peter's lifelong quest for a hookworm-anemia vaccine was helped by scientific grant support for his laboratory from both Rockefeller and Gates. Both individuals and the charities they created were game-changing for global health research, not only for Peter's laboratory, but for many other scientific enterprises as well.

Of course, the efforts of these plutocrats are not without criticism.[27] Even Peter has questioned some of the Gates Foundation's approaches to developing global health technologies, most notably their pivot in the last decade toward emphasizing the big-pharma companies over nonprofit product-development partnerships in order to achieve these objectives. And Mike has criticized Gates's messaging on climate solutions, which emphasizes unproven proposed new technology over practical efforts to decarbonize our energy infrastructure today.[28] So, once again, it's complicated.

But in the "up-is-down, black-is-white" world of today's political Right, Gates, Bloomberg, Zuckerberg, and their ilk, rather than being lauded for some of their enlightened philanthropy, have unfortunately instead been vilified as "controligarchs" who wield their wealth and power to advance a "globalist" (read: "Jewish") socialist new world order (NWO), where the "fake science" of climate change or COVID-19 vaccines is being used as a weapon to exert governmental control over individuals. Bill Gates's "work with" the infamous American financier and sex offender Jeffrey Epstein[29] has reinforced the conspiracy theory that Gates is part of some Jewish cabal. Look no further than the bizarre and risible conspiracy theory that Gates actually had Epstein murdered.[30] We see how virulent antisemitism creeps into these conspiracies. Not surprisingly, the climate and public health advocate Bloomberg, a prominent Jewish New Yorker, is a favorite target. Former House majority leader Kevin McCarthy once accused Bloomberg and progressive Jewish climate advocate and philanthropist Tom Steyer of trying to buy an election—in a tweet that McCarthy insisted was not an antisemitic dog whistle but took down nonetheless.[31] The preferred bogeyman of today's conspiracist Right, though, is almost certainly George Soros, a Hungarian American billionaire hedge-fund manager who has donated much of his fortune to progressive causes, including the groundbreaking Central European University. Born to a nonobserving Jewish family in

Budapest, he survived the Nazi occupation of Hungary, fleeing to the United Kingdom in 1947. According to University of Florida sociologist and biographer Armin Langer, Soros has become today's "perfect code word for conspiracy theories" of the Ultra Right linked to antisemitism.[32]

In his book *Jewish Space Lasers: The Rothschilds and 200 Years of Conspiracy Theories*, Michael Rothschild (no relation) traces the current conspiracies about Bloomberg, Soros, and other modern-day Jewish (or Jewish-friendly) plutocrats to those beginning with the Rothschild family of Europe. For centuries, the Rothschilds, wealthy, noble heirs to a Jewish banking fortune who rose to prominence in eighteenth-century Germany, have been front and center in numerous right-wing antisemitic conspiracy theories. Sadly, they persist and even infiltrate the US Congress. Consider the claim by far-right House Republican firebrand Representative Marjorie Taylor Greene that the Rothschilds used space lasers to ignite the historic California wildfires of 2018 in their effort to generate support for clean energy.[33] She later made similar, even more cryptic, claims about the massive Canadian wildfires of summer 2023 that blanketed East Coast cities such as New York in thick smoke.[34] Other conspiracy theories connect Gates to Soros, holding that *both* Soros and Gates planned to microchip people who are being tested for COVID-19.[35] Often, the annual Economic Forum held in Davos, Switzerland, is the alleged location of secretive meetings to implement these nefarious plans. Klaus Schwab, the Davos CEO, is reviled as the innkeeper of these events. The attacks on Gates, Soros, Bloomberg, Schwab, and others are part and parcel of a phony propaganda war waged against the so-called global elites by the Far Right (facilitated, ironically, by some of the billionaire tech bros).[36] However, in many cases even the term *elites* or *globalists* is little more than a thinly veiled code word for Jewish, long-standing antisemitic tropes. Both Peter and Mike (both of us have Jewish ancestry) are often accused of being part of this conspiracy, even though neither of has ever been invited to speak at Davos or hobnob with any of those global elites. Peter sometimes jokes that as a scientist, a vaccine scientist, and a Jewish vaccine scientist, he represents the *trifecta* for the conspiracists. The antisemitic attacks on Peter and Mike have historical roots that go back centuries.

The conspiracist attacks are a classic example of "projection" or "misdirection," honed over the decades by leading figures of the conservative movement such as Roy Cohn and Karl Rove and perfected by strategists of the modern Right like Steve Bannon and, of course, Donald Trump, who is known to have idolized Cohn. You simply accuse your opponent of the very thing that you are guilty of, confusing onlookers who no longer know whom to believe, while firing up your political base to attack your opponent and inoculating yourself against the underlying criticism. The claim that a shadowy cabal of progressive billionaires is in control of our politics today is intended to deflect your attention from one simple reality: such an effort *is* underway, only it's by the malevolent plutocrats of the authoritarian Right who are seeking to undermine democratic governance in the United States and abroad. Attacks on science and scientists are a key part of their plan.

MALEVOLENT PLUTOCRATS

Most relevant to our story are the abhorrent practices of a group of what we'll call *malevolent plutocrats* who promote a vigorous anti-science agenda. Some of them became wealthy as fossil-fuel barons, while others made their fortunes through more diversified businesses. Among the former—the oil and gas plutocrats—there is the Koch family, including Charles, who serves as chair and co-CEO of Koch Industries (his brother David passed away a few years ago), and a trio of Texas-based billionaires who made their fortune through fracking and related energy-business activities. Next, there is Rupert Murdoch and his son Lachlan, heir to his media empire, and finally the so-called billionaire tech bros, who amassed their wealth through high-tech computer industries. Several have become notorious for using their platforms to promote strong views, outside their range of expertise and poorly or not at all supported by the facts, and have inserted themselves into the public discourse over politics, social thought, and policy in an often toxic manner. Elon Musk, for example, openly and brazenly weaponized his wealth, power, and Twitter social media platform to influence the 2024 presidential election, even holding conversations with hostile enemies of the United States such as Vladimir Putin.[37] His efforts proved successful in reelecting Trump as president. Our focus here, however, is on the efforts by Musk and other plutocrats to target American science and the scientists themselves (us both included).

Several features of the malevolent plutocrats stand out to us. They include, variously, a fascination with cryptocurrency, ties to Vladimir Putin and Russia, and a penchant for infiltrating major think tanks and universities. It is not uncommon for cryptocurrency-proselytizing tech bros to hold antiscience viewpoints.[38] Why might this be the case? First, those who invest in cryptocurrency often distrust mainstream banking or economic systems, so it is not too much of a stretch to extend that distrust to government, scientific and academic institutions, and scientists themselves. The 2024 US presidential hopeful and billionaire biotech investor Vivek Ramaswamy is an exemplar of the subclass of tech bro known as the "cryptobro."[39] He proposed a plan to elevate the status of crypto and protect the industry and its developers, while scaling back the oversight of the US Security and Exchange Commission.[40] In September 2023, Ramaswamy publicly denounced his own decision to get vaccinated against COVID as he also called out Dr. Anthony Fauci and other US government officials for having "lied to us about what we knew about those vaccines before they were mandated."[41] Ramaswamy also promoted right-wing conspiracies related to the January 6 assault on the US Capitol.[42] And, of course, he's a climate-change denier too.[43] Still another irony about Ramaswamy and Musk, as well as RFK Jr., Governor Ron DeSantis, and J. D. Vance, is how they rail against "the elites" when their common denominator is a Harvard or Yale University education.

Meanwhile, state actors like Russia have used troll farms and bot armies to spread propaganda advantageous to their interests and to create an online army to advocate for their political agenda. This will be discussed further in Chapter 3, "The Petrostates," but given Russia's reliance almost exclusively on cryptocurrency to evade international sanctions against the country,[44] it is hardly surprising that they have sought to create a more cryptocurrency-friendly political environment by marshaling their online rabble for this cause. Nor is it surprising that tech bros who promote Russian propaganda,[45] like vaccine-conspiracy promoter Elon Musk and his huge online army of acolytes, would push crypto as well. In short, it's no real surprise that cryptocurrency and antiscience are common attributes of the tech-bro tribal identity.

Today's malevolent plutocrats frequently operate through a complicated web of entities as they wage war against science and scientists. They include political action committees, government lobbying organizations, local election campaigns, libertarian think tanks, conservative colleges, and even a few major universities. The Mercatus Center, a very large unit within George Mason University, for example, receives significant funding from the Kochs. The plutocrats often rely on recruiting and co-opting prominent university professors, writers, journalists, and other intellectuals to staff their think tanks, providing academic cover and a veneer of respectability. They have often targeted the science of both climate and pandemics, with the latter focused on denigrating vaccines or halting public health measures. The rationale for climate denialism seems straightforward because it offers a means to halt or slow government regulatory oversight of the fossil-fuel industry. However, the rationale for targeting pandemic science remains less obvious. In some cases, their passion to push back against pandemic prevention or vaccines fits into their greater antiregulation ideology, "Don't tell me what to do with my money or my body," although it must be said that this is selective reasoning at best—many of these individuals have no qualms supporting policies allowing the government to control women's reproductive options. In this view, attacking vaccines for their efficacy or safety represents collateral damage from halting vaccine mandates, a part of a larger libertarian campaign that champions the concepts of health freedom. In other words, in their zeal to stamp out the mandates, the plutocrats also wound up discrediting the effectiveness and safety of the actual vaccines. In addition, their efforts to derail public health recommendations may have been motivated by their concerns about stay-at-home orders (issued by governors) that slowed economies and reduced fossil-fuel consumption in the early part of the COVID-19 pandemic.

Attacking the experts is a key component of the antiscience agenda. Both of us have been on the front lines in the war on science and have seen this firsthand. Mike, for years, has been attacked by groups funded by the Kochs (and similarly minded conservative billionaires, such as the Scaifes and the Mercers) seeking to introduce doubt and suspicion around his climate research. He's been the target of conservative politicians and right-wing front groups attempting to obtain his personal emails in search of anything that might be used to embarrass or discredit him. When Mike served on the faculty at Penn State University, Republicans in the Pennsylvania General Assembly threatened to withhold Penn State's budget if punitive actions weren't taken against him for fabricated indiscretions. He's been subject to investigations by Republican politicians funded by the fossil-fuel industry and has endured threats against his life and that of his family through the tactic of "stochastic terrorism" (political violence induced through mass media, social media, or other forms of public discourse). At one point, as noted in the previous chapter, the FBI had to investigate a letter he had received laced with white powder.[46]

Similarly, antivaccine and antipandemic groups led or supported by plutocrats have attacked Peter, most notably in the summer of 2023. The episode involved a pile-on from Elon Musk, podcaster Joe Rogan, and RFK Jr., leading to a wave of personal stalking events and threats that required law enforcement. Mike and Peter have both at various times required law enforcement and even FBI protection. We discuss this further in Chapter 5, "The Propagandists." The focus here is on the leading malevolent plutocrats—who they are and what their agenda is.

BREAKING DOWN THE
MALEVOLENT PLUTOCRATS

A Koch and Some Bile

The archetypal malevolent plutocrats are the Koch brothers (technically now only the "Koch brother" Charles, since David has passed away). Koch Industries is the world's largest privately held fossil-fuel company, with an obvious financial interest in fossil fuels. The Kochs have funded groups with far-right ideologies, thereby linking extremist politics to antiscience. They have supported a constellation of think tanks and front groups (such as the "Heartland Institute" and "Competitive Enterprise Institute") whose names seem innocuous but whose primary role is to create doubt about the science behind human-caused climate change and to thwart the needed transition toward clean energy. They have also advanced a much broader libertarian, antiregulatory ideological agenda.

That's where the story intersects with vaccine denial, especially during the COVID pandemic. In the fall of 2020, the so-called Great

Barrington Declaration was drafted and promoted at a conference hosted by the American Institute of Economic Research, a libertarian think tank funded by the Charles Koch Foundation and located in Great Barrington, Massachusetts. The AIER president was previously vice president for research and policy at the Charles Koch Institute. This polemic promoted reckless disregard for public health and social distancing, together with the notion that herd immunity could be achieved through widespread natural infection, without resorting to COVID immunizations. In time, those promoting this approach joined forces with those who railed against vaccines. In their report for the Center for Media and Democracy titled "How the Koch Networked Hijacked the War on Covid," Walker Bragman and Alex Kotch describe how a "corporate-bankrolled campaign" undermined public health and "hijacked" governmental pandemic responses.[47] Stay-at-home measures issued by most of the US state governors in the first year of the pandemic, before COVID vaccines became available, adversely affected the oil and gas industry due to reduced demand—a threat to the Kochs' bottom line. Bragman and Kotch report on a press release from Americans for Prosperity, a Koch-founded far-right nonprofit: "We can achieve public health without depriving the people most in need of the products and services provided by businesses across the country." Once vaccines became available in 2021, the recovery was swift. But groups linked to Koch Industries launched an ambitious and aggressive, and ultimately dangerous and self-defeating, lobbying campaign to keep businesses open during the prevaccine phase. The danger of business as usual during this first phase of the pandemic lay in the fact that without stay-at-home measures prior to the availability of COVID vaccines, there would be widespread virus transmission resulting in huge surges of patients in hospital emergency rooms and intensive-care units that would overwhelm hospital staffing and cause deaths to skyrocket. During the first year of the pandemic, such surges led to high COVID mortality.[48]

Koch-supported nonprofits, policy think thanks, and lobbying firms, including Americans for Prosperity, the American Legislative Exchange Council (ALEC), and the Texas Public Policy Foundation, lobbied the Trump administration to avert shutdowns. They helped establish an alternative antiscientific reality, through a sophisticated network of academic entities and think tanks, wherein COVID mitigation policies were challenged as unsound science.[49] That is, incidentally, the precise modus operandi the Kochs and other climate-denying plutocrats have used for decades in their assault on climate science and climate scientists. The initial emphasis of the disinformation campaign was on disrupting stay-at-home orders, but it eventually expanded to other tactics.

A centerpiece of the aforementioned Great Barrington Declaration was the recommendation that governments avoid stay-at-home orders during the COVID-19 pandemic. Instead, it sought to create a two-tiered system in which young people would go about their normal activities, while older or otherwise vulnerable people would remain sequestered at home or in institutional facilities as a form of "focused protection."[50] The young would become infected to create a sufficient level of community immunity, also known as herd immunity, over some period of time, before older individuals could safely venture out. For all their vilification of academic elitism, the science-denying Right seems to exploit it for their own purposes whenever they can. Three professors at the most elite academic institutions, Stanford University, Harvard Medical School, and Oxford University, were chosen to lead the Great Barrington Declaration. The declaration gave academic cover to government leaders under pressure to avoid shutdowns even though it would lead to huge hospital surges and greater mortality.[51]

The flaws in this strategy are numerous and if implemented widely would create a public health disaster. First, it turns out that herd immunity was elusive due to the antigenic variation of the SARS-2 coronavirus and was never achieved despite the frequent and rosy predictions of the Great Barrington Declaration authors and their acolytes. Another major concern was the fact that COVID-19 in young people was often not a benign condition. COVID-19 was and still is a significant cause of death and severe illness in younger age groups and an important cause of chronic and debilitating long COVID. Last, focused protection of vulnerable groups was never realistic, and the sheer numbers of seriously ill middle-aged adults and seniors would overwhelm health systems and their emergency departments and intensive-care units. We learned fairly quickly in the COVID-19 pandemic that mortality rates skyrocketed once hospital ICUs and their personnel became overburdened or overwhelmed. The World Health Organization director general, Dr. Tedros, appropriately warned, "Herd immunity is achieved by protecting people from a virus, not by exposing them to it," and added, "Never in the history of public health has herd immunity been used as a strategy for responding to an outbreak, let alone a pandemic," calling such approaches "unethical."[52] Disturbingly, the declaration had the Harvard-Oxford-Stanford imprimatur that some politicians had sought to grant it a veneer of legitimacy. Unsurprisingly, President Trump invited the authors to the White House despite the objections of his own COVID-19 adviser Deborah Birx, who identified them as "a fringe group without grounding in epidemics, public health, or on-the-ground common sense experience."[53] One of the lead declaration authors, Jay Bhattacharya, was appointed by Donald Trump to lead the National Institutes of Health (NIH) in his second term.[54]

Trump embraced the misguided herd-immunity framing almost immediately. At a September 2020 ABC News town-hall event, Trump made a garbled statement about it: "And you'll develop herd, like a herd mentality. It's going to be—it's going to be herd-developed, and that's going to happen."[55] Then in November 2020, Florida governor Ron DeSantis ordered the resumption of most businesses, restaurants included, and blocked the enforcement of mask mandates and ultimately vaccine mandates. This anti-interventionist approach had negative public health consequences, with countries such as Sweden adopting similar policies suffering case fatality rates "many times that of its closest neighbors." One study found that almost 10 percent of deaths registered in Sweden could have been avoided if Sweden had issued mandatory stay-at-home orders.[56] Despite the failures of the Great Barrington Declaration, there was a concerted attempt to defend it through academic cover. Two of the Great Barrington authors became inaugural fellows at the newly created Academy for Science and Freedom at the Washington, DC, campus of Hillsdale College, "the Christian liberal-arts school at the heart of the culture wars."[57] However, according to Justin Feldman from Harvard University's FXB Center for Health & Human Rights, "They have no interest in science. . . . They have been wrong about the pandemic time and time again. They use their stature as 'experts' to push for policies that are indifferent to ongoing mass death."[58]

You Musk Be Joking

Elon Musk, the world's richest man, the tech bro of tech bros and cryptobro of cryptobros, has become a leading spreader of disinformation writ large. His own chatbot, "Grok," answered, in response to the query "Did Elon Musk spread misinformation to millions of people," "Yes, there is substantial evidence and analysis suggesting that Elon Musk has spread misinformation on various topics . . . to a very large audience through his social media platform, X (formerly known as Twitter)."[59] His promotion of antiscience disinformation largely emerged during the pandemic.[60] Prior to that, he did not appear to display any particular antipathy toward science. In fact, his holdings and business ventures such as Starlink, the satellite network, or Neuralink, a neurosciences biotech, were very much science and technology driven. His companies Tesla and Solar City were early critical drivers of the clean-energy revolution. Meanwhile, his foundation supported research on renewable energy, as well as pediatrics, space exploration, artificial intelligence (AI), and science and engineering education.[61] However, the goodwill he might have banked through these actions rapidly eroded as he transitioned dramatically toward extreme right-wing politics and science denial. Perhaps most critical of all, his purchase of Twitter—encouraged by venture capitalist and libertarian activist Peter Thiel,[62] an unapologetic opponent of democracy and proponent of climate and public health antiscience[63]—furthered a radical conservative political agenda. After purchasing it in late 2022, Musk converted Twitter into a forum for far-right extremism, earning him the title of "2023 Scoundrel of the Year" from the *New Republic*: "While he has mostly made headlines for his incompetence, he has unleashed and legitimized truly heinous forces on Twitter: He has welcomed back some of the world's most toxic people—Alex Jones, Donald Trump, innumerable Nazis and bigots—and has gone out of his way, again and again, to validate them. . . . He sees Twitter as a weapon, a way to not only push his agenda but to sic his army of loyalist losers on anyone he deems an enemy."[64]

Phil Markolin, a scientist and bioinformatics expert in Switzerland, looked closely at Musk's antiscience activities on Twitter, noting how the platform reinstated the most aggressive science-denying activists and how Twitter reconfigured the platform in a way that became toxic for many working scientists, especially climate scientists[65] and scientists combating the COVID-19 pandemic. Markolin quoted preeminent Australian COVID virologist Professor Eddie Holmes, who deleted his Twitter account after the social media site became what he described as "a post-Apocalyptic hell hole." Many other virologists and COVID scientists were driven out as well. Holmes declared, "The loss of science on Twitter is a contemporary mass extinction event."[66] Journalist Leo Hickman lamented "how toxic and alienating Twitter has become for people hoping to converse about the challenges of climate change in a nuanced, sensible way."[67] The evaporation of science from platforms like Twitter launched an exodus of science from the public conversation. We'll have more to say about that in Chapter 5, "The Propagandists," and we'll have more to say about Elon Musk and his dealings with Russia and Vladimir Putin, as well as the so-called Department of Government Efficiency, or DOGE—a legally questionable governmental entity that Trump has created for Musk to, among other things, slash science funding and eliminate or decimate scientific governmental agencies—in Chapter 3, "The Petrostates."

An Actual Bond Villain

No list of the leading malevolent plutocrats would be complete without Rupert Murdoch. Murdoch himself appears to be a climate denier. Back in February 2015 he tweeted, "Just flying over N Atlantic 300 miles of ice. Global warming!"[68] (the false implication being that if there is ice in winter, the planet can't be warming). He has also stated in an interview, "Climate change has been going on as long as the planet is here, and there will always be a little bit of it" (the false implication being that the impacts of fossil-fuel burning are mild and within the range of natural variability) and "If the sea level rises 6 inches...we can't mitigate that, we can't stop it. We've just got to stop building vast houses on seashores"[69] (the false implication being that we can just adapt to human-caused climate change, so there's no need to take action to mitigate it). Murdoch, on the other hand, does not appear to be a COVID vaccine denier. He famously took the vaccine himself as soon as he became eligible in late 2020, even as his network Fox News was promoting vaccine denial and misinformation through the Tucker Carlson show.[70]

The fact that he's not a COVID vaccine denier is really of no consequence, because regardless of what he may or may not personally believe, the media mogul—who has literally been portrayed as a Bond villain[71]—has wielded his international news empire as a weapon against science when it conflicts with his far-right ideology and financial bottom line. Murdoch's expansive media empire—which includes Fox News, the *Wall Street Journal*, and the *New York Post* in the United States; the *Sun* and the *Times* in the United Kingdom; and the *Australian*, *Sky News*, and the *Herald Sun* in Australia—is the greatest source of climate and COVID[72] misinformation in our news media today.

EVERYTHING'S BIGGER IN TEXAS— INCLUDING PLUTOCRACY

Texas is one of the hot spots of MAGA conservatism today. It is the center of the American oil empire, and it is run by the fossil-fuel industry. It also happens to be where Peter has lived and worked for more than a decade. As noted earlier, in 2011 Peter's vaccine-science and development scientific research group relocated to the renowned Texas Medical Center to develop new low-cost vaccines for COVID-19 and other neglected diseases. The Texas Medical Center is an enormous biomedical enterprise, the world's largest and first biomedicine city and high-tech life-sciences complex of steel, concrete, and glass roughly the size of downtown Los Angeles (Fig. 3).

Texas often likes to supersize everything, and on its website the Texas Medical Center calls itself "The World's Most Comprehensive Life Science Ecosystem." Relocating to Houston and the Texas Medical Center allowed Peter and his vaccine team (coheaded by his science partner of more than two decades, Dr. Maria Elena Bottazzi) to complete an ambitious and unprecedented undertaking, namely, to make low-cost and often patent-free vaccines for the world.[73]

Figure 3. Aerial view of the Texas Medical Center (original photo taken by Peter)

Construction of the Texas Medical Center began in the aftermath of World War II, through the support of Texas business community leaders who had a bold vision to make both Houston and Texas renowned in this area.[74] Central to making this happen was the enormous wealth Texas generated from the fossil-fuel industry, which supplied the oil and petrochemicals America needed to fuel its fighter planes, tanks, and battleships. The Texas fossil-fuel industry played a critical role in helping the United States defeat fascism during World War II. Now with the profits generated from that war effort, these Texas businessmen were prepared to give back and expand universities and health-care institutions. The Texas Medical Center was a major benefactor of that wealth. Overall, the Texas oilmen were a savvy, competitive, and, at times, hardscrabble group, who often labored outdoors in the harsh Texas climate, especially the extreme heat and sun of its unforgiving summers. Two especially important individuals included Monroe D. Anderson (M. D. Anderson), a cotton magnate (although later oil was discovered on his landholdings), and Hugh

Roy Cullen, who undertook high-risk yet successful drilling campaigns in East Texas.[75] Through such efforts, the M. D. Anderson Cancer Center became the largest and most successful research and treatment center devoted to cancer and its allied diseases, consistently ranked as the number-one cancer center by *U.S. News and World Report*. The Cullen family also became leading benefactors, providing the first large-scale finances for several Texas Medical Center institutions, including the Baylor College of Medicine, where Peter serves on the faculty. The National School of Tropical Medicine dean's office is located in the Cullen Building of Baylor College of Medicine, which was one of the original buildings of the Texas Medical Center.[76] Both the city of Houston and Baylor College of Medicine (and Peter!) are deeply grateful to the Cullen family.

Twentieth-century Texas oil barons known as "the big four"—Roy Cullen, Clint Murchison, Sid Richardson, and H. L. Hunt—also helped to promote the rise of post–World War II conservative Republican politics in the United States. Their motivations included a desire to expand Texas oil influence in Washington, DC, coupled with a strong aversion to communism spreading from the Soviet Union and China. Such fears explain why H. L. Hunt and other Texas oil magnates also backed Republican Wisconsin senator Joe McCarthy,[77] who led a vicious campaign of investigative activities through his House Un-American Activities Committee.[78] The embrace of this special form of American conservatism or anticommunism led to a conflicted feature of Texas philanthropy: the same benefactors who generously supported scientific and educational institutions also provided political support for campaigns, elected leaders, and political action committees that attacked science and scientists. During the era of McCarthyism that began in the late 1940s and continued into the 1950s, prominent US scientists were subjected to humiliating security investigations, denied visas or passports to travel abroad, and required to take loyalty oaths.[79] Among the McCarthy and HUAC targets was

Dr. J. Robert Oppenheimer, who led the science behind the Manhattan Project at the Los Alamos Laboratory but ultimately had his security clearances revoked.

Fast-forwarding to the twenty-first century, the financial transactions and political donations of a small but powerful group of ultraconservative Texas plutocrats from the oil fields of West Texas began to shift. Now, instead of supporting scientific institutions and scientists, these next-gen Texas oil barons pivoted to viewing vaccines or other aspects of science as potential threats to their updated vision of American conservatism. In 2022, CNN provided a detailed account of these new-generation plutocrats from Midland (Tim Dunn) and Cisco (Dan and Farris Wilks) in West-Central Texas, describing them as "billionaire oil and fracking magnates from the region" who use their wealth and influence to sway election outcomes and finance far-right PACs. CNN quotes Kel Seliger, a Texas GOP state senator: "It is a Russian-style oligarchy, pure and simple. . . . Really, really wealthy people who are willing to spend a lot of money to get policy made the way they want it—and they get it." CNN further adds: "The candidates Dunn and Wilks have supported have turned the state legislature into a laboratory for far-right policy that's starting to gain traction across the US."[80] The *Texas Monthly* calls Tim Dunn "a West Texas oilman" who is currently the "state's most powerful figure" and the "billionaire bully who wants to turn Texas into a Christian theocracy"; he is someone who has financed a "holy war against public education, renewable energy, and non-Christians"; meanwhile, Farris Wilks is a pastor in an ultraconservative church known as the Assembly of Yahweh 7th Day.[81]

Tim Dunn and the Wilkses have tapped their enormous wealth to support numerous political and business ventures. In 2019, the Public Accountability Initiative (PAI), a nonprofit watchdog group based in Buffalo, New York, did a deep dive into their projects, groups, and political candidates. For the Wilks brothers, they include Texans for

Vaccine Choice (TFVC), one of America's first antivaccine PACs that was created in 2015 to ensure the rights of parents to opt their children out of immunizations required for school entry or attendance and defeat candidates that promote school vaccine requirements.[82] Texans for Vaccine Choice actively promotes the tenets of health freedom and medical freedom, the preferred rallying cry for the antivaccine movement after the biomedical scientific community debunked the claim that vaccines cause autism.[83] On their website, Texans for Vaccine Choice notes that the organization "has matured into Texas' most influential medical liberty organization" and that "In a world where medical freedom is increasingly under attack, TFVC stands as a beacon." Although not in TFVC's specific mandate, Peter has noted how the tenets of health freedom or medical freedom these days often extend beyond vaccine choice. Aside from viewing school vaccine requirements as government overreach or money lining the pockets of the multinational pharma companies, some of the pros or propagandists in the health-freedom movement tout the benefits of natural immunity from acquiring the disease while downplaying its health risks.[84]

Through the lens of health freedom, the antivaccine lobby in America became inextricably linked to far-right Texas politics. According to both the *Texas Monthly* and the *Texas Observer*, the establishment of Texans for Vaccine Choice occurred in response to a state legislator who filed a bill in 2015 to eliminate nonmedical or conscientious-objector school vaccine exemptions. Effective lobbying by Texans for Vaccine Choice successfully defeated that bill, and ever since the group has used its support to back local candidates running on health- or medical-freedom platforms.[85]

Political support in Texas for antivaccine activism also goes beyond Texans for Vaccine Choice. Empower Texans, which ran from 2006 to 2020, was (according to PAI) a "conservative political project" funded by a small group of billionaires connected to the Texas oil and gas industry, including Dunn and the Wilks brothers, that "sought to push Texas politics further to the right." PAI also identifies private-equity firms that provide substantial funds, even hundreds of millions of dollars, to "prop up the business operations of some of the primary backers of Empower Texans," as well as a network of conservative and "evangelical Christian causes."[86] According to the *Houston Chronicle* and *Washington Post*, Empower Texans has also supported and advised Texans for Vaccine Choice.[87] The Central and West Texas oil plutocrats further finance ardent antivaccine activists in the Texas Legislature, such as former representative Jonathan Stickland (R-Bedford),[88] who has personally gone out of his way to publicly attack Peter on social media[89] and tweeted at him in 2019:

You are bought and paid for by the biggest special interest in politics. Do our state a favor and mind your own business. Parental rights mean more to us than your self-enriching 'science.' #txlege

Make the case for your sorcery to consumers on your own dime. Like every other business. Quit using the heavy hand of government to make your business profitable through mandates and immunity. It's disgusting.

It was especially disturbing to Peter that a sitting member of the Texas Legislature would accuse him (of all people) of making vaccines for financial gain or the benefit of the multinational companies when in fact the Texas Children's Hospital CVD had developed a COVID vaccine technology that was provided with no or minimal strings attached to vaccine producers in LMICs. Its hookworm-anemia vaccine is meant only for people living in extreme poverty who could never hope to pay for it.[90] As highlighted in the media, the Texas Children's Hospital scientists were pioneering efforts to bypass "big pharma" and make vaccines available to the world's poorest people and at the lowest possible cost.[91]

Among the consequences of this politically driven Texas antivaccine activism has been a steep rise in the number of children who do not receive their full complement of childhood immunizations. For the first time, almost one hundred thousand Texas schoolchildren were denied access to one or more lifesaving vaccines, almost a threefold increase in K–12 vaccine exemptions over the past decade.[92] And this number does not even include any of the more than seven hundred thousand homeschooled children.[93] We do not know the percentage of those kids that aren't receiving routine immunizations, and it is possible that hundreds of thousands of Texas schoolchildren do not receive the vaccines necessary to prevent measles, pertussis, and other life-threatening infectious illnesses.

But the worst was yet to come. The health-freedom and medical-freedom propaganda became weaponized during the COVID pandemic. In all, one hundred thousand Texans died because of COVID during the pandemic, making Texas among the worst-affected states in the nation in terms of total numbers of deaths. As Peter reported in 2022, almost one-half of those deaths (at least forty thousand) were unnecessary because so many Texans refused COVID immunizations that were extremely (90 percent) effective at preventing death and severe illness against the prevailing COVID variants.[94] Antivaccine activism became a leading cause of death in the state of Texas (and surrounding states), which had become an epicenter of the antivaccine movement thanks to the financing of antivaccine causes and PACs.[95] It did not help that the delta wave of COVID emerged in July 2021 just when the Conservative Political Action Conference was held in Dallas, Texas, where well-known antivaccine activists had been given prominent speaking roles. An article about that CPAC event and the cheering that erupted from the crowd after low COVID-vaccination rates were announced was titled, appropriately enough, "They Clapped for Death at CPAC."[96] If this sounds down-right dystopian, it's worth noting that an eerily similar scene occurs in the 1970s science-fiction movie *Logan's Run*. The film depicts a futuristic world where the population is strictly controlled. The people are treated to a public ceremony where all city dwellers turning thirty on that day compete in what is known as "carousel." Rising into the air in a rotating amusement park–like carousel, they are fired upon, to a cheering audience, earning "rebirth" if they make it all the way to the top alive. Spoiler alert: nobody ever does.

As the COVID pandemic began to subside in 2025, we saw how antivaccine sentiment spilled over to childhood immunizations. In February 2025, Texas experienced its first measles death in decades—among an unvaccinated school-age child—as a very large and dangerous epidemic lasting for weeks tore through several West Texas counties (and, eventually, dozens of other states), resulting in many hospitalizations among the unvaccinated. This is a tragedy that never needed to happen.

MERGER OF MENACE

A small group of malevolent plutocrats—the Kochs, Murdoch, Musk, and the Texas trio—have collectively impeded efforts to act on the climate crisis, the COVID-19 pandemic, or both. They have promoted climate denialism in the media and on social media, and they have blocked efforts to act on climate, setting us back decades. If policymakers had acted decades ago when the science was already conclusive, we could have decarbonized our economy more slowly, over decades, and avoided much damage, death and disruption. The opportunity cost of plutocrat-fueled inaction can be measured in billions of dollars, extreme weather disasters, and millions of lost lives.

During the COVID-19 pandemic, malevolent plutocrats contributed to a public health crisis—and hundreds of thousands of needless deaths—by financing and promoting an aggressive antiscience agenda, influencing government officials to enact highly flawed, dangerous, and unethical anti-interventionist strategies and shaping anti-vaccine and anti-stay-at-home policies. Either directly or indirectly, they intimidated prominent scientists, chased them off social media, and vilified them as enemies of the public.

In many cases, the plutocrats worked behind the scenes or through a complex web of libertarian think tanks, conservative policy institutes, PACs, and campaign contributions. Tracking the money trail is in itself a highly labor-intensive enterprise, especially when "dark money" is its main currency. By "dark," we mean its sources remain undisclosed and the funders unaccountable. A prime example is the so-called Donors Trust—essentially a money-laundering operation[97] that allows plutocrats like the Kochs to hide their political contributions.[98]

The plutocratic attacks on climate scientists and public health scientists have now merged in numerous ways. At the beginning of the pandemic, in early 2020, a dark-money group calling itself the "Center for American Greatness" penned an opinion article[99] attacking the hockey stick–like disease forecasts that leading epidemiologists had made of coronavirus cases by comparing those projections to Mike's famous climate hockey-stick curve (which, in their alternative science-denying world, had supposedly been "widely refuted").[100] The subtitle of the commentary—"There's Still Time to Find a Balance Between Public Health and the Economy"—invoked the specious premise, popular in the world of science denial, that there is somehow a trade-off between the health of our economy and the health of our people and planet.

More recently, as we were finalizing this book, the Heritage Foundation—an organization tied to the Koch brothers, the fossil-fuel industry, and Donors Trust, and a longtime promoter of climate-change denial[101]—used their Twitter account to tout a "review" of COVID science by the antiscience Brownstone Institute for Social and Economic Research. The Brownstone Institute was founded by former AIER editorial director Jeffrey Tucker to prevent "the recurrence of lockdowns." The review promoted the discredited idea that ivermectin—an antiparasitic treatment for human lymphatic filariasis (elephantiasis) or onchocerciasis (river blindness) in Africa and for deworming horses—is a viable treatment for COVID-19.[102] In Chapter 4, "The Pros," we go further down this dark avenue of substituting discredited or unproven treatments for proven life-saving interventions.

The Center for a Constructive Tomorrow (CFACT) is one of the key Koch-funded climate-disinformation outlets. In February 2022, it published a commentary by Joe Bastardi, a climate-change denier who has often butted heads with Mike (they were actually neighbors in the Pennsylvania college town of State College for several years). Titled "The Arrogance of Authority in Covid and Climate,"[103] the sad context for the screed was that Bastardi, who had lost both of his parents to COVID, was upset that hospital administrators were unwilling to entertain the use of "nonapproved therapeutics" to treat his parents, and this reinforced his disregard for scientific authority. Now, Mike may not see eye to eye with Bastardi, but he was deeply saddened that his personal loss and grief were so cynically weaponized by CFACT to advance their antiscientific agenda.

WHAT CAN WE DO?

Countering the plutocrats or their dark funding support won't be easy. To begin, we need to advocate for legislation to increase the transparency of funding sources or the activities of organizations that plutocrats endow or support. A few brave politicians are attempting to do that. Senator Sheldon Whitehouse (D-RI) has spoken out in congressional hearings, on television news programs, and in every other medium available to him to warn of the existential threat posed by dark money today.[104] He has cosponsored Senate legislation to end tax breaks for dark money[105] and the "DISCLOSE Act" that requires revelation of dark-money contributions.[106] But these efforts have been blocked by Republicans in Congress. We need far more Senator Whitehouses in Congress, and the only way that will happen is if people who actually care about these things turn out in droves to vote.

This did not come to pass in the 2024 presidential election. Supporters of Donald Trump, who ran on a proplutocrat platform ("Project 2025") of fossil-fuel extraction and climate inaction, turned out in greater numbers than those of Kamala Harris, who was committed to reining in polluters and plutocrats. Disappointingly, this even applied to youth voters (ages eighteen to twenty-nine),[107] a traditionally environmentally minded constituency. Mobilizing the youth vote is particularly important. We must convince younger voters that their future, the future of our planetary environment, and the survivability of our species truly do lie in the balance. If you're a parent of a youth, engage them in conversation about the threat posed by dark money and malevolent plutocrats who threaten our future with antiscience and disinformation.

But we also need help from others in the trenches in academia, the media, and nongovernmental organizations (NGOs) doing the hard investigative work of uncovering and exposing the hidden money trail. It takes entire organizations such as the Center for Media and Democracy to trace the dark money behind antiscience disinformation efforts. Simultaneously, we need to recruit allies from the business community who recognize the threat posed by antiscience. Most American business schools understand the critically important nature of business ethics, but the social threat posed by antiscience is a key new wrinkle that belongs in any twenty-first-century business curriculum. We must also reconsider the ethics training we currently provide graduate students and medical students and residents and extend this to climate and pandemic responses and preparedness. Finally, we must be willing to educate everyone around us—politicians, friends, classmates, students, colleagues—about the pernicious influence of plutocratic dark money. We should be having conversations about how Elon Musk has transformed Twitter—once a public square for the dissemination of scientific knowledge—into an antiscience cesspool of misinformation and disinformation. We should be talking about how the Murdoch media empire, through Fox News alone, spreads dangerous and deadly misinformation and disinformation to millions of viewers each evening. We should be communicating the pernicious and deadly impact of the antivaccine activism and climate disinformation funded by malevolent plutocrats. It constitutes a grave threat to us all and deserves far more attention than it is currently getting.

THE PETROSTATES

Continuous research by our best scientists . . . may be
made impossible by the creation of an atmosphere in which
no man feels safe against the public airing of unfounded
rumors, gossip, and vilification.

—President Harry S. Truman, September 13, 1948

A petrostate has multiple definitions and connotations, but most treat it as a state or country whose economy depends heavily on fossil fuels and carbon extraction, including revenue through oil, gas, and coal exports. The definition is easily extended, however, to a nation whose economy is primarily driven by the extraction of natural resources. True petrostates lack success in diversifying their economies and therefore experience economic boom-and-bust cycles depending on the prices of natural resources. Thus, countries such as Canada and the United States would not ordinarily be considered petrostates, even though carbon and fossil-fuel industries contribute significantly to their economies. However, some of the actions of the US government, involving the Republican Party in particular, very much resemble those of conventional petrostates, including a fossil fuel–driven policy agenda and the active promotion of climate denial and delay. We will discuss the "American petrostate" in some depth later.

Petrostates also tend toward huge disparities in income. Along with a dependence on fossil fuel–generated income, petrostates are often run by dictators, plutocrats, and oligarchs who acquire political or economic power (often without accountability) through the wealth they derive from extractive industries. An alternative, alliterative title for this chapter might instead be "Petrostates, Polluters, and Politicians," since we're really talking about a tripartite coalition of these three entities. Political corruption is indeed often rampant in petrostates, which frequently lean toward authoritarianism. The term *petro-authoritarianism* reflects an observation that "oil is an impediment to democracy."[1] Another phrase, *petro-masculinity*, pertains to strongmen or autocrats who promote fossil-fuel dependence.[2] Still another aspect of many petrostates is heavy investments in military power, with their government military expenditures rising or falling roughly with the price of their fossil-fuel reserves.[3] Examples of major petrostates include Russia, Azerbaijan, and Kazakhstan in Eurasia; Saudi Arabia, Iran, Kuwait, Libya, Oman, Qatar, the United Arab Emirates (UAE), and other Gulf countries in the Middle East and North Africa region; Cameroon, Chad, Equatorial Guinea, and Nigeria in Central or West Africa; Brunei and Indonesia in Southeast Asia; and Venezuela in the Americas. In addition, Brazil became increasingly petrostate-like under its past far-right populist president Jair Bolsonaro, who weakened climate measures, accelerated the deforestation and destruction of the Amazon rainforest, and worked to expand oil and liquefied natural-gas exports.[4]

Attacks on both science and scientists also constitute a signature feature of authoritarian regimes in general. We therefore use the term *petrostate* broadly in this book to include those authoritarian elements that might go beyond direct links to fossil fuels. This helps to explain why a petrostate might also be interested in targeting vaccines or claiming that the COVID-19 SARS-2 virus was made by scientists. There is a tragic historical legacy of authoritarian attacks on science and scientists, one that goes back to the early twentieth century when, for example, Stalin and Communist Russia viewed scientists (and other intellectuals) as threats to maintaining power and political control.[5] This thread has continued more or less uninterrupted over the past one hundred years. Today, many petrostates drive antiscience agendas, including present-day Russia, but also some regimes in the Middle East and increasingly Brazil, Venezuela, and low- and middle-income petrostates. And as mentioned above, even the United States does at times.

RUSSIA, RUSSIA, RUSSIA

Russia carries with it a historical legacy of antiscience dating back a century. Today its economy is almost entirely dependent on its continued extraction of the fossil fuels buried beneath its ground—fossil fuels that we cannot afford to burn if we are to avert catastrophic warming of the planet. And the nation is run by a brutal dictator, Vladimir Putin, who is intent not only on authoritarian domestic rule but on the destabilization of Western democracy itself. These factors combine in a perfect storm of consequences for the global spread of civilization-threatening antiscience.

Arguably more than any other single nation, Russia has gained notoriety for its horrific legacy of waging war on science and its scientists.[6] Twentieth-century Soviet Russia, in a span that began with the Bolshevik Revolution in 1917 and ended with the dissolution of the Soviet Union in 1991, was as violent as any nation in human history. The Soviet democide (murder by government) incurred tens of millions of victims during the Russian Civil War, the collectivization and Holodomor famine, and the so-called Great Purge during the first half of the twentieth century.[7] Now, more than two hundred thousand Russian soldiers may have died in Vladimir Putin's war against Ukraine.[8] Such overt disdain for human life tends to overshadow other Soviet-era horrors, including those that have occurred during the Putin regime, but it is in the USSR under Stalin when science and scientists—especially in the areas of theoretical physics and biomedicine—first came under organized, deliberate, and systematic attack.[9]

Stalin's distrust of scientists intensified during the 1930s, the period known as the Great Purge, during which he eliminated his rivals. He began by targeting and imprisoning physicists studying Einstein's concept of relativity, viewed by the Soviet leadership as antithetical to Communist ideologies.[10] Attacks on Mendelian genetics (named after Gregor Mendel, the founder of modern genetics) followed because it ran counter to the theories of Trofim Lysenko, an ambitious biologist with no formal scientific training who courted the favor of high-ranking Communist officials, including Stalin himself. The most acclaimed victim was Nikolai Vavilov, a prominent Soviet Mendelian geneticist. Vavilov openly feuded with Lysenko and did so on the grounds that his pseudoscientific ideas would eventually diminish all Russian science.[11] Vavilov's warnings proved sadly prophetic, but he paid dearly for his outspokenness and commitment to scientific truth. Vavilov was arrested, tossed into gulags, and ultimately died from starvation and exposure in a prison hospital in 1943 despite international appeals for his release.

Even after Stalin's death in 1953, Soviet scientists continued to be persecuted. The most famous was nuclear physicist and Nobel Peace Prize activist Andrei Sakharov, who became a dissident for promoting global nuclear disarmament and human rights.[12] In the mid-1980s, while Carl Sagan was raising concerns about the devastating planetary consequences of global thermonuclear war in the United States, Sakharov was doing the same in Russia. The impact of similar cautionary warnings from leading scientists in both countries helped provide an impetus for de-escalation in the nuclear arms race.

Russia's historical legacy of antiscience continues today through its assault on Western science, beginning with climate science. Fueling the attack is Russia's economic dependence on the extraction of oil and natural gas, together with its drive to undermine science as part of Russia's greater assault on Western democracy. Just prior to its offensive against Ukraine, fossil fuels provided almost one-half of Russia's total revenues and 60 percent of its export revenue.[13] That makes Vladimir Putin a highly motivated climate denier. He dismisses the notion that climate change is caused by carbon pollution, argues absurdly that we can easily adapt to it, and insists, rather disingenuously, that a warming planet would be a good thing for Russia.[14] He has done everything in his power to block any meaningful global action on climate, including tampering in the affairs of other countries.

That brings us to the final factor in the Russian antiscience perfect storm: Putin's antipathy toward Western democracy. Much like his Soviet predecessors, Putin, who attained power in 2012, has maintained an iron grip on Russia in some ways that resemble the stranglehold exercised by his Soviet predecessors. In 2024, Russian lawyer and opposition leader Alexei Navalny died in a freezing Russian Arctic prison at the hands of Putin,[15] suffering a fate similar to Vavilov's. However, Putin's approach to fostering antiscience is very different and slightly more nuanced than eliminating opposition. He simply exploits all the vulnerabilities present in the digital age. Putin has advanced his fossil-fuel agenda, for example, by exploiting cyberwarfare to block climate action in other nations and keep the rest of the world addicted to Russia's fossil fuels.

Let's start with the 2016 presidential election that saw Trump's ascendancy to the presidency. Putin viewed himself as benefiting from Hillary Clinton's defeat for both geopolitical and economic reasons.

A Trump presidency was critical to Russia's prospective partnership with fossil-fuel giant ExxonMobil. It would pave the way for a joint venture on an almost unprecedented scale between ExxonMobil and Rosneft, the Russian state oil company. Its overriding goal: to extract and monetize massive oil reserves from the Arctic, Siberia, and the Black Sea. Collectively, these reserves were valued at a half-trillion dollars.

In 2014, the Obama White House had imposed harsh economic sanctions on Russia in response to their invasion of the Crimean peninsula and subsequent annexation of the region belonging to Ukraine. This action blocked an attempted partnership between the two companies. Hillary Clinton would have maintained those sanctions, while Donald Trump clearly would not, as signaled by his appointment of Paul Manafort as his campaign manager. For years, Manafort had previously served as a lobbyist for Viktor Yanukovych, a past Ukrainian president who was tied to Putin and Russia. Among Manafort's first acts as Trump campaign manager was his alteration of the Republican platform at the July 2016 Republican National Convention, to strip language supporting Russian sanctions.[16]

Trump worked to do just that once in office. He appointed no less than Rex Tillerson—the CEO of ExxonMobil—to be his secretary of state and then began lifting the Russian sanctions. Fortunately, the Republican Party was not yet a wholly owned subsidiary of Trump, Inc., back then, and a few Senate Republicans joined with Democrats to block the effort. Past FBI director Robert Mueller would subsequently show that Russia had indeed sought to influence the election to elect Donald Trump, with complicity on the part of Trump's campaign. A half-trillion-dollar oil deal, it seems, helped to motivate the effort. Among other tactics, Russia manipulated social media, including Facebook and Twitter. Their attack plan included launching armies of bots and trolls and disseminating disinformation through government-sponsored news outlets, in an effort to divide Democratic voters during the 2016 election.[17] The Russian government aspired to siphon away supporters of Democratic candidate Hillary Clinton by convincing them to back Green Party candidate Jill Stein[18] (who at times appeared quite cozy with Putin and his inner circle)[19] or simply stay home and not vote.

Many of the attacks against Hillary Clinton during the 2016 election took the form of critical online commentary from ostensibly leftist or progressive climate advocates and activists. They criticized her climate policies, especially her position on "fracking" and natural-gas pipelines. We know now that many of the social media accounts involved were Russian trolls and bots. The idea was to drive a wedge within the environmental Left by convincing younger green progressives that there was little if any daylight between the two presidential candidates—Clinton and Trump—on the critical issue of climate action. Objectively, there was a *world* of difference between them, with only Clinton supporting meaningful global action on climate.[20] But the fact that many younger voters were misled into believing that Clinton and Trump were basically equivalent on this issue[21] almost certainly helped depress Democratic votes and turnout, handing the election to Trump—and advantaging Russia.

While we're on the topics of Russia, cyberwarfare, and climate, let's talk about a defining event at the intersection of these three things: "Climategate"—stolen emails that were misrepresented in order to manipulate the proceedings and outcome of the December 2009 Copenhagen climate summit. The perpetrators hacked into a computer server at the University of East Anglia in the United Kingdom, stole emails numbering in the thousands between the global community of climate scientists, and cherry-picked words and phrases from a small number of them, taking them out of context and misrepresenting them to suggest nefarious activity on the part of the scientists and to undermine public faith in climate science itself.[22] Mike was one of the central targets in the attack, along with UK climate scientists Phil Jones and Keith Briffa and US scientists Ben Santer, Kevin Trenberth, and Jonathan Overpeck. The common thread was that each of these scientists has played a prominent role in the Intergovernmental Panel on Climate Change.

With the hindsight we've gained, having witnessed Russia's tampering in the 2016 election, we can readily see Russia's fingerprints all over Climategate. Climategate may well have indeed constituted a test run for Russia's influence campaign in 2016 to elect Donald Trump. The same key player, Wikileaks, with its now well-established connections to Russian military intelligence,[23] helped disseminate purloined emails in both cases. Russia played a role in hosting and helping distribute the stolen emails, and recent evidence points to the computer hackers as having originated in Russia.[24] Russia of course had help. The emails were also hosted on Saudi Arabian servers. The Murdoch media—which had ties at the time to the Saudi royal family—helped amplify the various false claims to the point where mainstream print and television media eventually had no choice but to cover the faux scandal.[25]

Both Climategate and the 2016 election hack were based on the same modus operandi—for example, the use of stolen emails (from climate scientists rather than Hillary Clinton) to generate false controversies based on misrepresentations of their contents. The motive behind both Climategate and the 2016 election-influence campaign may have indeed been very much the same.[26] Like the effort to elect Donald Trump, Climategate advanced Russia's fossil-fuel agenda. It sought to undermine acceptance by the public and policymakers of the basic scientific evidence for human-caused warming at a critical moment, just days before the start of the all-important 2009 Copenhagen climate summit. The importance at the time of this summit can't be overstated. Climate advocates had branded it "Hopenhagen," for it was seen as the best opportunity yet for a meaningful global climate agreement at a time when the consequences of climate inaction were growing ever more clear and dire.

Russia has continued to hone these tactics, and the United States remains ground zero of their global assault. Given that the United States is the world's biggest legacy carbon polluter, it must demonstrate leadership. Halt climate progress in the United States, and in all likelihood, you halt meaningful global action on climate. A 2018 congressional report indicated that "Russian trolls used Facebook, Instagram, and Twitter to inflame U.S. political debate over energy policy and climate change, a finding that underscores how the Russian campaign of social media manipulation went well beyond the 2016 presidential election."[27]

Russia once again interfered in the 2024 US presidential election on Trump's behalf and arguably was the primary reason that he prevailed. Responding to the news that leading antivaccine activist RFK Jr. would be given control of public health agencies in the new administration, Democratic political strategist Simon Rosenberg issued a stern warning about the way that antiscience disinformation had been leveraged by Russia:

> MAGA/Trump/RFK/JDVance/Tucker are Russian-backed wrecking balls trying to destroy America from within. Undermining our public health systems has long been a focus of Russian info ops in the U.S. . . . sowing discord and confusion and causing more Americans to die—has been a central goal of Russian info ops in the US for years.[28]

A particular concern is the way that young male voters swung to Trump, defying their tendency in past elections to vote Democratic. According to *Wired* magazine, "Donald Trump owes at least part of his 2024 presidential election victory to the manosphere—the amorphous assortment of influencers who are mostly young, exclusively male, and increasingly the drivers of the remaining online monoculture."[29] Think petro-masculinity! *Inside Climate News* interviewed a number of these voters.[30] One individual they interviewed, Daniel Milani, a twenty-four-year-old in Chester County in Pennsylvania (one of the swing states won by Trump), said that he gets most of his news from YouTubers and specifically named the influential podcaster and YouTuber Tim Pool. Pool was recently exposed as one of the six right-wing influencers the Department of Justice found to have been unknowingly funded by Russia[31] to promote their disinformation.

It's easy to connect the dots to antiscientific disinformation here. *Inside Climate News* said that Milani indicated that "he doesn't have enough information to say whether or not human activities are the predominant factor causing global warming." If he's getting his information entirely from an online information ecosystem rife with science-denying Russian propaganda, it's understandable why he doesn't. The Washington, DC–based Center for Countering Digital Hate commissioned a survey of how new antiscience narratives pushed by online influencers are impacting younger audiences, and they found that a full 45 percent of male respondents believe that climate scientists are "manipulating data," and 41 percent agreed that climate change "is a hoax to control and oppress people."[32]

While the United States remains a major focus, Russia's efforts to divide and disrupt our virtual public space have gone global. Consider their interference, for example, in the 2016 UK elections.[33] They helped the climate change–denying[34] UK Independence Party successfully pass "Brexit," the withdrawal of the United Kingdom from the European Union, eroding the power of the EU to leverage international climate-policy efforts. They helped instigate the "Yellow Vest" revolts in France in 2018, sabotaging efforts to price carbon.[35] In that case, they incited protests and rioting in the streets by convincing French citizens that the proposed fuel tax would fall upon the working class and poor while benefiting multinational corporations.[36] In Canada, Russian bot farms were employed to convince progressives committed to environmental issues that Prime Minister Justin Trudeau, a supporter of carbon pricing, actually opposed climate action.[37] They stoked outrage against the Trudeau government during the 2019 Canadian election by focusing on contentious economic and immigration issues and, more to the point, climate policy.[38]

Russia's strategy here is clear. Their objective is to kill incipient policy efforts before they have a chance to take hold and succeed. They want to make carbon-pricing schemes politically toxic—associate them with social unrest, disruption, and economic pain. These efforts to sink global climate action, as we will see a bit later, are clearly bearing fruit.

If you guessed that precisely the same tactics are being used by Russia on the vaccine front, you guessed right. Prior to the pandemic, Russia launched both pro- and antivaccine messages to sow discord and create internal strife. The Putin government understands how vaccines have emerged as a key wedge issue in the United States (and other Western democracies) and recognizes that exacerbating the divide might further destabilize their enemies. Some of the disinformation originates from Russia Today (RT), a state-controlled news network, which also operates in multiple languages. Modern-day Russia has also promoted disinformation about the origins of serious viral infectious diseases, such as Ebola or HIV/AIDS, to suggest that these illnesses were created in American laboratories.[39] An objective in this case is to create distrust for the United States, especially among low- and middle-income countries. During the COVID-19 pandemic, there were similar Russian attempts to promote false rumors about COVID and to discredit the effectiveness or safety of American- or

European-made COVID vaccines in favor of Russia's homegrown Sputnik V vaccine.[40]

A tragic irony of this Russian-born disinformation campaign was its deadly impact on Russia itself. During the pandemic, vaccine hesitancy and refusal became widespread in Russia, with catastrophic consequences. By the fall of 2021, only 36 percent of Russia's population received at least a single shot of the COVID vaccine.[41] The results were predictable—the Russian Federation experienced some of the world's highest levels of COVID-19 mortality.[42]

In the meantime, Putin's science disinformation campaign remains as robust as ever. Trump and his supporters in the GOP have been called the "cult of Vladimir Putin"[43] by the gang on MSNBC's *Morning Joe*, for they appear to be working hand in glove with Putin to disrupt the democratic process through the promotion of antiscience. Their efforts have hindered action on climate as the crisis continues to escalate and have caused the United States to fall behind in future pandemic preparations. At the same time, Putin continues to pollute and destabilize democracies around the world on issues of climate and COVID-19 preparedness. Both Putin's hot war with Ukraine and his disinformation war on the United States and other Western democracies are mutually reinforcing. Antiscience could be considered front and center to Russian imperialism today. In the early twentieth century, the great Russian author Maxim Gorky said, "Without science, democracy has no future." Putin has taken that principle to heart, though not in the manner that Gorky likely intended.

SAUDI ARABIA AND THE MIDDLE EAST

If there is such a thing as a "marriage made in hell," surely it describes the relationship between Russian president Vladimir Putin and Saudi crown prince Mohammed bin Salman (also known as MBS). Both are brutal dictators who crush internal dissent and literally kill their critics.[44] And both, as we have already seen, have collaborated in the effort to block meaningful global climate progress. Like Russia, Saudi Arabia's primary economic asset is its fossil-fuel reserves. And like Russia, it views climate action as a threat to its economic interests and has promoted climate denial and delay, while working to block or water down international climate treaties.

The Kingdom of Saudi Arabia, like Russia, played a key role in the manufacturing of the 2009 Climategate pseudoscandal. At the time, the Saudi royal family was a major shareholder in two of the chief US media outlets (Fox News and the *Wall Street Journal*) used to promote the false Climategate claims.[45] Saudi web servers, like Russian web servers, posted the stolen materials,[46] and the Saudi government was first out of the starting gate to exploit the affair and demand an investigation of scientists.[47] Their lead climate-change negotiator, Mohammad Al-Sabban, then used it to undermine the December 2009 Copenhagen climate summit, asserting its "huge impact" on the negotiations. He further made the jaw-dropping assertion that "it appears from the details of the scandal that there is no relationship whatsoever between human activities and climate change."[48]

The collaboration between Russia and Saudi Arabia in the larger assault on climate action has been extensive. Consider what happened at the December 2023 COP28 UN Climate Summit in Dubai. Russia inserted a massive natural-gas loophole into the agreement,[49] while Saudi Arabia led the effort to oppose any agreement to phase out fossil fuels.[50] While Russian state-sponsored media like RT and Wikileaks have promoted climate denial far and wide, the Saudis have invested heavily in Rupert Murdoch's climate disinformation–promoting media empire. Both countries have engaged in cyberwar against other nations, weaponizing social media with antiscience-promoting trolls and bot armies to influence voters. The Saudi royal family[51] and Russian oligarchs[52] were among the major funders of Elon Musk's

Figure 4. Saudi crown prince Mohammed bin Salman and Russian president Vladimir Putin at the G20 leaders' summit in Buenos Aires, 2018. Photograph: Kevin Lamarque/Reuters

takeover of Twitter, a problematic development, as we have already seen, from the standpoint of climate denial and antiscience. And the cozy business relationships over the years between both Russian and Saudi companies and the Trump businesses has been well reported.[53]

The Saudis and Russia, along with oil producer Kuwait, have formed a bloc of climate-denying and climate action–impeding petrostates. With Trump as president, the United States has been part of that bloc. A headline in the *Guardian* on December 9, 2018, read, "US and Russia Ally with Saudi Arabia to Water Down Climate Pledge." The four countries—Saudi Arabia, Russia, Kuwait, and, embarrassingly, the United States—had organized opposition to a UN motion that supported the conclusions from an IPCC special report that a dramatic reduction in carbon emissions was needed to avert catastrophic climate change. This coalition of petrostates rejected the motion, agreeing only to "note" the findings.[54]

In more recent years, we have seen other Middle East petrostates effectively join the coalition, even if the corruptive nature of their role has at times been a bit more subtle. Consider the December 2023 COP28 international climate summit in Dubai, mentioned earlier. The host country, United Arab Emirates, was keen to use its role as host nation to burnish its purported green bona fides. And what transpired can only be described as a cynical exercise. An executive of the UAE's state oil firm, Sultan Al Jaber, was chosen to preside over the conference. He used that position to secure deals for the oil firm. In an interview at the time, he insisted that there is "no science" showing that a phaseout of fossil fuels is required to limit warming below dangerous levels. With such dismissive signaling from the host country, one can hardly be surprised at what happened subsequently: the UAE's petrostate ally Saudi Arabia was emboldened to join with Russia and China to oppose an agreement to "phase out" fossil fuels, and the oil cartel OPEC asked its members to block *any* deal to curb fossil-fuel dependence.

It had become entirely obvious at this point that petrostates must no longer be allowed to serve as host countries for this summit.[55] But alas, the United Nations then chose another petrostate—Azerbaijan—to host the subsequent (COP29) climate summit. And the result was all too predictable. As the Associated Press described it, Azerbaijan president Ilham Aliyev used his opening address to lash out at Western critics of his country's fossil-fuel industry: "As a president of COP29 of course, we will be a strong advocate for green transition, and we are doing it. But at the same time, we must be realistic." Aliyev called his country's fossil fuels a "gift from god."[56] Azerbaijan, like the UAE one year before, exploited its host status to promote fossil-fuel deals ahead of the climate summit.[57] Meanwhile, in what might be viewed as a cynical PR stunt, it launched the Baku Call on Climate Action for Peace, Relief, and Recovery to "support climate- and conflict-vulnerable nations."[58] This is like the arsonist promising he will buy you fire-resistant furniture.

It is worth noting that in the arena of biomedicine, the Saudi government (along with other Middle Eastern governments) has invested heavily in health care and the biotech sector and is working to make the Kingdom of Saudi Arabia a center of excellence for biomedicine in the region. Peter has interacted with Saudi medical schools and other institutions in the vaccine space, a collaboration that was solidified when he served as US science envoy for the Middle East and North Africa in 2015–2016. In addition to diversifying his economy through biotech and health-care investments, MBS has funded advances in artificial intelligence, including a new $40 billion investment in the spring of 2024. MBS and Saudi Arabia are thus somewhat of a paradox, providing support for areas of science and technology they find nonthreatening, while simultaneously working aggressively to advance antiscience about human-caused climate change and block meaningful global action on climate.

LATIN AMERICAN PETROSTATES: BRAZIL AND VENEZUELA

Over the decades whenever Peter has visited Latin American countries, he has been impressed with a near-total absence of antivaccine activism or antiscience beliefs around public health measures. He is often asked to speak to pediatric or infectious-disease societies in Mexico, Central American countries, Colombia, or Brazil; he has used such opportunities to praise their medical and scientific communities for holding the line against declining immunization rates that currently plague the United States and other Western nations. Unfortunately, things worsened following the presidential election of Jair Bolsonaro, a former military officer and far-right populist candidate elected in 2019. Aside from rolling back climate-change legislation and protections for the Amazon rainforest and its indigenous inhabitants, as COVID-19 spread across the country Bolsonaro espoused dubious health viewpoints similar to those of US President Donald Trump.

Bolsonaro's public statements downplayed the severity of the virus and the pandemic, even calling it a "fantasy" and comparing it to "the flu."[59] As COVID-19 spread, Brazil suffered from overcrowded hospitals and lacked access to adequate supplies for intensive-care units, while its leadership disparaged masks and other nonpharmaceutical interventions.[60] Bolsonaro publicly refused to take a COVID vaccine,[61] and even though Brazil has more capacity than any other Latin American country for producing its own vaccines, its two major vaccine producers failed to achieve timely production of their own indigenous COVID vaccines.[62] Political ideology and leanings strongly influenced personal vaccine decisions. Support for Bolsonaro in Brazil's 2018 and 2022 presidential elections was "significantly and inversely associated with COVID-19 vaccine uptake."[63] The nongovernmental organization Médecins sans Frontières declared Brazil's COVID response a humanitarian catastrophe,[64] one associated with the highest death tolls in the Americas.[65] In 2023, Peter was invited to speak to the Faculty of Medicine of the University of São Paulo, perhaps Latin America's top-ranked university, where he saw firsthand how many Brazilian professors and staff had become demoralized from unprecedented levels of vaccine refusal and hesitancy in their patient population. Something new and awful had happened.

The situation in Venezuela has some overlap with Brazil, but there are also important differences. Venezuela represents what some might call a "failed petrostate" due to a combination of oil dependence and declining oil production, insurmountable debt, economic instability and runaway inflation, and increasing authoritarian control under President Nicolás Maduro.[66] The consequences of this socioeconomic collapse are multifaceted and include a slowing or halting of public health measures to vaccinate the population or to implement proper insect vector–control programs.[67] As a result, there have been sharp rises in vaccine-preventable diseases such as measles, in addition to malaria, dengue, and other tropical diseases.[68] Certain indigenous populations living in the Amazon rainforest such as the Yanomami have suffered disproportionately, especially from measles, a situation with some resemblance to how measles and smallpox wiped out Native Americans in North America during European colonization.[69]

There has also been a significant exodus of the Venezuelan people and a disproportionate brain drain, meaning those with higher education have fled to other Latin American or European countries and urban centers in the United States, especially Florida and Texas. Many come to Houston, a gateway city for the Americas. Peter has met with scientists who have fled Venezuela, often for economic reasons, but also because they are under threat by the government.[70] They tell chilling stories of threats against them or their scientific colleagues, as well as thefts of scientific equipment and interruptions in pay, that make it impossible for scientists to provide for their families. In 2020, there were threats of government raids and arrests to intimidate scientists with the Venezuelan Academy of Physical, Mathematical, and Natural Sciences for accurately predicting the seriousness of the COVID pandemic and how it would affect Venezuela.[71] Such action hastened an already precipitous decline in Venezuelan scientific productivity. A 2023 report from the magazine *Science* reports that scientists who choose to stay in Venezuela are "marooned" and "mostly idle,"[72] while their laboratory equipment is stolen. Many who come to the United States have no choice but to take lower-level scientific positions as research associates or technicians even though they may have been professors in Venezuela. The Simón Bolívar Foundation is working to assist some of those scientists, including professionals who enroll in certificate programs at the National School of Tropical Medicine of Baylor College of Medicine where Peter serves as dean.[73]

THE AMERICAN PETROSTATE

Smaller mini-petrostates exist within sovereign nations. There is no better example than Texas, with its population of thirty million people and a $2.4 trillion GDP economy. Such numbers would place Texas in the top-ten global economies, ahead of Brazil, Italy, Russia, and South Korea.[74] Since the first gusher at the Spindletop oil fields near Beaumont in 1901, the "Big Rich" Texas now boasts a robust economy but one that is heavily dependent on fossil fuels. Some estimates find that as much as 35 percent of its multitrillion economy comes from oil and natural gas.[75] It is no coincidence that a new generation of Texas plutocrats linked to oil and natural gas now supports antiscience causes either directly or through PACs[76] and far-right elected officials who do their bidding. The Texas attacks on science and scientists thus couple directly to its petrostate status.

While Peter lives in the mini-petrostate of Texas, Mike lives in another—Pennsylvania. It is where oil was discovered in the United States—Titusville in the northwest corner of the state, in 1859. It is a state whose economy was built upon, and whose people derived their livelihoods from, two and a half centuries of coal mining. Today it is home to the largest natural-gas field in the country, the Marcellus Shale, recognized as such in the late 2000s.

Legacy to the state's deep history of fossil-fuel extraction, Penn State's College of Earth and Mineral Sciences—Mike's academic home for seventeen years prior to his taking a new position at the University of Pennsylvania in the fall of 2022—hosted, paradoxically it might seem, some of the world's leading climate scientists. But it also hosted leading fossil-fuel researchers. Nothing better encapsulates this seeming contradiction than the college's first-year seminars—small courses, capped at twenty students, taught by prominent faculty members. Back when he was at Penn State, Mike used to teach his seminar on Tuesday and Thursday late mornings just as geosciences colleague Terry Engelder was finishing up his own seminar, in the same room, on the first floor of the Deike Building. Often, they would chat briefly as Terry was cleaning up and Mike was setting up. Terry had become widely known for his prediction that as many as a half-quadrillion (that's a one with fifteen zeros after it) cubic feet of natural gas could be recovered from the Marcellus Shale, a rich deposit of shale gas centered beneath the surface of Pennsylvania but spanning a larger region from West Virginia to western New York. Mike's seminar focused on the science, impacts, and solutions to the climate crisis. Terry's seminar instead, one could argue, promoted the *cause* of the crisis; he touted the virtues of Pennsylvania's fossil-fuel resources. It was an awkward transition—literally—when he would finish teaching and Mike would start.

The tensions at Penn State reached a fever pitch in the wake of the manufactured pseudoscandal known as Climategate in late November 2009. Fossil-fuel interest groups had fabricated allegations of misconduct against Mike by misrepresenting the contents of the numerous stolen emails. Pennsylvania's Republican state legislature, coordinating with a front group (the Commonwealth Foundation) funded by plutocrat Richard Mellon Scaife, threatened to withhold all state funds from the university if it failed to take punitive actions against Mike. The university, feeling the pressure of these threats, initiated an inquiry into the various allegations of misconduct, followed by a formal investigation. A half-year later, in July 2010, they cleared Mike of allegations of impropriety. A year later, the National Science Foundation's Office of Inspector General—having been pressured by Republican senators to investigate Mike—would complete its own independent investigation, again finding no substance to the allegations. For the better part of two years, however, Mike remained under a dark cloud of suspicion, thanks to the efforts of fossil-fuel interest groups, their advocates in the Pennsylvania state government, and administrators feeling the pressure of fossil fuel–friendly politicians.[77]

The politics of the fossil-fuel industry have continued to impede climate policy in Pennsylvania, whether during the Republican administration of Tom Corbett from 2011 to 2015, where the governor continued to dismiss the science and promote the interests of the fossil-fuel industry,[78] or even the subsequent climate-friendlier Democratic administrations of Governors Tom Wolf and Josh Shapiro, wherein Republican state legislators and Republican-controlled courts have blocked executive efforts to regulate carbon pollution.[79] Of course, such climate obstructionism is hardly limited to Pennsylvania or Texas. Fossil-fuel interest groups and plutocrat-funded organizations like ALEC have conspired with Republican legislators across the country to promote fossil fuels and block climate policy.[80]

Indeed, the United States as a nation has flirted with petrostate status as GOP legislators and presidents have worked to codify the agenda of the fossil-fuel industry into law and oppose efforts to move toward clean energy. During the 1990s the fossil-fuel industry began to invest heavily in conservative policy groups, think tanks, and front groups advocating policies friendly to the fossil-fuel industry. Meanwhile, they funded conservative climate-denying politicians, most of who were from oil states.

Among them was James Inhofe (R-OK), who denigrated climate change as "the greatest hoax ever perpetrated on the American people" and was immortalized as "Senator Snowball" for brandishing a snowball on the Senate floor after a 2015 snowstorm, as if it were somehow disproof of human-caused warming. One of the largest recipients of fossil-fuel money in the US Senate, Inhofe used his chairmanship of the Senate Energy and Public Works Committee for years to promote climate denial, boost fossil fuels, and block any climate legislation. His antipathy toward science-based policy also extended to COVID-19. He was one of only ten senators who voted against the Families First Coronavirus Response Act, which mandated paid leave for COVID-related health issues. He also voted against the American

Rescue Plan, which provided economic relief from the pandemic, stimulus checks, and support for a national vaccination program. In a cruel twist of irony, Inhofe passed away on July 9, 2024, from a stroke after a long struggle with long COVID, which had forced his retirement in January 2023.[81]

Then there was Joe Barton (R-TX), powerful chair of the House Energy and Commerce Committee from 2004 to 2007. Among the largest recipients of money from polluters in the US House of Representatives (where he earned the nickname "Smokey Joe"), Barton used his chairmanship to promote the fossil-fuel industry, block renewable-energy projects, and attack climate science and climate scientists (we'll hear more about this later). He once cited the biblical flood myth as purported evidence against human-caused climate change.[82]

Consider as a more recent example former congressman Lamar Smith (R-TX)—yes, also from Texas. Smith was chair of the House Committee on Science, Space, and Technology from 2012 to 2018, during which he attempted to slash government funding for climate science and sought to directly insert Congress into the funding decision-making process at the National Science Foundation. He threatened investigations against climate scientists whose work he didn't like. He also subjected scientists at the National Oceanographic and Atmospheric Administration to a subpoena for their personal emails simply because he disliked a study they had published that refuted a favorite talking point of climate deniers (the false claim that global warming had ended).[83]

The White House itself was weaponized against climate action during the George W. Bush years. The *New York Times* uncovered an internal memo from the American Petroleum Institute, a fossil fuel–industry trade group, revealing it was seeking to "maximize the impact of scientific views consistent with [theirs] with Congress, the media, and other key audiences."[84] They got their wish. Bush himself actually had reasonable views on climate. He appointed a moderate proenvironment Republican, former New Jersey governor Christine Todd Whitman, to run the Environmental Protection Agency (EPA). She signaled that she intended to regulate carbon emissions under the Clean Air Act. That's when the far more senior Vice President Dick Cheney, with his close ties to the fossil-fuel industry and its front groups like the Competitive Enterprise Institute (CEI), swooped in. Cheney, working hand in glove with CEI, had Whitman removed from the EPA and secured control of the White House Center for Environmental Quality, where he installed a lobbyist from the American Petroleum Institute, Phil Cooney, as director. From this perch, Cooney would implement a fossil fuel–friendly policy agenda from within the White House. The White House, under Cheney's direction, would work to dismantle climate policies, while distancing itself from consensus statements on climate change drafted by authoritative scientific bodies such as the IPCC. The veritable fox was now guarding the henhouse.[85]

As Mike details in *The Hockey Stick and the Climate Wars*, this is the moment when the attacks on the hockey stick escalated. *New York Times* science reporter Andrew Revkin published a series of exposés showing how Cooney had been unilaterally editing government reports to downplay the seriousness of climate change. Cooney also led an effort, within the White House, to attack and discredit the hockey stick, which had become a powerful symbol of the climate threat. The attack was part and parcel of the Republican-led assault on climate science and climate action, with the hockey stick and Mike now the central targets.

With the election of Barack Obama in 2008, however, there was a dramatic improvement in the American political climate for climate. During the eight years of the Obama administration, despite concerted resistance to any congressional climate bills from Republicans, there was substantial climate progress in the form of executive presidential action. That included implementation of an EPA clean-energy plan to reduce carbon emissions in the power sector, adoption of stricter fuel-efficiency standards, the successful negotiation of a bilateral agreement between the world's two largest carbon polluters—the United States and China—and, in part as a result of that development, a critical international agreement in Paris committing the nations of the world to substantially reduce their carbon emissions.

All of that progress was threatened, however, with the most extreme instance of climate-policy whiplash to date—the 2016 election of Donald Trump as president, thanks in substantial part, as we already now know, to Vladimir Putin. The incoming Trump administration outsourced its energy and environmental policy to fossil-fuel interests and plutocrats in an unspoken agreement that was reached to secure their support. A 2020 Brookings Institution report counted more than seventy actions to "weaken environmental protections," including backpedaling on climate-change policies and lifting regulatory restrictions on the fossil-fuel industry.[86] Trump, meanwhile, announced his intention to withdraw from the 2015 Paris climate accords.[87] Yet it could have been far worse. There was massive incompetence on the part of the administration, with an almost comical lack of leadership and a near-total absence of damage-control capacity. Fossil-fuel toadies that had been handpicked by plutocrats like the Koch brothers, including Scott Pruitt at EPA and Ryan Zinke at Interior, were all forced to step down amid scandal and embarrassment. The harm done, while very real, could largely be undone if the administration was held to a single term—which they were.

Among the first actions of the new Biden administration in February 2021 was to rejoin the Paris Agreement,[88] signaling a new era of American leadership and international engagement on climate. The White House soon signed the Inflation Reduction Act into law, with its nearly half-billion-dollar investment in renewable energy, along with other legislation (the 2021 Infrastructure Investment and Jobs Act and 2022 CHIPS and Science Act) that helped accelerate the transition toward clean energy. Independent policy analysts such as the Rhodium Group estimate that these efforts put the United States on a path close to, if not quite meeting, their international commitment at COP26 of a 50 percent reduction in carbon emissions by 2030.[89]

All that progress, however, is now gravely threatened by a second Trump presidency. A recent study estimated that a second Trump presidency means an extra four billion tons of carbon pollution by 2030.[90] However, that is an incomplete assessment of the true impact. Trump's victory in the 2024 election signals to the world an end to American leadership on climate during the critical next half-decade during which emissions must be dramatically reduced. It is possible, on the other hand, that the impact will be mitigated by the fact that even fossil-fuel companies like ExxonMobil are uncomfortable with Trump's threat to pull out of global climate agreements.[91] And China appears ready to step up to the plate and help fill the vacuum created by lost American leadership.[92]

Republicans have promised even worse, however, now that they've regained control of Congress. A Republican 2024 Climate Strategy that has come to be known as "Project 2025" was effectively written by the fossil-fuel industry. It proposes dismantling climate policy across the US government in concert with the Trump administration. The plan annihilates the EPA, charged with enforcing environmental policies, while reversing the EPA's 2009 "endangerment finding" that classifies carbon dioxide as a Clean Air Act–regulated pollutant. The plan also seeks the elimination of the National Oceanic and Atmospheric Administration, which not only monitors our changing climate but also runs the National Hurricane Center. It would seek to repeal the aforementioned Inflation Reduction Act (though that might meet with more resistance than anticipated, given that much of the funding goes to red states, and Republicans will not have a filibuster-proof majority). The Project 2025 agenda also eliminates clean-energy initiatives within the Department of Energy, and it strikes climate—which is recognized by military experts as ranking among our greatest national security threats—from the agenda of the National Security Council. It supports additional Arctic oil drilling and codifies that the US government has an "obligation to develop vast oil and gas and coal resources."[93] In other words, full Republican control of our federal government is now game over for American climate action in the near term. As this book went to press in the early months of 2025, the implementation of this agenda by the Trump administration, his "DOGE" appointee Elon Musk, and the Republican-led Congress was well underway.[94]

Lest you think it's only climate action that is under assault by the GOP, rest assured they have in fact—as is almost always the case for authoritarian governments—pursued a carte blanche agenda of ideologically motivated antiscience. The Republican Party is the chief promoter of COVID-19 antiscience in the United States. During the 2010s, antivaccine activism morphed into a libertarian and right-leaning health-freedom movement that first spread through Texas and sought to discredit school requirements for childhood immunizations. These new political trends took an especially dark turn during the COVID pandemic, when the libertarian philosophy concerning childhood immunizations was adopted by the Far Right and soon extended to COVID mRNA vaccinations. Now elected GOP leaders actively sought to discredit COVID vaccines. The attacks on vaccinations peaked during the delta-variant wave in the last half of 2021, right after COVID mRNA vaccines had become freely and widely available. It is important to remember that these vaccines were approximately 90 percent protective against severe illness and deaths caused by the delta and BA.1 omicron variants.[95] Therefore, overwhelmingly, the deaths that commenced in the summer of 2021 were avoidable and occurred because so many Americans across southern states like Texas, Florida, and all those in between fell victim to the antivaccine rhetoric and claims and refused COVID immunizations.[96] In many cases, Americans had replaced their COVID-19 vaccines with ivermectin or hydroxychloroquine, believing such interventions were equivalent, even though both drugs were shown to be ineffective either as a treatment or when used as chemoprophylaxis.

A follow-up analysis revealed that COVID deaths occurred along partisan lines, with those living in conservative or red states experiencing far-lower immunization rates and far-higher death rates.[97] A *New York Times* columnist and reporter pronounced it "red Covid."[98] The details of this great red-COVID American tragedy unfolded as follows. Beginning in the spring and summer of 2021, despite the overwhelming evidence that mRNA COVID-19 vaccines were highly protective, the political Right in the United States worked to discredit COVID vaccine mandates. Sadly, for the country, these GOP-led attacks went even further, disparaging the effectiveness and safety of COVID vaccines. A predatory disinformation campaign began in the spring and summer of 2021, as mRNA COVID vaccines became freely and widely available, and when a new delta-variant wave was just starting to gain momentum in US southern states. The timing could not have been worse. As COVID vaccination rates began to stall in the spring of 2021 and the delta variant first began to accelerate, the Conservative Political Action Conference commenced in Dallas, Texas. Prominent antivaccine activists and Republican elected leaders came together in Dallas to boast about how the GOP had drawn a line in the sand and would discourage Americans from accepting COVID immunizations. According to one observer, "The [CPAC] audience upon hearing that a [vaccination] goal of President Biden's has not been met, simply erupts into enthusiastic applause. This roomful of adult human beings puts its hands together for more death."[99]

CPAC signaled the beginning of one of the great human tragedies in modern American history. Americans across the US southern states died in prodigious numbers by declining COVID immunizations during the delta wave, which lasted for most of the last half of 2021 and into 2022. Elected leaders in the US Congress disparaged their importance, safety, or effectiveness.[100] At CPAC, House Freedom Caucus members were outspoken in their antivaccine sentiments. Representative Madison Cawthorn (R-NC) compared immunizations to confiscating guns and Bibles,[101] and Representative Lauren Boebert (R-CO) echoed such defiance.[102] Just prior to CPAC, Representative Marjorie Taylor Greene (R-GA) labeled vaccinators as "medical brown shirts,"[103] using a Nazi analogy common among antivaccine activists who cynically exploit the worst genocide in world history as they compare vaccinations to the Holocaust.[104] Other House Freedom Caucus members, including Representatives Jim Jordan (R-OH), Mo Brooks (R-AL), and Paul Gosar (R-AZ), brazenly espoused antivaccine rhetoric during the delta wave, as did two US senators.[105] Senator Ron Johnson (R-WI) hosted a vaccine-injury roundtable, promoted hydroxychloroquine, and even put forward the ridiculous notion that COVID vaccines could cause HIV/AIDS; Senator Rand Paul (R-KY) also piled on with antivaccine commentary and rhetoric.[106] All these members of Congress (unsurprisingly) are also climate-change deniers.[107]

Peter became a central target of their attacks. Both Senators Johnson and Paul maligned him on social media and elsewhere.[108] Johnson accused him of being "a card-carrying member of the Covid Cartel,"[109] although how making low-cost neglected-disease vaccines for global health constitutes a cartel is mystifying. Paul stated that Peter "fears his connection to Chinese military scientists will be exposed" when, of course, he has no such ties and has never met (or corresponded with) anyone from the Chinese military.[110] Paul continued this line of conspiracy-laden ridiculousness in his book,[111] published by Regnery, a favorite book publisher of the Far Right.[112] For Peter, the fact that he was never employed by the US government and has no relationship with either the state of Wisconsin or the state of Kentucky yet was being subject to attack by their elected representatives in the US Senate was profoundly disturbing. It signaled something very dark and foreboding about the use of state power to attack scientists in America, especially given the subsequent full takeover of our federal government by the GOP in early 2025.

Earlier in the pandemic, Republican governor Ron DeSantis (R-FL) downplayed COVID on Fox News and attacked Peter after he had correctly predicted that his state of Florida would get slammed by the delta wave.[113] He boasted of having "defeated Fauci-ism."[114] As we will see in Chapter 6, "The Press," the antivaccine agenda of these elected officials was amplified on Fox News constantly during the pandemic.

These attacks had grave consequences. Multiple studies, including those led by Charles Gaba who heads ACA Signups, provided evidence for the lowest COVID immunization rates and corresponding highest death rates occurring in Republican-majority or "red states."[115] In all, Peter estimated that approximately two hundred thousand Americans died from "red Covid" because they refused to get vaccinated during the delta-variant wave in 2021 and BA.1 omicron variant wave.[116] Similar estimates were derived by scientists from the Harvard T. H. Chan School of Public Health.[117] The deaths that occurred in these waves were closely linked to political-party affiliations and sentiments. The Kaiser Family Foundation, a health-policy think tank tracking COVID immunizations, found that "an unvaccinated person is three times as likely to lean Republican as they are to lean Democrat."[118] Of course, that's only part of the story. The ill-advised rhetoric from Trump and GOP members opposing masking, social distancing, and other measures advised by public health experts certainly added to the toll. By some estimates, as many as four hundred thousand lives might have been lost due to Trump and the GOP's antiscience policies and messaging during the pandemic.[119]

In 2023, the GOP doubled down in their attempts to revise history and manufacture a troubling alternate reality.[120] First, prominent members of the GOP attempted to blame immunizations for the COVID deaths. This helps us to understand the ludicrous statements of the Florida surgeon general highlighted in Chapter 4, "The Pros," claiming COVID vaccines constitute the "anti-Christ," or efforts by the Texas attorney general to threaten lawsuits against Pfizer for allegedly misleading the public about mRNA vaccines. Representative Marjorie Taylor Greene even proposed congressional efforts to block the funding for vaccine research,[121] despite an analysis from the Yale School of Public Health finding that mRNA vaccines saved two to three million American lives.[122]

Then, just when you thought this campaign of disinformation and doubt couldn't sink to any lower depths, a House select subcommittee on the COVID pandemic asserted that US scientists invented the COVID virus—either through their own virologic research conducted in the United States or through NIH funds that were allegedly sent to the Wuhan Institute of Virology.[123] On its official Twitter site, the subcommittee announced in June 2023: "Get your popcorn ready folks . . . ,"[124] not even hiding their intentions to generate political theater and sound bites for Fox News. The subcommittee proceeded to parade prominent American virologists such as Kristian Andersen (Scripps Institute) and Robert Garry (Tulane University) in front of C-SPAN cameras in an attempt to discredit or humiliate them. Unsurprisingly, Taylor Greene was among the more outspoken members. According to *Newsweek*, in follow-up closed-door hearings in January 2024 with former NIAID director Anthony Fauci, Greene "accused Fauci of 'enhancing viruses' to create vaccines to treat them and called for the doctor to be jailed following his testimony before the panel."[125]

Later, at a follow-up subcommittee hearing, Taylor Greene told Fauci, "You should be prosecuted for crimes against humanity. You belong in prison, Dr. Fauci." Senator Rand Paul repeated Greene's taunt, asserting that Fauci "should go to prison."[126] Eventually, Peter too gained the attention of the subcommittee and was interviewed, even though he develops vaccines and has no involvement in research into COVID origins or fundamental studies on virus genetics and pathogenicity. In case there is any doubt about the GOP's intentions with these subcommittee hearings, Representative Ronny Jackson (R-TX), a subcommittee member who had formerly served as Donald Trump's White House physician,[127] promised, "There will be a public hearing eventually," adding "That will be a lot more theater."[128]

All this political theater was in cynical defiance of scientific reality. There is now overwhelming scientific evidence—published in our top peer-reviewed scientific journals, including *Cell* and *Science*—for the natural origins of COVID-19. It occurred through zoonotic spillover from animals, as highlighted in Chapter 1, with the emergence of two related SARS-2 virus strains in Wuhan animal markets; this is similar to how the original SARS emerged in 2002.[129] Even the US director of national intelligence (DNI) assessed that "SARS-CoV-2, the virus that causes COVID-19, probably emerged and infected humans through an initial small-scale exposure that occurred no later than November 2019 with the first known cluster of COVID-19 cases arising in Wuhan, China in December 2019."[130]

Of course, other senior elected officials, including former president Donald Trump, have been more than happy to pile on as the congressional witch hunts have ensued. During the 2024 presidential campaign, Trump repeated his commitment to "dismantle the deep state" by reclassifying the positions of government scientists who previously kept their jobs regardless of which political party was in power.[131] Converting scientist positions to political appointments would eliminate a cadre of dedicated professional scientists and therefore further limit the ability of the US government to fight pandemics, the climate crisis, and other global threats to the nation.

Trump in his first term in office went out of his way to downplay the severity of the COVID-19 threat and pandemic, while promoting the unproven use of hydroxychloroquine to avoid implementing necessary stay-at-home orders before vaccines were made available. We learned, through a book published by Bob Woodward in late 2024, that Trump secretly sent COVID-19 tests to Vladimir Putin during the pandemic when there was a shortage in the United States.[132] In his reelection bid in 2024, Trump threatened to defund public schools that required student vaccinations.[133] And in 2025, Trump appointed RFK Jr. to a major health post—director of the Department of Health and Human Services (HHS)—in his new administration. He was confirmed by the Republican-majority Senate.[134] Almost as if to underscore the depravity of this appointment, RFK Jr. has indicated that he may try to criminalize vaccine science he doesn't like, possibly now using his office to pursue fraud cases against scientific journals for publishing articles that support mainstream findings regarding the efficacy of vaccines.[135] The spirit of Lysenko, it appears, is alive and well—in America.

It's worthy of note that, when it comes to Donald Trump, the fruit doesn't fall far from the tree. Starting in 2024, the former president's son Donald Trump Jr. began promoting the sales of ivermectin and other spectacular medical interventions through the "Wellness Company."[136] A prominent antivaccine activist serves as their chief scientific officer; the *Daily Mail* refers to their product line as "MAGA Goop," promoting vaccine spike-protein detox, in addition to ivermectin. He, meanwhile, attempts to discredit vaccine scientists, retweeting a far-right conspiracy Twitter post calling Peter a "charlatan."[137]

In the first hundred days of the second Trump administration, as this book went to press, the antiscience attacks massively intensified. Christina Pagel, a professor of operational research at University College London, detailed thirty-five distinct actions that the Trump administration took in its first six weeks in what she characterizes as "a deliberate war on science and academic freedom" in which they have "slashed research budgets, purged health and scientific agencies, censored research, and threatened universities."[138]

A legally dubious and Orwellian new "Department of Government Efficiency" was established, installing as its point person none other than Elon Musk. Its antiscience agenda became increasingly evident in early 2025 when it began to populate its website with climate-denial propaganda from the Koch-funded Competitive Enterprise Institute.[139] DOGE is attempting to make massive cuts to the NSF and NIH budgets. It seeks to slash funding for universities, government research labs, academic health centers, and biomedical research institutions, forcing significant layoffs in our scientific workforce.[140]

Trump, Musk, and DOGE are canceling grants at the NIH and National Science Foundation and seeking to radically reduce the amount of overhead that universities and science laboratories receive to cover the costs of administering research-related activities. They have sought to defund the National Oceanic and Atmospheric Administration and Environmental Protection Agency. They have ended all funding by the Department of Defense for the study of climate change, extremism, or disinformation,[141] and they have eliminated the Office of the Chief Scientist at NASA (which was held by a Biden-appointed climate specialist).[142] They have even outlawed any reference by the federal government agencies to "climate."[143] "This is an utter disaster," said Michael S. Lubell, a professor of physics at the City College of New York and former spokesman for the American Physical Society, the world's largest group of physicists. "Climate science is dead. God knows what's going to happen to biomedicine. This marks the beginning of the decline of the golden age of American science."[144]

The most pernicious aspect, from our standpoint, is the devastating message these actions send to young and aspiring scientists: don't look to the US government to support American science. Mike saw the sadness, fear, and desperation in the eyes of the brave early-career scientists who organized the "Stand Up for Science" rally in Washington, DC, in early March 2025 at the foot of the Lincoln Memorial.[145] As Sam Stein of the *Bulwark* put it, "It's not just jobs cut and agencies gutted. It's the talent that will be lost for generations to come."[146]

Although federal judges have issued injunctions or restraining orders against many of these actions, it is presently unclear how this will play out in the courts. And this may be just the beginning of even greater cuts to US science to come over the next couple of years. Exactly what fills the void remains a big unknown. Even greater support through the industrial sector or from the venture-capital or philanthropic community is unlikely to come close to addressing the funding shortfalls. America, a country literally built on a foundation of enlightenment and science (think Benjamin Franklin and Thomas Jefferson; both were scientists, after all), may experience an unprecedented brain drain of its scientific talent to other countries—our loss, their gain.

There is, unquestionably, a coordinated, concerted attack on science by today's Republican Party—the American petrostate, if you will—with climate and biomedicine as focal points of the assault. Much of it involves party elites and corporate funders weaponizing the Republican base to serve as ground forces in their war on science. They do so by convincing their rank-and-file that evil scientists want to take away their liberty and freedom. Climate advocates, they say, will take away your hamburgers and your plane ticket to see grandma during the holidays. At the 2021 CPAC in Dallas, party faithful were warned that public health scientists will first vaccinate you and then take away your guns and Bibles.

These attacks endanger us all. Our ability to conduct virologic research is threatened, as is our ability to assess and mitigate climate risk. Beyond beating back antiscience through the electoral process and voting out politicians who promote antiscience, we will need champions in the executive, legislative, and judicial branches of government to counter antiscientific framing. Even the Biden administration was timid in this regard. There was mostly silence from those who need to be speaking out, including the White House Office of Science and Technology Policy, President's Council of Advisors on Science and Technology, and the Democratic leadership in the House and Senate. Nina Jankowicz, the former director of the US Department of Homeland Security's Disinformation Governance Board, has written about her experiences after being forced to resign due to political pressure from the GOP.[147] She has lamented how the Biden administration failed to oppose in any meaningful way the effort from the GOP to dismantle her office and board. Martin Luther King Jr. famously stated: "In the end, we will remember not the words of our enemies, but the silence of our friends." So far, the silence has—in some instances—been deafening.

The bottom line is that so far, the leadership of the Democratic Party has not prioritized standing up to the antiscience machine of the GOP and their malevolent plutocrat allies. At a practical level, this unfortunate reality means that the defense of science in America falls upon the scientists themselves. Some scientists, the two of us included, are willing to take this on. However, we are ill-equipped to face off against a well-honed machine with the financial and political support of a major American political party, tech-bro billionaires, and, as we will see in Chapters 4–6, a cadre of pros, propagandists, and a press has often been less than helpful. Ultimately, defeating this antiscience alliance cannot fall to the scientists alone or even the scientists working with their universities and scientific institutions. Such an effort will require significant political backing and financial support, possibly from science-supporting major donors, working together to create an organized counterforce for the defense and promotion of science. We will return to this matter in Chapter 7, "The Path Forward."

A TOLKIENESQUE FIGHT

It is said that those who fail to learn from history are doomed to repeat it. And that maxim undoubtedly holds with the current ideologically driven assault on science by bad state actors and compromised politicians. We hear ever more clearly the reverberations of warnings past: the inquisition of Galileo by the Roman Catholic Church in the early seventeenth century, the rise of Lysenkoism in the Soviet Union, and the attacks on Einstein's theory of relativity in Weimar Germany during the 1930s. And, closer to home, there was Joseph McCarthy and the House Un-American Activities Committee in the mid-twentieth century tasked with investigating scientists suspected of communist ties or leanings. That ultimately led to the removal of the security clearance of physicists who had served on the World War II–era Manhattan Project; we will return to that episode in a bit.

Let us revisit, in this context, the matter of the House COVID-19 pandemic subcommittee formed by House Republicans in 2023 in their attempts to politicize COVID-19 and discredit American science and scientists, for it bore an eerie resemblance to this earlier episode in American political history. The message sent from the GOP-led House subcommittee, as well as some members of the House Freedom Caucus and Senators Rand Paul and Ron Johnson, after all, was quite clear: biomedical scientists, like Peter, who worked to save lives during the COVID-19 pandemic were to be seen as public enemies. Such sentiment belongs in the same category as the political conspiracies of QAnon. It is dangerous for our future and has the potential to undermine American science for years to come. The GOP antiscience conspiracies could eventually discourage the next generation of young people from embarking on careers in scientific research—all for the political expediency of a handful of plutocrats and unhinged political extremists. As Peter pointed out in a 2023 opinion piece in the *Los Angeles Times*:

> The U.S. must also recognize how anti-science rhetoric has emerged as a new lethal force and find mechanisms to halt its advance. Pseudoscience carved a path of destruction in the U.S.S.R. almost 100 years ago, and now it is happening again. . . . We must find ways to preserve our achievements in biomedicine and support scientists, even if that means both the scientists and those in positions of power engage political leaders and challenge ideologues to reject their anti-science rhetoric and agenda. Otherwise, almost a century of America's preeminence in science will soon decline, our democratic values will erode, and our global stature will fall.

America is a nation built on science and technology, and the conspiracies promoted by House COVID subcommittee rank, ironically, among the most truly un-American activities in the recent history of the US Congress.[148]

While Peter has found himself at the receiving end of GOP-fueled attacks on COVID science (including a very tense closed-door meeting with Republican staffers and lawyers in the fall of 2024), Mike has experienced congressional assaults on his science. Back in 2005, he testified at a hostile hearing held by the aforementioned congressman Joe Barton (R-TX). The hearing was part of a larger concerted effort by fossil fuel–funded Republicans to discredit him and his research that yielded the iconic hockey-stick curve. In early 2010, the newly elected far-right Republican attorney general of Virginia, Ken Cuccinelli, engaged in what the *Washington Post* editorial board characterized as a "witch hunt" against him and the University of Virginia, where he had formerly been a faculty member.[149] Their editorial cartoonist Tom Toles ridiculed Cuccinelli's actions as a modern-day Galilean inquisition (in ways not too different from how the attacks on Peter were portrayed in a more recent Dave Whamond cartoon—Figures 5 and 6).

In his 2012 book, *The Hockey Stick and the Climate Wars*, Mike recounts his experiences—and there have been many—at the center of the various politically motivated attacks against him and the "hockey stick." He invokes an earlier episode from the late 1940s McCarthy era, previously detailed by Carl Sagan in *The Demon-Haunted World*.

The episode involved former US president Harry S. Truman and American physicist Edward U. Condon. Condon was a distinguished scientist administrator who served as both the director of the National Bureau of Standards and the president of the American Physical Society. He had *also* been a participant in the Manhattan Project, headed up by J. Robert Oppenheimer, to develop an atomic bomb. Of note, Condon has a minor role in the 2023 Academy Award–winning film *Oppenheimer*, depicted by Finnish actor Olli Haaskivi.

Oppenheimer was required to testify before HUAC in 1949. He admitted past ties to the American Communist Party but denied being involved in espionage. No punitive actions were taken against him (though his security clearance would famously be revoked four years later by a review panel). Condon, however, faced withering attacks by HUAC. He testified in 1948 before committee chair Representative J. Parnell Thomas (R-NJ). Thomas disparaged Condon as "Dr. *Condom*," calling him American security's "weakest link," or, worse, a "missing link." In subsequent testimony, a Republican congressman later chastised Condon: "It says here that you have been at the forefront of a revolutionary movement in physics called [read slowly] 'quantum mechanics.' It strikes this hearing that if you could

Figure 5 (*above*) and Figure 6 (*below*). Political cartoons comparing modern-day attacks on science to the Roman Inquisition attacking Galileo (*above* from Tom Toles; *below* from Dave Whamond)

be at the forefront of one revolutionary movement, you could be at the forefront of another."

President Harry Truman delivered the remarks quoted at the beginning of this chapter at an annual meeting of the American Association for the Advancement of Science, while at Condon's side on September 13, 1948. Condon was eventually exonerated, while Thomas wound up serving time in federal prison on a fraud conviction (unrelated to the events described here).

Truman's words resonate profoundly today, more than three-quarters of a century later. The fact that antiscience has been embraced so fully by one of the two major parties in the United States is grave cause for concern. Today's Republican Party is an authoritarian, anti-democratic political entity that opposes, for purely ideological reasons, all measures to address the climate crisis or deal with pandemic threats. If they continue with their efforts to dismantle and slash key science agencies within the government, staff remaining positions with antiscience ideologues, and engage in attacks against scientists whose findings they don't like, America's already limited ability to respond to emerging health- and climate-related catastrophes will be mortally threatened.

What happens in America doesn't, of course, stay in America. With assistance from Vladimir Putin, American antiscience campaigns have spread to other countries. Look no further than neighbor to the north Canada and the dangerous Freedom Convoy movement, especially prominent—it is no coincidence—in the oil-soaked province of Alberta.[150] US-style antivaccine sentiments are embraced by far-right groups in Europe. Antivaccine activism from the United States is contaminating low- and middle-income countries on the African continent and elsewhere.[151] Italian prime minister Giorgia Meloni expresses dismissiveness about climate change, even in the face of blistering heat and devastating wildfires.[152] Right-wing prime minister of Hungary Victor Orbán does too.[153] And Saudi Arabia and other Middle Eastern petrostates that have helped block global climate action are now reaping what they've sown, as increasingly searing, deadly heat in the region is thwarting efforts by these nations to diversify their economies by branding themselves as meccas for twenty-first-century tourism.[154] As authoritarianism spreads, so too do science denial and the disastrous impacts of that denial.

Given the critical importance of American leadership in any global policy efforts to address the existential climate and pandemic crises, we face a stark realty: the Republican Party now represents a very real threat to human civilization itself. It pains us to have to make such dramatic statements, but we are not alone. Noam Chomsky called the GOP "the most dangerous organization in human history,"[155] and for good reason. Unfortunately, the field of battle today is littered with the carcasses of moderate Republicans who opposed the Trump/MAGA march toward authoritarianism and fascism. According to early-eighteenth-century poet Alexander Pope, "hope springs eternal." So you'll forgive us for noting that historically the Republican Party has supported science. After all, the National Academy of Sciences began during the presidency of Abraham Lincoln, while President Dwight Eisenhower launched NASA. There is not much evidence that today's Republican Party shares such an outlook on the importance of science.

However, the history of American politics is one of significant change. Consider the collapse of the conservative Whig Party (the "Whigs") in the mid-nineteenth century following the repeal of the Missouri Compromise in 1854. This reopening of the North-South debate over slavery formed a schism between northern and southern Whigs, with the northern Whigs leaving to join the antislavery Republican Party. What emerged thereafter were two major parties, one, the Republicans, representing antislavery northerners, and the other, the Democrats, reflecting proslavery southerners. History, of course, has a distinct sense of irony. Another realignment, a century later, resulting from emerging Democratic support for civil rights in the 1940s would eventually result in a thorough reversal, with civil rights–averse southerners leaving the Democratic Party for the Republican Party.

Arguably, what we see today is the logical continuation of that trend. Today's Republican Party is led by a nativist demagogue, assisted by a right-wing media echo chamber that feeds its voter base with disinformation, much of it tinged with toxic race-baiting, culture-war rhetoric. The result, to wax Tolkienesque, is a massive army of "MAGA" Orcs who do battle (that is to say, turn out in elections to vote for Republicans) for their masters (the plutocrats who fund the Republican Party candidates and machinery) and their agenda (a deregulatory environment—for fossil fuels especially).

At least part of the GOP base has been misled through years now of pro-Trump propaganda into supporting autocratic rule. From a philosophic standpoint, it is the very opposite of the "limited government" traditionally espoused by the Republican Party, but it's convenient to plutocrats as long as their guy remains in charge. Where the support of top-down policy implementation ends, of course, is at the national boundary. That's not a coincidence, either. Dismissing organizations promoting international climate policy as dangerous "new world order" schemes designed to reward "globalists" (read: "Jews") once again exploits the carefully cultivated nativism and bigotry in service—big surprise—of petrostates and fossil-fuel interests. In the absence of coordinated intergovernmental policy, despots, autocrats, and tyrants run free. A world that lacks enforcement mechanisms to ensure democratic governance is precisely what petrostates like Russia want.

And that is precisely what Trump and Republicans have given them. Pete Hegseth, Trump's secretary of defense, in early 2025 ordered the US Cyber Command to stand down on all countermeasures to deal with Russian cyberwar and cyber-influence campaigns.[156]

Around the same time, the Department of Defense cut programs to defend the United States against chemical, biological, and nuclear arms at a time when Russia is expanding their own capabilities in these areas.[157] It is difficult to imagine how US policies would look different if they were literally dictated by Putin himself. Trump and congressional Republicans, in short, have chosen unilateral disarmament in the current battle with Russia and other petrostate bad actors.

So where does that leave us? To return to our *Lord of the Rings* metaphor, much as the Orcs could not be reasoned with or turned toward the side of good, the forces of antiscience, who are the product of years of disinformative propaganda, will not easily be deprogrammed. Any venture that seeks to marshal them in the battle against disinformation and antiscience is at best a heavy lift and likely doomed to failure. The battle for Middle Earth wasn't won by appeasing the dark lord and his forces. It was won by defeating them. The way forward today doesn't involve winning over the hearts and minds of those entrenched in antiscience. It involves political organizing and mobilizing the majority of Americans who still do believe in informed policymaking, international cooperation, and the preservation of our civilization and our planet. We will have more to say about this later.

4
—

THE PROS

[Bjorn] Lomborg's arguments . . . understate the potential economic impacts of climate change and exaggerate the costs of cutting greenhouse gases. And he has promoted them apparently secure in the knowledge that they will not be fact-checked by book publishers or newspaper comment editors.

—Bob Ward, policy and communications director,
London School of Economics

[Robert] Kennedy [Jr.] insists he's not "anti-vaccine," but many of his debunked arguments are straight from the anti-vaccine playbook, which he and his nonprofit have helped write. . . . He conveniently ignores the scientific literature—often vast, and of higher quality—that runs counter to his beliefs.

—Jessica McDonald, FactCheck.org,
August 2023

As we learned in previous chapters, a new group of plutocrats—some linked to the fossil-fuel industry, others seeking to promote a far-right deregulatory agenda—today funds and influences an orchestrated, well-financed, and predatory antiscience campaign. They collude, as we've seen, with politicians and polluters, especially those linked to petrostates and related authoritarian regimes. Arguably the best-known plutocrat today is Elon Musk. He wields extraordinary power not only through his massive wealth and ownership of the Twitter social media platform, but also through the unprecedented political influence he's been granted in the form of the Trump administration's legally questionable "DOGE." The plutocracy operates through a complex array of libertarian think tanks, policy institutes, conservative colleges or universities, PACs, and political campaigns. Many of these entities depend on dark money flowing through multiple streams and mechanisms. Together, the plutocrats and petrostates (and the polluters and politicians with whom they collaborate) shape or finance the key policies, legislative actions, disinformation campaigns, and miscellaneous activities that underlie the antiscience ecosystem.

That system begins with the professionals ("the pros") who generate tangible disinformation content (deliverables) legitimized by their professional reputations, including academic, licensing, certifications, or other credentials. The pros, in turn, work closely with the propagandists and the press to create a pervasive rhetorical infrastructure built on deflection, division, and disinformation (Fig. 7).

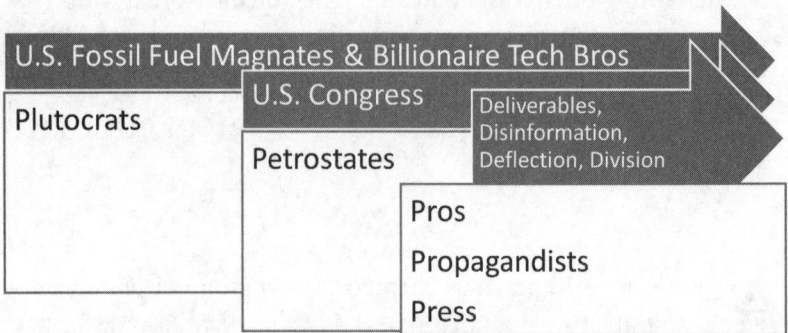

Figure 7. The antiscience ecosystem and the interrelatedness of the five *p*'s

In this chapter we focus on the pros who generate much of the content or key antiscience talking points. As a group, the pros include credentialed experts, oftentimes holding doctoral degrees, postgraduate certifications, and in some cases (current or former) major university or academic appointments.

In the climate arena, the pros include individuals with scientific credentials who have been financially lured by polluters and plutocrats and weaponized into a force to attack mainstream science and scientists. There are also the paid propagandists with no scientific credentials but plenty of media savvy and access to wide platforms.

In biomedicine there are also professional antivaccine activists and science deniers, with many pushing alternative medicines or supplements of unproven value or those that can cause actual harm. Some generate enormous revenue through the sales of such snake oil. Antiscience merch—miracle cures, nutritional and mineral supplements, and conspiracy books on Amazon.com—has become big business. Some pros hold university or hospital appointments, including prestigious academic health centers where they push the false tenets of herd immunity during the COVID pandemic, promote lab-leak conspiracies, or seek to discredit vaccines. Often, they will begin their antiscience attacks in one arena—say, promoting herd immunity—but then enlarge their remit to disparage vaccines and push the false hope of miracle cures or nutritional supplements.

The pros often have a big platform, appearing as expert talking heads on Fox News or other conservative media (discussed further in Chapter 6, "The Press") or major podcasts. Some even run for political office. Several have been tapped for leadership positions in the second Trump administration. They are known for an aggressive offense, as they spearhead attacks against scientists, working overtime to dehumanize and discredit them. They are our most aggressive and unrelenting detractors and are typically well remunerated for their efforts. In some cases, they are worth tens or even hundreds of millions of dollars.

THE DENIERS CLUB

In the climate arena, a handful of industry-funded or -connected individuals have played an outsized role in the spread of misinformation and disinformation. Some came up through the ranks of climate science itself and encountered reinforcement for their contrarian views and assertions by proindustry front groups and antiregulatory organizations and funders. Some are lobbyists who are compensated to serve as attack dogs. Yet others are ostensible "experts" who have no formal scientific training but have ingratiated themselves into the modern mass media culture, where they are often presented as experts by partisan outlets. Meet some of the key members of *The Deniers Club*.[1]

Fred Singer

The granddaddy of industry-funded climate denialism was surely S. Fred Singer. We encountered Singer earlier in Chapter 1 in the context of the climate wars of the 1990s and his attacks on climate scientist Ben Santer, whose work helped establish, by the mid-1990s, a detectable impact of fossil-fuel burning on the climate.

Carbon dioxide is not a pollutant. On the contrary, it makes crops and forests grow faster.

—S. Fred Singer

Figure 8. Tom Toles / *Washington Post*

Singer was a Cold War physicist with libertarian leanings. His ideologically motivated assaults on scientists stretched back decades. He was a fierce critic of Carl Sagan and his work on nuclear winter. He wrote that "Sagan's scenario may well be correct, but the range of uncertainty is so great that the prediction is not particularly useful."[2] Note how scientific uncertainty was used here as a pretext for inaction, when the principle of precaution actually suggests just the opposite. The uncertainty trope has been used repeatedly by special interests, partisans, and ideologues, like Singer, in seeking to undermine efforts to tackle environmental crises.

Formerly a faculty member in the Department of Environmental Sciences at the University of Virginia, Singer left the university (and academia) in 1990 and launched the Orwellian "Science and Environmental Policy Project," funded by corporate interests, including tobacco, pesticides, and fossil-fuel companies. The mission? To undermine acceptance by the public and policymakers of the science of ozone depletion, tobacco, and—last but certainly not least—climate change.[3] Singer would in fact focus almost exclusively on debunking climate science in subsequent decades, his attacks advancing the agenda of powerful corporate fossil-fuel interests.

In March 2008, Singer, with support from the fossil fuel–industry funded Heartland Institute, published a denialist response to the newly published fourth assessment report of the IPCC. Named "Climate Change Reconsidered: Nongovernmental International Panel on Climate Change (NIPCC)," it was not only titled but formatted to mimic the actual report. Embracing the "uncertainty" trope, Singer not only denied the science linking fossil-fuel burning to the warming planet but also challenged the notion that the consequences of ongoing fossil-fuel burning, as projected by models, would be detrimental. The "summary for policymakers" (titled "Nature, Not Human Activity, Rules the Climate"—itself a fallacy) claimed that "a warmer world will be a safer and healthier world for humans and wildlife alike."

ABC News dismissed the report as "fabricated nonsense."[4] Singer passed away in early 2020, but he continued to promote climate denialism to the bitter end. In 2018, he wrote an op-ed in the *Wall Street Journal* with the risible title "The Sea Is Rising, but Not Because of Climate Change."[5] Treating this absurd suggestion with the derision it deserved, Mike coauthored a rebuttal in the *Wall Street Journal* days later that began with the rhetorical question: "Would the Journal Run the Op-ed 'Objects Are Falling, but Not Because of Gravity'?"[6]

Steve Milloy

Next in our lineup is Steven Milloy. Milloy is not a scientist, but that doesn't stop him from playing one on TV. On Fox News, he is presented as an environmental science expert, while he accepts payments—and talking points—from fossil-fuel and agrichemical companies. Milloy is a self-proclaimed "junkman," calling out the purported "junk science" indicating harm from tobacco products, pesticides, and fossil fuels. He has used his perch at Fox News for years to promote climate-denialist claims and memes. You might call him a denier for hire. When a group of more than two hundred scientists (Mike included) wrote an open letter demanding that the American Museum of Natural History cease in accepting money from climate-denying plutocrat Rebekah Mercer, Milloy drafted his own letter (many of the cosignatories of which were tied to the Mercer Family Foundation) that laughably insisted, among other things, that "it is quite likely that future generations will benefit from the enrichment of Earth's atmosphere with more carbon dioxide."[7]

Mike has in fact had several run-ins over the years with Milloy. Milloy was instrumental in spreading smears about climate scientists during the 2009 Climategate episode where, as readers might recall, scientists' emails had been stolen and then quoted out of context in an effort to manufacture doubt and controversy about climate science in advance of the 2009 Copenhagen climate summit. Milloy even sponsored a Google ad ridiculing Mike that would pop up during internet searches on his name or "climate change."[8] In 2012, days before Mike was scheduled to deliver a high-profile lecture at the University of California at Los Angeles (UCLA), Milloy offered a $500 cash award to anyone willing to ask him a question "debunking" his science.[9]

We don't agree ... that man-made emissions of carbon dioxide (CO_2) and other greenhouse gases are having either detectable or predictable effects on climate.

—Steve Milloy

Figure 9. Tom Toles / *Washington Post*

Pursuing character-assassination campaigns against individuals whose work threatens his industry clients is standard operating procedure for Milloy. One classic example is Tyrone Hayes, of UC Berkeley, who spent years studying the health impacts of a weed killer called Atrazine, produced by Milloy's client Syngenta. Hayes had shown that Atrazine leads to sex abnormalities in frogs, causing them to become hermaphrodites (developing both ovaries and testes) even at doses below the current the EPA limits. Milloy went to work. He denounced Hayes with charges of scientific dishonesty, writing a commentary, "Freaky-Frog Fraud," that accused Hayes of being "determined to scare the public about atrazine."[10] Even the pope hasn't escaped Milloy's smears. When Pope Francis issued his famous papal encyclical *Laudato si'* in 2015, calling for international action on climate, Milloy immediately lashed out. He labeled him the "red Pope" and accused him of waging a "cultural revolution."[11] They say you can learn a lot about someone from who their role models are. Milloy's, by his own account, include Joe McCarthy and Richard Nixon.[12]

Marc Morano

Next, we have Marc Morano. He helped launch the original widely discredited "swift-boat" campaign deriding John F. Kerry's Vietnam military service while he worked as a reporter for the "Conservative News Service," funded by ExxonMobil. Subsequently, Morano would go on to serve as communications director for notorious climate-change denier Senator James M. Inhofe (R-OK), bringing his political smear tactics (which ironically are now sometimes generically referred to as "swift boating") to climate science. Among others, Morano has attacked James Hansen, former head of the NASA Goddard Institute for Space Studies—and one of the most distinguished climate scientists in the modern era—as a "wannabe Unabomber" intent on potential nefarious practices, including "blowing up cities," "razing cities," or "ridding the world of industrial civilization." Mike has also endured attacks from Morano, who has referred to him as a "charlatan" or someone responsible for "the best science that politics can manufacture."[13] Pot calling the kettle black, perhaps?

Climate scientists "deserve to be publicly flogged."

—Marc Morano

Figure 10. Tom Toles / *Washington Post*

Today, Morano runs the anti-climate-science website ClimateDepot.com, tied to the front group CFACT (Center for a Constructive Tomorrow) that has been funded by a combination of fossil-fuel interests and conservative plutocrats such as the Scaife family.[14] Through his website, and on right-wing news networks like Fox News, Morano produces a steady stream of untruthful claims about climate science and climate scientists. Truth and consistency matter little to him. That's why one week he's on Fox News insisting that President Joe Biden is "failing to fulfill his green agenda,"[15] while just weeks later he's lambasting Biden for his climate-forward actions, which include "forcing" an energy transition[16] and "halting" liquefied natural gas.[17] In his recent Fox News appearances, Morano has insisted that universities are trying to scare young people about climate change, that climate policy—rather than climate impacts—poses the real risk to society, that phasing out fossil fuels is an "insane goal," and that renewable energy isn't necessary and doesn't work. All in a day's work!

Bjorn Lomborg

Bjorn Lomborg calls himself the "skeptical environmentalist." Of course, he's neither. A true skeptic subjects both sides of a proposition to scrutiny without favor, critically evaluating and weighing the evidence, rather than indiscriminately rejecting findings that don't comport with their ideologically driven preconceptions. And brandishing a Greenpeace T-shirt, as he's wont to do, hardly establishes your environmental bona fides when you're penning commentaries for the *Wall Street Journal*, the *New York Times*, and *USA Today* dismissing the impacts of climate change, downplaying the role of renewable energy, and promoting the supposed virtues of fossil fuels. Lomborg's "Copenhagen Consensus Center" does not disclose its funding or that its address is a parcel service based in Lowell, Massachusetts (in other words a virtual entity).

Lomborg perfectly reflects an insidious new form of climate denialism that has taken precedence as the reality of human-caused warming becomes clear to the person on the street and very difficult to assail. He doesn't dismiss human-caused climate change outright. Instead, he denies the seriousness of the threat and the need to take meaningful action. As Mike puts it: "With a smile and a professed concern for the environment and the poor, he scolds those who would misguidedly wean us off fossil fuels and promote clean energy."[18] His claims might seem credible, but he understates climate projections and potential impacts, damages, and costs. In an opinion piece for *Project Syndicate*, he insisted that "a 20-foot rise in sea levels . . . would inundate about 16,000 square miles of coastline, where more than 400 million people live. That's a lot of people, to be sure, but hardly all of mankind. In fact, it amounts to less than 6% of the world's population—which is to say that 94% of the population would not be inundated."[19] The plight of four hundred million people, after all, is minuscule compared to the potential billions of dollars in lost fossil fuel–industry profits.

On average, global warming is not going to harm the developing world.

—Bjorn Lomborg

Figure 11. Tom Toles / *Washington Post*

A classic example of Lomborg's modus operandi relates to the Canadian wildfires of summer 2023 that blanketed major East Coast cities with smoke, including Mike's home city of Philadelphia, which had the worst air quality in the world for several days. The wildfires were part of a trend toward more widespread, dangerous, and deadly wildfires in North America tied to warming and drying. Have no fear, Lomborg was quick to announce to *Wall Street Journal* readers in a full-length commentary on August 1 that climate change had nothing to do with it. Mike published a letter of response in the *Journal* days later, noting, among other things, that Lomborg's dismissive claims had been contradicted on the news pages of the *Wall Street Journal* itself just days earlier in a news article titled "The World Bakes Under Extreme Heat" that stated, "Punishing heat this summer has helped stoke wildfires in places like Canada, Southern California, Spain and...Greece."[20] Lomborg, as Mike noted, had also misstated his own views in the commentary, falsely claiming that Mike had asserted that climate action is the "only way" to reduce vulnerability to wildfires. In the interview Lomborg was quoting, Mike had in fact emphasized the importance of adaptation, in addition to mitigation (reducing carbon emissions), in reducing our vulnerability to extreme weather but noted that there are thresholds beyond which we will exceed our adaptive capacity. Mike noted in his response that this was in fact the real lesson *Wall Street Journal* readers should take away from the unprecedented summer we had just experienced.

THE WELLNESS AND HEALTH-FREEDOM EMPIRE

Today's cadre of professional propagandists seeking to discredit vaccines or other mainstream aspects of public health or biomedicine can trace their roots to fringe elements of alternative medicine that go back two hundred years or more.

Origins of Health Freedom: Botanicals and Eclectic Medicine

In their efforts to expel the British from America, the leaders of the founding colonies acquired an understandable antagonism toward licensing restrictions, because they bore a resemblance to the Crown's trade monopolies for imports and exports. The Boston Tea Party protests against the British East India Company serve as an example.

According to health-law expert Dr. Lewis Grossman, such attitudes spilled over to an aversion to medical licensing or other forms of government oversight. Practically speaking, this meant that individuals could self-identify as medical professionals, even in the absence of formal training, and promote unproven cures and treatments. Taking advantage of this milieu was Samuel Thomson, an entrepreneurial herbalist and healer from New England who arguably became the founder of the health-freedom movement in America. He collected and organized botanical remedies of uncertain value (especially in the South and the Midwest) and transformed them into America's first health and wellness empire. According to Grossman, by 1839 approximately three million Americans, or 20 percent of the US population, bought into Thomson's system of proprietary cures (for twenty dollars), bolstered by an aggressive sales force, a network of botanic societies and conventions, and multiple editions of Thomson's book *New Guide to Health*, which was a bestseller.[21] Thomson scattered his sales force across rural America where educational attainment was low, making it easy to convince the inhabitants to share his contempt for "elitist medical practices" found in the more populated cities and towns in the Northeast.[22]

During this period in the early 1830s, French political philosopher Alexis de Tocqueville noted in his travels across the heartland of the United States that "Americans are more addicted to practical than to theoretical science," and Thomson's writings and practices fitted nicely into this viewpoint, while harmonizing with the populist presidency of Andrew Jackson.[23] Thomson's botanical empire and system of propaganda exhibited many attributes of today's pros in the public health space who rail against vaccines and condemn the elites, while presenting hydroxychloroquine, ivermectin, or nutritional supplements as spectacular cures. It is also interesting and relevant, but perhaps not surprising, that Thomson opposed smallpox immunizations;

in time, those promoting unproven cures would consistently wage a war against vaccines and vaccinations.[24]

By the middle of the nineteenth century, promoting unproven and spectacular medical cures had become a signature feature of American medicine. Those pushing these medicines represented the first generation of public health pros, and they became ingrained in the American psyche. As some might say, they became "a thing." Even though Thomson's botanical health-freedom empire declined following his death in 1843, it resurfaced through schools of botanical or eclectic medicine started by Dr. Wooster Beach,[25] until they too were finally closed after an extensive fact-finding trip published in the landmark Flexner Report.[26] The Flexner Report (the actual title was *Medical Education in the United States and Canada*, written by Abraham Flexner) called for modernizing American medical education and demanding adherence to scientific principles and foundations. However, after Flexner, other health-freedom organizations subsequently gained traction, including a group calling itself the National League for Medical Freedom, founded by a libertarian activist and magazine editor,[27] which closed in 1916, and later in the 1950s through a new group calling itself the National Health Federation. The NHF promoted laetrile, a chemical extracted from certain plants including apricot seeds, as a spectacular cure for cancer, which was later adopted in the 1970s by the far-right John Birch Society.[28]

Both groups contributed to the modern paradigm of ultraconservatives touting unproven cures and treatments. During the 1990s the NHF successfully lobbied heavily for the inclusion of specific language in the Dietary Supplement Health and Education Act (DSHEA). The language reduced the regulatory oversight of the US Food and Drug Administration (FDA) regarding dietary supplements, allowing significant exemptions to the stringent requirements for drugs or vaccines. That meant that the American public could be subject to predation by unscrupulous health promoters, including those from sectarian medical movements. Americans remained vulnerable to the next generation of cranks, ideologues, and grifters pushing unproven or ineffective medications, nutritional supplements, or remedies. Their sales would halt only if the FDA had evidence that they were harmful.[29] The parade of preposterous miracle-cure ads you're forced to watch on late-night television containing a fleeting small-print disclaimer that their claims have not actually been evaluated by any legitimate health-science body? You can thank libertarian activists for that.

Even after DSHEA, there were multiple attempts led by Representative Ron Paul (R-TX), the father of Senator Rand Paul (R-KY), to sponsor or introduce bills that might solidify and enhance the health- or medical-freedom agenda.[30] Such actions helped to set the stage for a new generation of antiscience professionals. But it needed a torch. It got two. The first was autism. The second was the COVID-19 pandemic.

Autism

Some benchmark the origins of the modern antivaccine movement to a 1998 paper in the prominent British medical journal the *Lancet* written by Andrew Wakefield and collaborators. It purported to show that live viruses that were attenuated in the measles-mumps-rubella (MMR) vaccine had the ability to replicate in the lymphoid tissues of the colons of twelve children with pervasive developmental disorder,[31] now known as autism. The implication was that the MMR vaccine could cause autism, explaining what some perceived to be a rise in diagnosed autism cases.[32] The article was ultimately retracted by the *Lancet* editors, after an investigation by the British General Medical Council revealed multiple study biases in how the children were selected for participation, in addition to financial and ethical conflicts, and a "callous disregard" for the safety of the children who received unnecessary or invasive tests.[33] The *British Medical Journal* deemed the study "fraudulent,"[34] and the British General Medical Council removed Dr. Wakefield from the medical register—halting his ability to practice medicine in the United Kingdom—for "serious and wide-ranging findings against him."[35] Following publication of the *Lancet* article, multiple epidemiological studies showed that children who received the MMR vaccine were no more likely to acquire autism than those who did not. However, the claims about links between vaccines and autism persisted, in part because they coincided with the beginnings of social media with rapidly evolving platforms such as Myspace (2003), Facebook (2004), YouTube (2005), and Twitter (2006) in the early 2000s. The false claims kept coming even though the specific vaccine or vaccine ingredient alleged to cause autism was constantly changing. New antivaccine groups formed, and a parade of unqualified individuals stepped forward, making increasingly outrageous claims. The biomedical community was now regularly called upon to debunk the latest phony assertion, a sort of scientific "whack-a-mole."[36]

In 2005 we got what you might call version 2.0 of the antiscience vaccine and autism campaign. It came in the form of an article by environmental attorney (and now head of HHS) Robert F. Kennedy Jr., simultaneously published in both *Salon* and *Rolling Stone*. RFK Jr. argued that a mercury-based thimerosal preservative (used in some vaccines) was causing autism and other neurodevelopmental conditions; this followed a medical hypothesis that made a similar claim put forward in a medical journal.[37] However, the *Salon* article was retracted by the editors because, as they put it, "subsequent critics, including most recently, Seth Mnookin in his book, *The Panic Virus*, further eroded any faith we had in the story's value."[38] Once again, the biomedical scientific community conducted well-designed epidemiologic studies to convincingly show there was no association with autism.[39] There was also a published nonhuman primate study conducted in infant rhesus macaques for mercury-thimerosal to confirm there was no link with autism.[40] Even then, a growing antivaccine lobby persisted in claiming that if it wasn't MMR or thimerosal, then perhaps it was spacing vaccines too close together or adjuvants containing aluminum found in some vaccines.[41] The point was there was a committed group of individuals and organizations that steadfastly refused to back away from their beliefs regarding a link between vaccines and autism, despite mountains of scientific evidence to the contrary—a striking example of the increasingly widespread antiscience or pseudoscience fallacy known as "motivated reasoning."

In parallel to the epidemiologic studies, an important body of evidence was building on the true causes of autism and other neurodevelopmental conditions. A more plausible and far more compelling picture emerged based on research conducted by America's premier scientific research organizations such as the Broad Institute at Harvard and the Massachusetts Institute of Technology (MIT) that identified specific autism genes involved in early fetal brain development.[42] As a result, we could now unravel the genetics of autism and how genes alter neurodevelopment. Further studies from Stanford University's medical school led by Dr. Sergiu Pasca show how these autism genes alter the migrations of neurons contained within brain assembloids or organoids—collections of brain cells that assemble into mini-brain-like structures in the labs.[43] With such a full and rich scientific framework for explaining the genetic basis of autism and its impact on fetal brain development, there was no need to invoke vaccines as a cause of autism. In fact, we now have strong evidence to the contrary and an alternative body of evidence to understand how the neurologic processes leading to autism begin well before birth.[44]

Peter became involved in this public conversation about vaccines and autism, because he is not just a vaccine scientist and pediatrician, but also the parent of a child, Rachel, who has autism and intellectual disabilities. After extensive discussions with his family, including Rachel, and several leaders in the bioethics community, including New York University (NYU) professor Arthur Caplan, Peter wrote his 2018 book, *Vaccines Did Not Cause Rachel's Autism: My Journey as a Vaccine Scientist, Pediatrician, and Autism Dad*, with a foreword by Dr. Caplan.[45] This book summarizes the compelling evidence showing there was no vaccine-autism link and explained how autism begins before an infant is born or receives vaccines, describing the genes affecting autism and other aspects of neurodevelopment. The book also reports on efforts by Baylor College of Medicine to conduct whole exome genomic sequencing to isolate Rachel's autism gene, which it turned out was similar to some of the others previously identified by the Broad Institute. The response from the scientific community to the book was overwhelmingly positive, but for a growing antivaccine lobby, Peter was labeled a public enemy and was subjected to wave after wave of harassment and threats online through emails and social media and several stalking episodes.[46] Many of those attacking him turned out to be pushing their own dubious autism cures or nutritional supplements. In some cases (such as therapies to attempt binding or absorbing heavy metals from the blood, a process known as chelation), the proposed treatments were toxic or even lethal. Nonetheless, they have become the new practices of the modern twenty-first-century health-freedom and medical-freedom pros.

Disinformation Dozen

In the beginning it was difficult to make sense of the individuals and groups who were targeting Peter or questioning vaccine-autism links. They appeared independent of each other, at least on the surface, but were united in their commitment to promoting antivaccine disinformation. However, a new nonprofit disinformation watchdog group, the Center for Countering Digital Hate, established in the United States and United Kingdom, found common threads among the antivaccine groups.[47] The CCDH's mission is to conduct research on groups spreading hate and disinformation across different platforms. It then works with the public and advertisers to alter or halt the harmful practices of tech platforms and social media companies that are amplifying the disinformation. The CCDH also works to shape public policy, and one of their major focus areas is addressing both climate and vaccine disinformation.[48] The CCDH identified a group of twelve individuals or groups responsible for 65 percent of the antivaccine disinformation and named them the "Disinformation Dozen."[49] They pointed out how antivaccine activists have professionalized, formed organizations that successfully monetize the internet, and taken in annual revenue in the tens of millions of dollars (while gaining tens of millions of followers). Collectively, the CCDH notes, the "anti-vaxx industry" "is worth up to $1.1 billion to Big Tech with 62 million followers across their platforms."[50] As the CCDH also points out, "Anti-vaxx is big business. But lies cost lives."

Among the pros in the public health arena who have promoted and spread the disinformation are those selling nutritional supplements, facilitated by the reduced FDA oversight in the wake of DSHEA. The CCDH has also identified those monetizing the internet through their *Substack* and newsletter sales, especially two prominent antivaccine activists, Alex Berenson and Joseph Mercola.[51] Berenson, a former news reporter, journalist, and novelist who downplayed the dangers of the COVID pandemic on Fox News and other news outlets before becoming an ardent antivaccine activist, was a featured speaker at the highly visible "they clapped for death" 2021 CPAC conference of conservatives in Dallas, discussed in Chapter 3, "The Petrostates," while Mercola is a Florida-based osteopathic physician who was reported by the *New York Times* to be "the most influential spreader of coronavirus misinformation online."[52]

The *New York Times* has reported that "Dr. Mercola has built a vast operation to push natural health cures, disseminate anti-vaccination content and profit from all of it" and cited his own affidavit "claiming his net worth was 'in excess of $100 million.'"[53] In a subsequent documentary titled *Superspreader*, the *New York Times* reported on headlines from Mercola's website, such as "Learn How Homeopathy Cured a Boy of Autism," "Vitamin D: The Silver Bullet for Cancer," and "Your Flu Shot Contains a Dangerous Neurotoxin."[54] Such headlines align with health-freedom rhetoric and activities of the botanic and eclectic movements in the nineteenth century. According to the CCDH, "Mercola has used his newsletter to push claims that 'More children have died from COVID shot than from COVID' while Berenson declared that 'Vaccines don't stop Covid hospitalizations or deaths.'"[55]

The assertions from the public health "pros" often aggregate into three major themes or modern-day health-freedom tenets. These were highlighted previously in Peter's book *The Deadly Rise of Anti-science* and include the unacceptability of vaccine mandates; together with the concept that choice represents an essential core American value, choice extending to unproven cures (think hydroxychloroquine or ivermectin) and nutritional supplements, and prioritizing "natural immunity" from acquired infections together with nutritional supplements, often while disparaging vaccinations; and finally demonizing or promoting mistrust in the big-pharma companies.

What Else Do We Have to Say? RFK (Jr.)

Unsurprisingly, RFK Jr. also made the Disinformation Dozen list. He is, as we already know, a prominent antivaccine activist. But when he announced his intention to run for the US presidency and oppose the reelection of President Joe Biden in 2023, he suddenly had an even larger national stage for advancing his antiscientific views on vaccines and health.[56] This trend continued after the 2024 US presidential election, when Donald Trump nominated RFK Jr. to become HHS secretary, with a mandate to reorganize its key agencies, including the FDA, Centers for Disease Control (CDC), and NIH. Confirmed in February 2025, RFK Jr., working with Elon Musk and DOGE, is seeking to gut the NIH, firing thousands of employees.[57]

Both Mike and Peter have interacted with RFK Jr. in the past. For Mike, this was in his role as a climate scientist engaged in efforts to communicate the science and its implications to the public. Mike had done several interviews with RFK Jr. over the years and was always impressed by his framing of the topic, including his affirmative stance on the underlying science and his urgency in calling for action. As a former environmental lawyer for the National Resources Defense Council, there was no "bothsidesism" in his characterization of the problem. No false balance, none of the equivocation that is all too common among media interviewers. In fact, the topic of Mike's most recent (2015) interview with him was about the latest "smackdown on climate deniers."[58] How disappointingly ironic that RFK Jr., years later, would be hiring climate deniers rather than smacking them down.[59]

Peter first became connected to RFK Jr. in January 2017 at the request of the NIH after Kennedy had been asked by Trump's transition team to advise them about vaccines. At a meeting in Trump Tower in New York, RFK Jr. told reporters (prior to Trump's inauguration) that he had been asked to chair a commission investigating "vaccine safety and scientific integrity." The Trump team subsequently appeared to soften this commitment, instead indicating they were "exploring the possibility of forming a committee on Autism, which affects many families," while adding that "no decisions have been made at this time."[60] During this period, Peter received a call and request from the NIH leadership, both NIH director Dr. Francis Collins and Dr. Anthony Fauci, the NIAID director of the NIH, to engage in a series of discussions with RFK Jr. Their rationale was that Peter, as both a vaccine scientist and the parent of Rachel, his daughter with autism and intellectual disabilities, was perhaps best positioned to help RFK Jr. understand the falsehood of any purported vaccine-autism association.

Peter's interactions with RFK Jr. began as a series of extensive telephone calls, mediated by a member of the Kennedy family, although subsequently the calls switched to email. Both the calls and the emails, while interesting, were also unproductive. Peter felt that RFK Jr. was deeply dug in and often made conspiratorial claims about the multinational pharma companies working in the vaccine space. He also conveniently ignored scientific findings to suit his own false narrative. In his emails with Peter, RFK Jr. eventually began to copy several of his colleagues who, unbeknownst to Peter at the time, were some of the most prominent antivaccine activists. Meanwhile, RFK Jr. globalized his antivaccine advocacy, potentially contributing to American Samoa's worst measles epidemic in modern times when he "flooded the area with misinformation" and Samoa's MMR immunization rates decreased to a "dangerously low level." A subsequent devastating measles epidemic resulted in fifty-seven hundred measles cases and eighty-three deaths, mostly among young children.[61] Then in 2019, RFK Jr. and his associates organized an antivaccine rally on the Upper West Side of New York at the Riverside Church, one of the iconic churches in the history of the US civil rights movement. This served potentially to sow distrust of vaccines among the African American community.[62] These activities also set the stage for an explosion of antivaccine activism during the COVID-19 pandemic.[63]

Propped up by his flawed claim of a linkage between mercury and vaccines, RFK Jr. founded the World Mercury Project, which in 2018 expanded to become today's Children's Health Defense, one of the most influential antivaccine organizations there is.[64] He previously served as chairman and chief litigation counsel. RFK Jr. also promotes conspiracies and scientifically unproven theories regarding other environmental causes of chronic illnesses, including Wi-Fi and 5G radiation as putative causes of cancer, antidepressants, and school shootings.[65] According to Jessica McDonald from the Penn Annenberg Public Policy Center's FactCheck.org:

> An undercurrent to many of Kennedy's science-based claims is that he is uniquely positioned to understand the science, whereas actual scientists are not. "I don't necessarily believe all the scientists, because I can read science myself," he told the *New Yorker* in July, just after misrepresenting the science of the COVID-19 vaccines. "That's what I do for a living. I read science critically."
>
> But time and time again, a review of the evidence contradicts Kennedy's views. He misrepresents major conclusions from papers and gets other details wrong. He conveniently ignores the scientific literature—often vast, and of higher quality—that runs counter to his beliefs. He misleads on vaccine law and misunderstands key governmental programs, consistently viewing them through a lens of conspiracy and corruption.[66]

An illustration of how he cherry-picks facts and factoids to produce fantasy narratives is provided by his comments about Peter in his 2021 book, *The Real Anthony Fauci: Bill Gates, Big Pharma, and the Global War on Democracy and Public Health*, published by his Children's Health Defense organization and Skyhorse Publishing.[67] In the book, he misrepresents Peter's actions, dismissing him as a "CNN television doctor" and risibly claiming he has sought to make criticism of Fauci a felony and that Fauci and Bill Gates have made some sort of secret deal to hand him millions of dollars. He has absurdly sought to link Peter to big pharma.

Debunking RFK Jr.'s statements is straightforward. As one of the few scientists in America who had been developing coronavirus vaccines for more than a decade, Peter was a frequently invited guest on the cable news channels to discuss COVID-19 during the pandemic. Without receiving any payment for any interview, he appeared regularly on MSNBC, CNN, *PBS NewsHour*, CBS, ABC, NBC, and even Fox News. He also appeared regularly on local news outlets in Texas. RFK Jr.'s characterization of Peter as a "CNN television doctor" diminishes his role as a scientist who co-led the development of a low-cost COVID vaccine for LMICs (as highlighted in Chapter 1, "The 1-2-3 Punch"). His comment about Peter regarding hate-crimes protection is similarly disingenuous—Peter's actual comments were in regard to protecting professional scientists from political attacks or threats to themselves and their families.[68]

Regarding the millions of dollars he's allegedly taken from the NIAID and the NIH: The NIAID is one of twenty-seven research institutes and centers of the NIH and is the world's largest supporter of infectious-disease scientific research. Almost every major scientist in America who studies microbes and infectious diseases applies for research grants from the NIAID. These grants are peer-reviewed by an external advisory committee of experts who are selected from universities and research institutes across the country. The grants are then scored and ranked and are considered highly competitive. An NIH institute director typically has no role in influencing this decision. In terms of support from Bill Gates, Peter's university has indeed received previous support from the Bill & Melinda Gates Foundation to develop a hookworm-anemia vaccine (as noted in Chapter 2, "The Plutocrats"), which is now completing clinical trials. This would be an extraordinary technology for addressing disease and poverty in LMICs and would become the second parasitic-disease vaccine in clinical use after the malaria vaccine.[69] Finally, for the implication that Peter is connected with "big pharma," the reality is quite the opposite, as Peter and his team of scientists have received international acclaim for finding ways to develop vaccines without involving the multinational pharma companies.[70]

In 2023, two journalists investigated the financial support of RFK Jr.'s Children's Health Defense to uncover a complex web of finance organizations with no obvious subject-matter interest in Children's Health Defense's antivaccine mission. These included Fidelity Charitable, a 501(c)(3) public charity that donated $1 million to Children's Health Defense in 2021–2022, as well as $235,000 to Del Bigtree's Informed Consent Action Network.[71] Bigtree is a prominent antivaccine activist and was asked to serve as RFK Jr.'s communication director for his 2024 US presidential bid.[72] Ultimately, Bragman and Kotch identified more than $15 million in donor-advised funds provided to antivaccine activist groups. DAFs can operate as pass-through organizations, allowing wealthy individuals or foundations to distribute funds quietly, or "under the radar": "DAFs are a common vehicle for funding entities like hate groups—or Covid-misinformation operations,"[73] and serve as one form of what is commonly referred to as "dark money."

Next-Gen Pros

The health-freedom movement today extends beyond Joseph Mercola, RFK Jr., and the other members of the Disinformation Dozen to include doctors associated with alternative or contrarian medical professional societies and groups. Besides the central tenets of health freedom, these organizations lean heavily on promoting spectacular cures for COVID-19 and other conditions, including ivermectin and hydroxychloroquine, despite the overwhelming evidence that these have no pharmacological benefits for COVID-19,[74] and evidence that hydroxychloroquine itself is toxic and could lead to mortality.[75] Among these organizations is America's Frontline Doctors (AFLD), whose website home page boasts "Protecting Your Medical

Freedom," while touting itself as "the nation's premier Civil Liberties Organization." *Time* characterized it as a "right-wing anti-vaccine group that claimed the U.S. government was suppressing effective treatments, hospitals were killing COVID patients, and vaccines for the virus caused cancer."[76] According to the *Intercept*, AFLD has made millions of dollars through online consultations and hydroxychloroquine or ivermectin prescriptions. The *Intercept* states, "America's Frontline Doctors . . . is working in tandem with a small network of health care companies to sow distrust in the Covid-19 vaccine, dupe tens of thousands of people into seeking ineffective treatments for the disease, and then sell consultations and millions of dollars' worth of those medications."[77] In a disturbing article, "What Price Was My Father's Life Worth?" interviewing Julie Moore about her father's choice to take ivermectin shipped to his home after a $90 telemedicine consultation with AFLD, *Time* writes:

In his final days, Moore could only watch through the glass as her dad battled delirium, trying to tear off his oxygen mask in a panic. "The worst patients we've seen are the ones that delayed treatment because they were self-medicating through ivermectin," Moore says a nurse told her. "You wouldn't believe how many people we've treated who have done this." When Moore asked if any of those patients left the hospital, the nurses shook their heads no.

The extent of what Moore calls her dad's "death by deception" only became clear after he died. In his office, she found emails and documents from AFLD outlining their "COVID protocol," printed out and annotated with characteristic meticulousness. "My dad was a highly intelligent man. He was a fantastic pharmacist," Moore says. "But he was also 82. They were very convincing, and they were lying."[78]

One individual with past connections to AFLD is Dr. Joseph Ladapo, the current Florida surgeon general under Governor Ron DeSantis,[79] who in January 2024 warned the public in Florida against taking mRNA vaccines for COVID, claiming they could modify an individual's DNA or be linked to cancer.[80] Ladapo publicly denounced mRNA vaccines as "the anti-Christ of all products,"[81] an act that was widely condemned by the mainstream scientific community.[82] Peter went on several news broadcasts to provide the scientific context for understanding why Ladapo was wrong.[83] So we increasingly saw during the pandemic how health-freedom propaganda took a two-fisted approach—both disparaging vaccines and promoting spectacular cures of unproven value, most notably ivermectin. In November 2023, the AFLD board chair and founder, Dr. Simone Gold, took to Twitter to declare Peter a member of that organization's "Deadly Disinformation Dozen" list—an unimaginative and cynical effort at projection.[84] In 2022, Gold was sentenced to a sixty-day prison term and ordered to pay a fine for storming the Capitol on January 6, 2020. She was granted a full pardon, however, by newly reelected President Trump in January 2025.[85]

Beyond AFLD, another US-based health-freedom organization that promotes the use of ivermectin is the Front Line COVID-19 Critical Care Alliance (FLCCC), led by two founding presidents, Drs. Paul Marik and Pierre Kory, both internists and critical-care physicians. According to *STAT News* and their investigative reporters: "Ivermectin is highly lucrative for the doctors who sell it. Members of Front Line Covid-19 Critical Care Alliance and America's Frontline Doctors offer access to the drug online via telemedicine, for high fees, and often without insurance coverage. . . . Kory has launched his own 'advanced Covid-19 care center,' with a 'specialized focus' on long Covid consultations. An online appointment with a member of his team costs $1,250, which includes an initial video visit and two follow-ups, while meeting with Kory himself costs $1,650."[86]

In the meantime, the FLCCC has received DAFs from Vanguard Charitable and Fidelity Charitable.[87] According to news reports in January 2025, the American Board of Internal Medicine revoked the board certifications of both Drs. Marik and Kory, and possibly other antivaccine physicians who promote ivermectin or other unproven treatments for COVID.[88] Along similar lines is the Wellness Company, described by McGill University as "a striking example of the very lucrative libertarian medical movement that claims to stand against the profit-motivated pharmaceutical industry while replacing drugs with expensive dietary supplement."[89] Some of the physicians linked to these organizations make a practice of denouncing Peter on social media and media platforms. Often, the pros seek to call attention to themselves by demonizing mainstream scientists or by turning them into soft targets. Stoking the faux-outrage machine to build a base around an esprit de corps, even when it is based on pseudoscience, is another commonly used tactic. In an interesting twist, the FLCCC changed its name in January 2025 to the Independent Medical Alliance, specifying Dr. Joseph Varon as its president and chief medical officer. Dr. Varon is a Houston-area critical-care physician and someone Peter has known and interacted with, maintaining a friendly and collegial relationship, during the COVID pandemic. Possibly, it is a positive sign. One can hope, anyway.

The Brownstone Institute is yet another dark money–funded[90] antiscience organization that generates health-freedom disinformation. It has been accused of "waging information warfare on public health efforts to tackle the COVID-19 pandemic." Their founder and president (previous editorial director of AIER) has worked closely with the Great Barrington Declaration coauthors.[91] Dr. David Gorski, a cancer surgeon and physician-scientist at Wayne State University's medical school, has reported on their litany of all too familiar conspiracy theories about vaccines and COVID origins and their touting of unproven cures.⁹² On its website the Brownstone Institute lists the Independent Medical Alliance as one of the "Friends of Brownstone," along with the *Epoch Times*, a far-right newspaper linked to the Falun Gong new religious movement. Their rhetoric is highly conspiratorial in nature, "suggesting that information about [vaccine] side-effects was being suppressed."⁹³ Among the titles of the articles on their website are "Conspiracy Theorists Were Right About Climate Lockdowns" and "Contaminated: We've Been Their Lab Rats All Along."⁹⁴

There is a persistent trend of antivaccine activism and COVID misinformation that is linked to the lucrative health-freedom, wellness, and nutritional-supplement industry with historical ties that go back to early America. The movement taps into something much older and more ingrained in the American psyche, but has evolved from the time of Samuel Thomson to become a sophisticated network of shadowy groups that promote antivaccine conspiracies and magical and unproven cures and treatments for COVID-19.

The prescient 2011 film *Contagion* imagined a chillingly COVID-like global outbreak, involving a novel zoonotic respiratory virus spread through multiple spillover events from bats to pigs and then to humans. Public health experts work against the clock to develop a vaccine for the virus, which threatens to kill hundreds of millions of people around the world. The key antagonist is antiscience conspiracy theorist and blogger Alan Krumwiede, who profits from the online sale of bogus homeopathic cures derived from forsythia, which he falsely claims to have used to cure himself of the virus. Hordes of people seeking forsythia swarm and overwhelm the pharmacies. The lead protagonist, meanwhile, is CDC scientist Dr. Ally Hextall, who develops a successful vaccine and saves the day.

Today's antivax pros (with considerable help from our media, as we'll learn in Chapter 6, "The Press") have managed to perversely flip this script, making public health scientists like Tony Fauci and Peter the villains, while presenting themselves as the heroes. They include members of the Disinformation Dozen and groups such as AFLD and FLCCC, with the Brownstone Institute providing them with quasi-academic and think-tank cover. Social science research has revealed the outsized ability of physicians and other credentialed experts to damage public health owing to the trust the public has placed in them. What furthers the damage today is the way the pros are able to make public health misinformation go viral, if you'll forgive the pun, in the era of social media.[95] The flow of dark money, including DAFs, is a pernicious new ingredient as well. Walker Bragman and his colleague Alex Kotch deserve enormous credit for tracking the dark-money transfers. One important way to reduce the influence and spread of disinformation by the pros is to better regulate DAFs, demand greater transparency, or, ideally, halt the flow of funds to organizations committed to damaging public health or attacking scientists. However, as Bragman and Kotch point out: "Cutting off the funding spigot could help . . . but doing so is no easy task. Although DAF sponsors have full legal control over the funds they distribute, getting them to stop funding anti-vaccine and anti-public health organizations may prove difficult."[96]

DEFEATING THE DOGS OF WAR

As we have seen with both climate and COVID-19, there are rogues' galleries of professional disinformers—the pros we speak of—who have promoted antiscience aimed at undermining public understanding of the underlying science. In later chapters, we will see how this new professionalization of antiscience couples with inauthentic social media accounts, podcasters, an "intellectual dark web," and the mainstream media to create a dangerous force that targets both science and scientists.

In the case of climate, the players are invariably connected with fossil-fuel interests whose business model is fundamentally incompatible with the needed decarbonization of our economy. With COVID-19, they tend to be funded by a snake-oil industry of pseudoscientific cures marketed to the vaccine averse, exploiting distrust in mainstream public health science and vaccines. The pros and plutocrats are thus seen to be well-enmeshed cogs in a finely tuned science-denial machine. Indeed, this synergy challenges simple solutions to the problem. One could try to cut off their funding source, but, as we saw in Chapter 2, "The Plutocrats," the dark-money apparatus put in place by billionaire plutocrats makes this essentially impossible to do without eliminating dark money itself. We can't just target one head of this antiscience hydra. We have to target the entire beast.

While fundamental political change will ultimately be needed to accomplish that task—and we speak to this later—there are nonetheless things we can do meanwhile in an attempt to stem the tide of antiscience disinformation pushed by the pros. Professional credentialing organizations and university administrators can clamp down on those who leverage their academic positions or credentials to promote disinformation. Recently, state licensing boards for the practice of medicine have begun to suspend or revoke medical licenses. In addition, the American Board of Internal Medicine, the organization that oversees board certification in the specialty of internal medicine in the United States, has successfully revoked this credential from some of the worst offenders. However, because university faculty and professors are—and should be—afforded a high degree of academic freedom, reining in those who make unsupportable claims is challenging. Academic freedom goes only so far, however. Where human mortality is involved, such as during a pandemic or as a result of deadly climate change–amplified weather extremes, we need a broader ethical framework of professionalism and ethics. Such efforts could help rein in the pros. But what about another class of bad actors—*the propagandists*? That's the topic of our next chapter.

THE PROPAGANDISTS

Twitter, for all of its many flaws, was once a vital breaking news service. It is not that now. It's not entirely clear *what* it is, beyond a toxic cesspool increasingly made in the twisted image of its deeply unwell owner. Changes to its verified user system, Musk's decision to open the floodgates to bigots and trolls, and his own presence on the site have destroyed any utility it once had. It is now a source of endless misinformation and propaganda.

—Alan Shepherd, *New Republic*, December 2023

B y *propagandists*, we mean the online trolls, bot armies, and cadre of fake experts who collaborate with, or are employed by, or *deployed* by, plutocrats and petrostates (and polluters) to spread misinformation and disinformation, primarily online. The distinction between the *propagandists* and the "pros" is somewhat subtle and, at times, a bit blurry. The pros tend to be credentialed (and sometimes fallen or contrarian) experts who are tasked with generating much of the disinformation content, while the propagandists' role is to disseminate and spread disinformation. The propagandists often rely heavily or exclusively on creating viral content in the newer media environments of podcasts and what we term *antisocial media*—social media platforms weaponized by bad actors to promote ideologically motivated antiscience. But the pros and propagandists typically work synergistically. The pros bring an air of authority to a particular talking point, meme, or framing, while the propagandists help make sure it goes viral. Further blurring the distinction between pros and propagandists is their tendency to tag-team, wherein the pros make a false, although seemingly authoritative, statement, then feed it to propagandists who unleash it on an unsuspecting public. And if at times one is confused as to who is a pro versus a propagandist, well that's partly intentional.

Since Elon Musk took over Twitter, rebranded it "X," and invited back many of the antiscience trolls who were previously kicked out, this social media platform has become a weapon of choice in the attack against both science and scientists. For some, it is a "a toxic cesspool increasingly made in the twisted image of its deeply unwell owner," Elon Musk.[1] In the antiscience space, Musk maintains a dual role, in fact, as both a plutocrat and a propagandist.

Other propagandists run ideologically themed podcasts, which are dogmatic and often uncompromising, and use excessive or inflammatory language. Many are ideologues who are skilled in the tools and techniques of propaganda. Alex Jones is a prime example, a quintessential propagandist who, through his *InfoWars* podcasts, promotes antiscience views on both climate and COVID. He has engaged in frequent attacks against both Mike and Peter. In 2024 a Texas bankruptcy judge ordered Jones—a notorious Sandy Hook school-shooting denier—to liquidate his assets, including *InfoWars*, after a successful lawsuit was brought against him by the Sandy Hook families.

Jordan Peterson is another example. Peterson has reported ties to the Heritage Foundation and fracking billionaires and has promoted an agenda that aligns with Project 2025 while dismissing climate science as a "pseudo-religion."[2] Some might classify him as a pro because he is an emeritus psychology professor at the University of Toronto and fashions himself (or many consider him) as a public intellectual. But he has no credentials in climate or biomedical science and is better characterized as a propagandist for the antiscience disinformation he promotes.

These days Joe Rogan, a former martial-arts color commentator and entertainer turned podcaster, ranks among the most influential of the lot. Originally rather mainstream in his views, by the last half of 2021 or early 2022, Rogan had become a full-on equal-opportunity promoter of climate denial and COVID antiscience. He has invited each of us on his podcast to debate science deniers—a practice we don't engage in because it grants them unearned stature. Peter laments Rogan's pivot toward antiscience since he once considered him a friend and even appeared twice on his show prior to Joe's descent and pivot to antiscience propaganda.

Swiss science blogger and computational biology and bioinformatics expert Phil Markolin has categorized antiscience propagandists into at least eighteen different archetypes divided in three different categories, including the merchants of doubt, merchants of ambiguity, and merchants of confusion.[3] The term *merchants of doubt* refers to the classic book of that title by science historians Naomi Oreskes and Erik M. Conway.[4] Published in 2010, the book reported on the practices used by earlier antiscience campaigns to discredit studies on the health dangers of tobacco, air pollution, environmental degradation, and of course climate change.

Sadly, these approaches have since been fine-tuned to land us where we are today with the attacks on pandemic threats and climate science and nearly every matter where science and ideology might collide—be it the rights of women to reproductive choice, the gun-violence epidemic, or even the study of disinformation itself. Our colleague Renée DiResta, formerly of the Stanford Internet Observatory (SIO), has written extensively about the rise of disinformation and online propaganda. She is also someone who has found herself in the crosshairs of the antiscience attack machine. After she went on Joe Rogan's podcast in March 2019 to talk about Russian bots and troll farms, she found herself besieged, ironically, by Russian bots and troll farms. She has also been attacked by right-wing media and, like Mike and Peter, Republican politicians and front groups tied to the Kochs and dark money–laundering operations such as Donors Trust.[5]

For the propagandists, a key feature of the antiscience playbook is its reliance on conspiracy theories or conspiracy-filled rhetoric. A social psychology research group at the University of Paris notes how low critical-thinking abilities, in many cases related to low educational attainment, may be useful predictors of those who are susceptible to conspiratorial ideation.[6] In this sense, the propagandist attacks on science and scientists represent a form of predation targeting individuals or groups lacking education. In 2021, the Kaiser Family Foundation noted that the less educated were more likely to refuse COVID immunizations,[7] while a 2018 Pew Research Center survey found that less educated people are less likely to see climate change as a threat.[8] Together, propaganda and conspiracy theories constitute two essential elements of the propagandist's powerful disinformation arsenal.

CONSPIRACY THEORIES

Over the years, both Mike and Peter have come across many well-intentioned people who nonetheless succumb to false conspiracy beliefs. There is no question in our minds that the COVID-19 pandemic was an accelerant of such attitudes. Social scientists will no doubt be studying the precise reasons for a long time, but it is plausible they include social isolation, feelings of loss of control, and other forms of stress. Once conspiracy beliefs take hold and critical-thinking skills decline, individuals become easily weaponized as vectors of misinformation. Add grievance and anger, mixed with bigotry and white rage, sexism, homophobia, and antisemitism, and you create an incendiary brew that is easily ignited. The victims of misinformation campaigns are readily marshaled to propagate conspiracy theories through their social networks. A subset of them may coalesce into an army of angry, misinformed, and even violent individuals (think January 6).

Much of the conspiratorial ideation today feels uniquely crafted to feed narratives of climate or pandemic denial and inaction—so much so, in fact, that it was the topic of a course Mike taught with his University of Pennsylvania colleague Kathleen Hall Jamieson, director of Penn's Annenberg Public Policy Center and cofounder of FactCheck.org. The graduate students in the class were tasked with identifying and debunking the leading climate conspiracy theories encountered on social media today. It is challenging to even keep up with the major conspiracy theories in cyberspace, and of course as soon as you think you have a handle on them, another one soon arises. Here we offer a typology of the major conspiratorial theories. While many of them began in the realm of climate denial, several have expanded to embrace pandemic and COVID conspiracies in recent years.

New World Order

A common theme is how powerful actors—be it the pope, the United Nations, the World Economic Forum, the "deep state," and, especially, "*globalists*"—have fabricated the climate crisis in an effort to further their agenda of a socialist new world order. Today, Alex Jones is a leading promoter of this conspiratorial worldview (he simply abbreviates it "NWO" in his social media postings and podcasts).

Peter remembers the first time Jones accused him of being a major actor in the NWO and doing a Google search in order to find out what it actually meant.

Space Lasers, HAARP, Chemtrails, and Cloud Seeding

We learned earlier of MAGA Republican and Representative Marjorie Taylor Greene's infamous Facebook post claiming that the antisemitic Right's eternal boogeyman, the Rothschilds, used space lasers to ignite the historic California wildfires of 2018. But several other popular conspiracy theories feed into this same basic narrative. They include the belief that a nebulous web of governments and nongovernmental organizations has managed to hide huge natural reservoirs of water to manufacture water scarcity in various regions of the globe[9] (this idea actually entered into popular culture as a key plot device in the 2008 James Bond movie *Quantum of Solace*). Or that such groups and individuals are impacting the environment by deploying novel technology such as "HAARP" (high-power radio-frequency transmission used to study the structure of the ionosphere) and "chemtrails," mind-controlling chemicals sprayed into the atmosphere, as ostensibly evidenced by the white streaks seen in the wake of aircraft. Those are actually "contrails"—small cloudlike structures formed from the water vapor released by the combustion of jet fuel. Conspiracy theories about HAARP and "space lasers" have their origins in Russian-ministry and state-owned media propaganda dating back two decades.[10]

A new variant on this theme was on display in mid-April 2024. A massive flooding event occurred in Dubai, located in the heart of the Saharan desert. The claim that it was the result of "cloud seeding" by the UAE government went viral on social media. Judith Curry, a former professor of earth sciences at Georgia Tech University who now, among other things, produces videos[11] downplaying the threat of climate change for "Prager U"—an ultra-right Potemkin university—took to Twitter to promote this false framing: "Torrential rains in Dubai—unintended consequence of cloud seeding?"[12] Her tweet garnered hundreds of retweets and more than seventy-five thousand views. As for the absurdity that cloud seeding was behind this event, the Associated Press noted that "one way to know for certain that it was not caused by tinkering with clouds is that it was forecast days in advance," with computer models having forecast several inches of rain (a year's worth of rainfall for that region) a full six days earlier.[13] The false "cloud-seeding" framing provided a convenient distraction from any discussion of what is *actually* causing increasingly extreme rainfall events—a warmer, more moisture-laden atmosphere, and a "stuck" jet-stream pattern, favored by human-caused climate change, that keeps deep extratropical cyclones like the one that impacted Dubai in place for days on end.[14]

In October 2024, during the homestretch of the 2024 presidential election, Russia helped amplify and spread false claims about a pair of damaging and deadly landfalling hurricanes—Helene and Milton—in their attempt to help elect Donald Trump.[15] While most of the false and misleading claims centered on the purported lack of response by the Biden administration, one of the more viral conspiracy theories maintained that the hurricanes had been generated by Democrats in an effort to target "red" states like Florida. And yes, Congresswoman Marjorie Taylor Greene was one of the main spreaders of this ridiculous claim.[16] As if we needed another example of hypocrisy and irony, the fact that the very same people who dismiss human-caused climate change (which scientific studies demonstrate to have substantially worsened the impacts of the two storms)[17] are convinced that humans can nonetheless manufacture weather phenomena out of thin air takes it to a whole new level. And if you're looking for more irony still, Florida's Republican governor Ron DeSantis, rather than confronting the monumental threat posed to his state head-on, has demanded that climate-change references be removed from Florida textbooks[18] and rejected $350 million in federal funds for climate resilience that would have helped Floridians deal with the devastating onslaught of catastrophic hurricanes they've experienced over the past two years.[19]

Agenda 2030 vs. Project 2025

While climate denial is a dominant theme in these conspiracy theories, a broader war on science is at play. Unsurprisingly, some of the leading climate-denial protagonists embrace numerous conspiracy theories. Consider the example of Tony Heller, a climate denier who was banned from Twitter for promoting COVID-19 misinformation. Heller's Twitter account was resurrected when Elon Musk took over the company and restored numerous previously banned antiscience-promoting accounts in late 2022.[20] Today Heller has a healthy 75,000 followers on Twitter and another 125,000 subscribers to his YouTube channel. Not only does he promote climate denial and COVID-19 disinformation, but—rather offensively—he also spews conspiracy theories about the 2012 Sandy Hook elementary school massacre in which twenty first graders were murdered by a lone gunman with an assault weapon.[21] There are peer-reviewed studies in the social science literature that demonstrate that climate denialists tend to traffic in generally distasteful messaging. Misogynist and authoritarian attitudes correlate with opposition to climate policy.[22]

An overarching conspiracy theory involves the 2030 Agenda for Sustainable Development, simply referred to as "Agenda 2030."[23] Launched in New York at the UN General Assembly annual meeting in late September 2015, it proposes an ambitious vision of "a world of universal respect for human rights and human dignity, the rule of law, justice, equality and non-discrimination." An anodyne program to foster a set of global sustainable development goals, Agenda 2030 nonetheless possesses all the elements for a far-right conspiracy theory. Like the reviled IPCC, it operates under the auspices of the United Nations, feeding classic "one world government" or "new world order" fears. And its goals of poverty reduction, alleviating hunger, global public health, and universal equity play into plaintive cries of "socialism." Many of the attacks on Agenda 2030 start with the right-wing Heritage Foundation,[24] filtering down into social media and other avenues, where the propagandists operate.

It is worth appreciating the masterful if deeply cynical projection on display here by the architects of this particular conspiracy theory, as "Project 2025"—which they have helped create—is everything they falsely claim "Agenda 2030" to be. While Agenda 2030 represents an agreement among all UN member states, Project 2025—a right-wing plan to turn over the reins of the US government to polluters—was hatched by the aforementioned plutocrat-funded Heritage Foundation, written by a shadowy group of a hundred former members of the Trump administration and other conservative activists, and funded by plutocrat dark money.[25] It's not a conspiracy theory but an actual conspiracy. Whereas Agenda 2030 supports universal goals of equity, social justice, and environmental sustainability, the similar-sounding Project 2025 promotes inequality, injustice, and environmental devastation.[26]

The Great Reset

A closely related topic of conspiratorial ideation is the "Great Reset" initiative, a global economic recovery plan launched in June 2020 in response to the COVID-19 pandemic. Sponsored by the World Economic Forum (WEF) and promoted by Britain's (then) Prince Charles, the objective was to not just rebuild the global economy in the wake of the economically devastating pandemic but do so in a manner that prioritizes sustainable development. Like Agenda 2030, though, the Great Reset called for the coordinated actions of large multinational organizations and emphasized principles of environmental sustainability. All the usual right-wing conspiracy trigger words were there: The WEF! Prince Charles! COVID-19! Sustainable development! And we can see how distrust of public health science and vaccines might readily be woven into conspiratorial narratives. Were COVID-19 and the vaccinations and lockdowns just a pretext for a radical clean-energy agenda? The propagandists certainly wanted us to think so.

You might recall from Chapter 4 ("The Pros") the Brownstone Institute, a group funded by dark money. It provides a prime example of the potential blurring between pros and propagandists, since both operate in the Brownstone space. The Brownstone Institute has played a key role in the promotion of antivaccine messaging, including the infamous Great Barrington Declaration. It has also played a role in the promotion of climate denial. And it has sought to tie these two things together. In a 2023 web post on the Brownstone Institute's website titled "Covid Emergency, Climate Emergency: Same Thing," Brownstone affiliate W. Aaron Vandiver warned that a climate-emergency declaration "could infringe on civil liberties and human rights" and warned of the threat of "climate lockdowns."[27]

This was not a one-off. In April 2024, Mike received an award from the EcoHealth Alliance, a nongovernmental organization that advocates for a "one health" approach to addressing the health of people and the planet. For instance, the EcoHealth Alliance tracks the emergence of high-pathogen zoonotic viruses from bats, such as coronaviruses or the Ebola virus, to identify processes by which humans become infected to ignite pandemics. The previous year's recipient was Peter. An individual named Toby Rogers, who identifies himself as a fellow of the Brownstone Institute in his Twitter profile, tweeted about the development: "Wait, hold the f*ck up. Michael Mann (creator of the global warming 'hockey stick graph') spoke at an

EcoHealth Alliance conference this week!? The same EcoHealth Alliance that laundered the money to the Wuhan Institute of Virology to create SARS-CoV-2. And Hotez reposted this!??"[28] It had hundreds of retweets and more than sixty thousand views. A week later, Rogers posted that Peter and Mike are in cahoots with "corporate actors with malevolent intentions" (irony is indeed dead), insisting that "global warming" has become a cover for the proliferation of the "biowarfare industrial economy."[29]

This conspiratorial narrative—that COVID-19 policy and climate policy are part of an interconnected plot to take away freedom and liberty—has been heavily promoted by the pros, so it's hardly a surprise we encounter it from the propagandists, to whom they throw forward passes. Earlier, we mentioned Marc Morano, an industry-funded attack dog and all-purpose denier for hire. On April 21, 2024, the eve of Earth Day, he went on Fox Business to say: "If Joe Biden declared a national climate emergency, he would have COVID-like powers under that emergency" and that he could "bypass democracy and impose the Green New Deal on America without a single vote of Congress," adding that "this is truly a Halloween story, not a story for Earth Day. This is a truly frightening story."[30] Note the use of fear messaging designed to incite conservative viewers by convincing them that climate action is a thinly disguised excuse for perpetual lockdowns like those they were forced to endure during the pandemic.

Fifteen-Minute Cities and Low-Traffic Neighborhoods (LTNs)

We wouldn't be finished discussing conspiracy theories if we didn't talk about "fifteen-minute cities." The concept was introduced by Carlos Moreno, a professor at the Sorbonne in Paris, as a sustainability-oriented way of rethinking the structure of cities and urban environments. The "fifteen minutes" embraces the concept that one ought to be able to access all of the critical services and resources one needs within a short potentially walkable or cycling distance from one's place of residence. Several cities, including Paris, had adopted the idea by the time it became the focus of conspiracy theorists in late 2022.[31]

Of course, this is actually the way communities were constructed prior to the advent of modern highways and transportation systems and widespread personal ownership of vehicles. But it has more recently been reinvented as a way to minimize reliance on fossil fuel–driven transportation systems and to maximize the cohesion of communities. So here we have the brainchild of a Parisian professor (an "elite!"), favoring connected communities ("socialism!"), which minimizes our reliance on fossil fuel–driven transportation systems ("loss of freedom!")—all of the requisite ingredients for a conservative conspiracy theory. As Daniel Zuidijk and Olivia Rudgard of *Bloomberg News* put it, "A coalition of anti-vaxxers, conspiracy theorists and far-right influencers have positioned the 15-minute city concept as a totalitarian plot" intended to keep people trapped in their homes.[32]

How do the conspiracy theorists connect fifteen-minute cities to COVID? That requires one more motivated leap of logic, equating fifteen-minute cities with low-traffic neighborhoods. LTNs, which involve blocking off streets for exclusive use by pedestrians, became widespread in the United Kingdom during the pandemic. Fifteen-minute cities, LTNs, Agenda 2030, the Great Reset, and "lockdowns" are seen as various interlocking attributes of a clandestine master plan hatched by global elites to advance socialist agendas under the guise of sustainability and societal progress. "Scotty the Kid" is a social media influencer with thousands of *Substack* subscribers and more than 135,000 Instagram followers. His social media posts are replete with nativism, Trump advocacy, and antivax propaganda. He insisted in a viral January 28, 2023, Instagram post titled "Climate Change Lockdowns disguised as 15 Minute Cities under the UN Agenda 2030," "Your Government is pushing ahead with plans to bring 15-minute cities to a location near you. They are a brainchild of the UN's Agenda 2030 and are in effect Climate Change lockdowns."[33]

———

Conspiracy theorists are quick to pigeonhole any development into their narrative, regardless of how far-fetched. A case in point is the February 2023 East Palestine, Ohio, train derailment, which spilled toxic chemicals into the air and water, requiring the evacuation of residents within a one-mile radius of the spill. This disaster was a direct consequence of deregulatory excess by congressional Republicans. As Biden White House spokesperson Andrew Bates explained, "Congressional Republicans and former Trump Administration officials owe East Palestine an apology for selling them out to rail industry lobbyists when they dismantled Obama-Biden rail safety protections as well as EPA powers to rapidly contain spills."[34] But that's hardly a helpful narrative for polluters. So instead, we got stories about how the train derailment was actually a training ground for the implementation of fifteen-minute cities. Let's meet some of the individuals behind this canard. First, there's John Sabal. Otherwise known as "QAnon John," Sabal runs the Patriot Voice website. In 2021 QAnon John and his partner, "QueenAnon Amy," hosted a convention at the Dallas Omni Hotel, where speakers, according to the *Dallas Observer*, "spewed homophobia, transphobia, misinformation about the pandemic (repeatedly referring to it as the 'plandemic') and false claims about an alleged pedophile cabal in the U.S."[35] Meanwhile, Sabal uses his website to promote his for-profit QAnon-themed conferences. For a $50,000 online contribution, you have the option of becoming one of his "Patriot Platinum" sponsors.[36] Sabal is also a social media influencer who promotes QAnon theories to his sizable Twitter following of more than one hundred thousand. In the wake of the East Palestine train wreck, he tweeted to those followers that "Cleveland Ohio wants to be the FIRST '15 minute smart city' in the United States . . . Guess what?? East Palestine is approx 90 miles away from Cleveland. Guess what the US Govt can do with 'contaminated' & 'toxic' land . . . Evict the residents FORCEFULLY. I can't think of a better way to MAKE people relocate to a HEAV-ILY surveilled, and controlled WEF prison city." Sherri Tenpenny, an antivaxxer who also spreads COVID-19 misinformation online (and is listed among the CCDH's Disinformation Dozen), tweeted: "Can they make it any more obvious? They use disasters to bring in and force ppl into their 15 minute cities."[37]

TRANSITION RISK

Does this behavior represent the spontaneous, organic uprising of aggrieved citizens? Or is something more orchestrated and sinister afoot? Who might truly be threatened by the Great Reset or fifteen-minute cities? To attempt to answer that question, let's talk about "transition risk." It's a term used in the world of investment to describe the exposure a company or entire industry faces in the event of developments, technological or political, that lead to a large-scale shift away from reliance on their services or products. Policies favoring a rapid transition away from fossil fuels toward clean energy fundamentally threaten the world's wealthiest and most powerful industry, the fossil-fuel industry. They stand to lose more than a trillion dollars in proven oil, gas, and coal reserves—stranded assets—that they will be unable to monetize.[38]

One can imagine how concerned they might have been following the inaugural WEF virtual meeting on the Great Reset initiative on June 3, 2020. International Monetary Fund managing director Kristalina Georgieva explained how with the pandemic recovery we could shift away from fossil fuels toward clean energy by altering the economic incentives and, in particular, eliminating subsidies for the fossil-fuel industry. None other than fossil-fuel giant British Petroleum's own CEO, Bernard Looney, conceded that fossil-fuel subsidies must end and expressed support for the green investment policies that the European Union was beginning to enact. Referring to their investments in alternative energy, Looney said. "We all know there is a carbon budget. It is finite, it is running out." He spoke to the moral imperative of quickly transitioning off fossil fuels: "Given the choice, I would choose my grandchildren every time."[39] His words must have sent shockwaves throughout the industry.

What might have sounded a bit like "game over" for the fossil-fuel industry, however, was in fact "game on" for polluters and petrostates, who engaged in a concerted campaign to halt these incipient efforts. Writing to then president Donald Trump, Archbishop Carlo Maria Viganò—coincidentally, a veritable cheerleader for Russian president Vladimir Putin and his political agenda[40]—warned Trump that "a global plan called the Great Reset is underway. . . . Its architect is a global élite that wants to subdue all of humanity, imposing coercive measures with which to drastically limit individual freedoms and those of entire populations."[41]

The conservative noise machine, meanwhile, kicked into high gear. The key players were a virtual who's who of alt-right influencers and media personalities. Among them were former Fox News host Tucker Carlson.[42] In November 2020, Carlson went on a rant on his Fox News show in a segment titled "The Elites Want COVID-19 Lockdowns to Usher in a 'Great Reset' and That Should Terrify You."[43]

Then there's disgraced alt-right Sandy Hook–denying, conspiracy theory–promoting radio host Alex Jones.[44] Jones did a segment, "The Great Reset's End Game Is Total Extermination of All Humans on Earth," in which he stated that the goal of the Great Reset is the "manipulation of citizens and nations through ignorance and fear." He has also stated that its true intent is to "teach us to not have money and to be poor, and that's how you save yourself is not having a car, going to a job, then we're going to dictate how you live your life now."[45] Australia's far-right One Nation party leader Pauline Hanson called the Great Reset a "socialist left Marxist view of the world,"[46] while conservative UK writer and podcaster James Delingpole called it a "global communist takeover plan."[47]

We have seen the crucial role played by social media in the spread of these conspiracy theories. British journalist Ben Sixsmith has noted, for example, that "Great Reset" disinformation was spread by the "fringes of Right-Wing Twitter."[48] As noted in the last chapter, petrostates such as Russia have weaponized social media, using troll farms and bot armies to spread antiscience disinformation that furthers their political agenda. A congressional report from 2018 found that "Russian trolls used Facebook, Instagram and Twitter to inflame U.S. political debate over energy policy and climate change, a finding that underscores how the Russian campaign of social media manipulation went beyond the 2016 president election."[49] Increasingly, the propagandists have become critical allies of Russia and an extension of its disinformation machine.

And it's not only petrostate bad actors. Fossil-fuel companies and the various front groups they support use similar methods to manipulate public opinion.[50] They weaponize social media to advance their economic agenda, creating memes or messages, and amplifying them using troll farms and bot armies. That lures genuine individuals into the fray, and we get pile-ons and gang tackles. We'll look at some examples now.

ANTISOCIAL MEDIA AND BOTS OF WAR

As we've noted, a favored approach of the pros and propagandists is to seed a prospective online discussion and then use trolls and bots to amplify a preferred narrative. This tactic has long been employed in attacks on climate science and gained strength during the COVID-19 pandemic. As reported by the *Guardian* back in 2020, "The social media conversation over the climate crisis is being reshaped by an army of automated Twitter bots, with a new analysis finding that a quarter of all tweets about climate on an average day are produced by bots. . . . The stunning levels of Twitter bot activity on topics related to global heating and the climate crisis is distorting the online discourse to include far more climate science denialism than it would otherwise."[51]

Since then, of course, this virtual cancer—with help from Elon Musk—has metastasized to afflict our entire body politic, including COVID-19, vaccines, and the emerging climate-denial/antivax nexus. One of the primary modes of attack involves portraying the scientists as villains. While we encountered recent examples of this with public health scientists in the last chapter, it has been going on for decades in the climate arena. In the wake of the manufactured late 2009–2010 Climategate affair, Mike and many of his climate-scientist colleagues reported threatening emails.[52] Kevin Trenberth, a scientist at the National Center for Atmospheric Research, has a "9-page document of 'extremely foul, nasty, abusive' e-mails" sent to him over a period of a few months. Other scientists received death threats and had to have security details when they attended conferences. Consistent with the historical linkage between antiscience and antisemitism that we documented earlier, Mike and other climate scientists were accused of being part of a Jewish conspiracy to defraud the American people through perpetration of a "climate-change hoax." Their pictures with the words *Jew* next to them were posted[53] on the website of a white supremacist organization, stormfront.org. A comment on the site stated that the "global warming scam has always borne the stench of the same old Jewish liars, thieves, swindlers and murderers."[54] Such things don't happen in a vacuum. They are a manifestation of the sort of stochastic terrorism discussed in Chapter 2, "The Plutocrats." Stormfront.org posters have murdered nearly one hundred people in recent shootings of Jews and Muslims.[55] The pros and propagandists manufacture storylines that intentionally play into preexisting con-spiracist narratives, including those that are rooted in centuries-old antisemitic tropes, with the knowledge that this carries with it the threat of physical harm and even murder.

Let's now meet Paul Driessen. Driessen has been employed by a constellation of industry-funded antiscience groups, including the Center for a Constructive Tomorrow, the Frontiers of Freedom, the Atlas Economic Research Foundation, and the Center for the Defense of Free Enterprise.[56] He rails against "eco-imperialism,"[57] and, like Bjorn Lomborg, one of the pros we encountered earlier, he likes to brandish his ostensible environmental bona fides. He claims to have been a "former member of environmental organizations such as the Sierra Club and Zero Population Growth." But he purports to have "abandoned their cause when he recognized that the environ-mental movement had become intolerant in its views, inflexible in its demands, unwilling to recognize our tremendous strides in protect-ing the environment, and insensitive to the needs of billions of people who lack the food, electricity, safe water, healthcare and other basic necessities that we take for granted."[58]

In mid-October 2009, just weeks—curiously enough—before the release of the stolen Climategate emails, Driessen published the commentary "None Dare Call It Fraud: The 'Science' Driving Global Warming Policy" on right-wing websites such as the Post Chronicle and Townhall.com,[59] where he denounced the "inconvenient truth" behind "global warming hysteria," insisting it is a scheme to "enrich Al Gore" and "alarmist scientists." He encouraged followers to "get on your telephone or computer and tell your legislators and local media this nonsense has got to stop." He closed by writing: "It may be that none dare call it fraud—but it comes perilously close."

The repeated reference to "fraud" by industry-funded actors like Driessen is not coincidental. It is intended to stoke hate and resentment among his readers. Mike and his colleagues—you are supposed to believe—were engaged in a *fraud* that would compensate "alarmist scientists who get the next $89 billion in US government research money" and "financial institutions that process trillion$$ in carbon trades." Even though *ExxonMobil* was making unprecedented corporate profits, somehow, we're supposed to believe it is the *climate scientists* who are somehow benefiting financially. And—as we're told by posters on Stormfront.org—they are "Jews." It is noteworthy that the antivaccine crowd has tried to do the same with Peter. Even though he makes low-cost and often patent-free vaccines for the world's poor, they accuse him of secretly taking "big-pharma" money, while conveniently ignoring the millions of dollars raked in through their nutritional-supplement, ivermectin, and hydroxychloroquine business lines.

As Mike comments at length in *The New Climate War*, climate denial is untenable today with the vast majority of our population because they are witnessing profound impacts already playing out. In midsummer 2023, Peter suffered through a rough Houston summer, with an unprecedented forty-five days of temperatures above 100°F that began in June, including Houston's highest-recorded temperature ever—109°F on two different days in August. A few months earlier in May 2024, dozens of people had just died of heat exposure in a massive early heat wave with temperatures soaring to 110–114°F over large parts of South and Southeast Asia in what one historian characterized as the most extreme climate-driven event to date.[60] So the polluters and petrostates and the individuals and organizations that advocate for them can, in general, no longer deny the existence of climate change. They have instead largely turned to other tactics—delay, deflection, division, and so forth. However, in today's fractured media environment, microtargeting is easy to do. You don't have to choose just one tactic. On Twitter, for example, people tend to become trapped within bubbles of like-minded individuals with similar interests, worries, or aspirations.

It's thus possible today for bad actors to pursue a multipronged strategy that simultaneously seeks to reinforce climate denial among conservatives, while breeding despair, doom, and disillusionment among progressives. The objective is to undermine any collective sense of either urgency or agency. As we know, the use of trolls and bots to infect social media discourse is a key component of ongoing disinformation campaigns. And while we can still witness troll and bot armies deployed to promote denialist framing, far greater emphasis today seems to be placed on divisive messaging aimed at disengaging climate activists and advocates.

DIVIDE AND CONQUER

A preferred strategy is "divide and conquer." It involves creating wedges to divide climate advocates by manufacturing rifts between climate activists and mainstream climate communicators. Mike and his colleague Katharine Hayhoe have been on the receiving end of vitriol from ostensible climate activists that undoubtedly has its origins in agents provocateurs stirring the pot in an effort to weaponize climate advocates to do their dirty work for them. Consider the attack on Hayhoe by Anthony Watts, a climate-change denier who has ties to the Koch-funded Heartland Institute.[61] Watts published an article by Eric Worral on his blog that falsely claimed that Hayhoe had criticized youth climate activist Greta Thunberg's "Climate 'Shaming' Crusade."[62] Hayhoe responded on Twitter, "I don't normally bother to call out liars on

Twitter but I will here, as they're trying to invent a disagreement to drive a wedge between us. @GretaThunberg is not personally shaming anyone: she is acting according to her principles."[63]

Responding to criticism she is subjected to regarding her *own* lifestyle choices—for example, why she flies to conferences and why she has chosen to have children—Hayhoe responds: "Flying and eating and having children is often framed as a purity test. It's like, 'So you say you care about climate change. But if you have a child or eat meat or, heaven forbid, have ever stepped in an airplane, then you are not one of our allies. You are one of our enemies."[64]

Hayhoe notes that the sorts of attacks that once came only from climate-change deniers now arise from "people who are not only concerned but are part of the fight." Jonathan Foley, executive director of the climate advocacy group Project Drawdown, has aptly characterized this fractiousness as "the climate movement eating its own."[65] In fact, a seemingly paradoxical nexus of doomers, dividers, and deniers has now emerged. This odd confluence was noted by Katharine in a post on Bluesky:

I've long suspected it, but it was officially confirmed today. Doomerism/personal guilters have wrapped so far around, they are now co-posting with climate dismissives. The below are responses to a post by a doomer ridiculing my newsletter. Michael Mann, I bet you have similar proof.[66]

Doomism produces viral social media content—what's been termed "climate doom porn," marked by dramatic but unsupported claims of collapsing ice sheets, runaway warming, and imminent extinction. Doom porn sells, and it has surely borne fruit for the polluters, petrostates, and plutocrats who are fanning its flames. Consider the vitriol directed at Katharine Hayhoe and Mike by ostensible climate advocates who insist it's too late to act and dismiss our messaging on urgency and efficacy as "hopium," the implication being that we are selling "hope" in the way, say, junkies on the street might sell drugs. It's the sort of smear you might expect from climate deniers, but instead it comes from those who ostensibly are on the side of climate action. "I loathe Mann & Hayhoe," tweets Eliot Jacobson, a self-avowed "doomer"[67] with a substantial Twitter following (seventy-five thousand), who derides us as "hopium addicts." "Mann (like Hayhoe) is a serial blocker for anyone who challenges his hopium. Gimme someone else," says another doomer on Twitter.[68]

These are just a few examples. Twitter is rife with such accusations against prominent climate scientists and climate communicators. From the standpoint of bad actors opposed to climate action, the attacks constitute a "twofer." The first, and most obvious, is that *doom-mongering* convinces many would-be climate advocates that climate action is a hopeless cause. But the blistering attacks against mainstream climate science and scientists advance an agenda of *division*, dividing rank-and-file climate activists and leading climate science communicators.

This divisive battle has been carefully nurtured by bots and trolls, with others joining in the fray, unwittingly allowing themselves to be weaponized for this agenda. Not everyone falls for it, of course. But the doomers have risen from relative obscurity to prominence in a political economy where extreme claims and vitriolic attacks go viral and create huge, almost cultlike followings that are indeed—as we will see shortly—readily monetized.

Some of the friendly fire comes from fellow scientists who have gone down the path of doomism or at least what we might call "soft doomism," that is, emissions reductions alone are not adequate to prevent catastrophic warming. An example is Kevin Anderson, a perfectly well-respected British climate scientist. Anderson has accused the mainstream climate researchers of *understating* the climate-change threat to secure grant money: "The overall framing is firmly set in a politically dogmatic stone with academia and much of the climate community running scared of questioning this for fear of loss of funding, prestige, etc."[69] The accusation is disturbingly similar to the (opposite) accusation by climate deniers—an example of which we saw earlier this chapter—that climate scientists are *overstating* the climate threat to secure grant money. One wonders which it is. Are climate researchers understating or overstating the threat? Logic dictates it can't be both.

Even revered climate scientist James Hansen, whose early predictions of warming proved prophetic, has gotten sucked into the vortex of soft doomism. As we learned in Chapter 1, the scientific consensus is that we can still avert a catastrophic planetary warming of 1.5°C (3°F) if we rapidly reduce carbon emissions this decade. Hansen has claimed that the climate-research community has underestimated the sensitivity of the climate to carbon dioxide emissions and that sustained carbon emissions will cause us to unavoidably cross that threshold.[70] His rhetoric has grown increasingly heated and conspiratorial in nature, including vitriolic attacks on mainstream science and scientists, such as tweeting in late 2023: "The United Nations and COP28 are lying. They know the 1.5°C and 2°C global warming targets are dead."[71] Hansen has argued that we should instead turn to potentially very dangerous "geoengineering" schemes—proposed technofixes like shooting reflective chemicals into the stratosphere to reflect back sunlight or dumping iron particulates into the oceans to fertilize the natural uptake of carbon by algae.

There are several troubling issues here. First, Hansen is conflating his dour assumptions about policy intervention with assumptions about climate physics.[72] Second, Hansen uses this misleading framing to argue for potentially dangerous geoengineering technofixes. Such interventions suffer both from potential unintended consequences (shooting chemical particulates into the stratosphere to block out sunlight could have adverse and unpredictable impacts on our atmosphere and our climate) and from what's known as "moral hazard" (the belief that there is a simple technofix that we can employ in the future provides an excuse for continued fossil-fuel burning today).

In the end, polluters and petrostates are the ones who benefit from high-profile climate scientists being pitted against each other. They would like nothing more than for us to accept the supposed inevitability of a fossil-fuel future, fed by this overall framing. So, we get division and deflection, with doomism in the mix. A feeding frenzy ensues. It starts with the journalists and the scientists they quote. The articles are posted on social media and provide fodder for divisive trolls and bots. Authentic users soon get entrained into the fracas and join in on the pile-on. As a result, climate Twitter today is filled with toxic doomist messaging and assaults on leading climate communicators who are subject to an endless onslaught of "hopium" accusations from ostensible climate advocates anytime we dare claim that it's not too late to do something about the climate crisis.[73] This may be the most successful gambit yet in the assault on climate action.

Bad actors may sow the seeds of division here, but there are self-reinforcing mechanisms that cause it to deepen. One's following—the number of followers one has—is crucial currency in the social media world. It directly influences reach and impact. But equally important is virality, the potential for social media posts to get massively reposted. Indeed, both the number of followers and the virality of one's posts work together to determine influence, impact, and—if that's what one is seeking—the potential for remuneration. Extreme claims and, especially, *conflict* have far greater capacity to go viral given prevailing social media algorithms. These basic attributes of social media work together to create an incentive for outrageous and inflammatory claims, divisive discourse, and, consequently, framing that feeds on such ingredients. So, we see a move toward antiscience messaging on the part of many social media influencers seeking to maximize their potential audience. As noted earlier, we saw such a transition several years ago with celebrity podcaster Joe Rogan, who was friendly with Peter until Rogan fully embraced COVID antiscience during the pandemic. But there are other examples of this phenomenon.

Indeed, Mike had a very similar experience with the famous author Naomi Wolf. Wolf was a leading voice in the so-called third-wave feminism movement of the 1980s and 1990s, with her break-through 1991 bestseller, *The Beauty Myth*, which spoke to female empowerment and the constrictions imposed on women by a patri-archal society's obsession with female appearance. Feminists such as Gloria Steinem and Betty Friedan praised her work. A few years ago, she expressed support for Mike on Twitter over attacks he'd been subjected to on the medium, bemoaning similar attacks she had suf-fered.[74] Familiar mostly with her earlier scholarship, Mike and she became "Twitter friends." She even promoted his book *The New Climate War*.[75]

Admittedly, Wolf had shown some signs of contrarianism in the past. For example, she appeared to subscribe to certain climate con-spiracy theories that we encountered earlier in this chapter regarding clouds and "chemtrails."[76] Moreover, as some observers who have sub-jected her earlier work to greater scrutiny have noted, her intellectual reasoning was sometimes suspect and her views not nearly as progres-sive as they might have seemed on the surface. In the *New Repub-lic*, Liza Featherstone delivers a blistering critique of *The Beauty Myth* that indicts Wolf's entire philosophical framing as "positively neolib-eral," consisting of "appeals to the individual, not to society," that are "almost anti-political" in nature.[77] Such appeals are indeed strikingly reminiscent of the tactic of "deflection" used by fossil-fuel interests to thwart governmental regulation by focusing on the individual car-bon footprint, rather than the emissions by corporate polluters. Mike discusses the tactic in great depth in *The New Climate War*—the very book of his that Wolf had ironically lauded and promoted.

You probably have some idea where this is going. As with Rogan, a switch seems to have been flipped during the pandemic. "The Beauty Myth Author Has Gone from Being a Feminist Icon to an Anti-vaxxer Banned by Twitter" read the subtitle of Featherstone's article (adding "But she's always struggled with the truth"). Here's how Rebecca Onion put it in *Slate*:

> Naomi Wolf is a COVID truther. The bestselling author spent the pandemic doing things like celebrating indoor restaurant meals and declaring that children are losing the reflex to smile because of masks (when asked for evidence, she stated: "The children I see around me is the citation"). She went on Tucker Carlson in February—not long after an actual coup attempt—to warn that the United States was "moving into a coup situation" because of COVID restrictions. Just this week, she shared a 1944 photo of a Jewish couple in the Budapest ghetto wearing stars on their jackets, with the caption "Biden: 'Show me your papers.'"[78]

Wolf has successfully remade herself as a leading COVID conspiracy theorist (who frequently attacks Peter on social media). Her 2022 book, *The Bodies of Others: The New Authoritarians, COVID-19 and the War Against the Human*, is 336 pages of antiscientific drivel. But it was a bestseller, thanks both to the huge audience that exists for COVID antiscience and to laudatory blurbs from the likes of Steve Bannon and Tucker Carlson.

Finally, let's consider the role of the "soft doomers." Led in substantial part, regrettably, by James Hansen, they believe that decarbonization is inadequate for addressing the climate crisis, and so we must do something else. With Hansen leading the charge, they see geoengineering as the solution, something that has wide appeal to tech bros given its connection to novel ostensible technology and the opportunity for business ventures and profit. This was even mocked in the movie *Don't Look Up*, featuring a tech billionaire villain who insisted he could stop the impending planet-destroying comet with his own proprietary technology that, conveniently enough, would also make him huge amounts of money. Spoiler alert—it didn't work.

Some prominent climate doomers, like Jem Bendell in the United Kingdom, are also antivaxxers.[79] If that seems odd, keep in mind that both positions reflect a rejection of authority, expertise, and the "scientific establishment." Just like the pros in the pandemic space who use their platform to market bogus cures for COVID-19, there are soft doomers in the poser/pretender/propagandist arena who are monetizing *their* platform. Consider Silicon Valley tech investor Dan Miller. Mike first encountered him on Twitter back in early 2020 when he weighed in on a tweet Mike had posted about carbon pricing.[80] He continued to comment on Mike's posts, typically contesting the science therein, despite any expertise in climate science. Mostly this involved disputing the mainstream scientific findings Mike reported on regarding the potential to avert warming targets such as 1.5°C and 2°C through rapid societal decarbonization[81] (this pattern of behavior was so irritating that Mike was eventually compelled to block him). Miller, notably, is the managing director of a for-profit venture capital firm, the Roda Group, that has substantial investments in geoengineering (atmospheric carbon-capture) technology.[82]

PROPAGANDISTS PIVOT TO PANDEMICS

Over time, the same playbook we saw in the climate arena of pros and propagandists weaponizing social media and public forums has readily been adapted in the public health domain. In the COVID and pandemic space that Peter operates in, many of the pros—as we've seen—are credentialed individuals, with doctoral degrees or even current or former university appointments. They push health disinformation while monetizing their operations through selling nutritional supplements and fake autism cures or, during the pandemic, promoting ivermectin and other spectacular COVID cures (which don't work) such as azithromycin and hydroxychloroquine. They sometimes participate in online medical visits or write antivaccine and pandemic conspiracy books sold on Amazon.com. The pros include the "Disinformation Dozen" and some of the health-care providers connected to AFLD and FLCCC. The propagandists help amplify the pros' messages on social media or podcasts. With the podcast example, the propagandist hosts and brands the podcast and conducts the interview, while the pro is the one being interviewed. As with climate science, the distinction can be blurry, but the propagandists tend to specialize in social media and podcasts rather than conventional media.

During the COVID-19 pandemic, the propagandists did irreparable harm to public health in the United States and elsewhere. The damage was so great that the US surgeon general, Dr. Vivek Murthy, made countering health misinformation on social media one of his top priorities.[83] He emphasized how misinformation on social media was leading many Americans to reject COVID immunizations and other public health interventions or choose hydroxychloroquine or ivermectin in lieu of vaccinations. He also highlighted the negative impact of even brief exposure to antivaccine misinformation and how social media and other forms of misinformation were weaponized to harass or threaten health-care workers.[84] A few months later in early 2022, the Stanford Internet Observatory further detailed the rapid spread of antivaccine conspiracies on social media, noting how the same propagandists keep resurfacing to promote them.[85] Unfortunately, even the US surgeon general's ability to issue social media advisories came under attack, with the antiscience lobby attempting to use the courts to hand-tie the US government. In the case of *Murthy v. Missouri*, the

Missouri attorney general, later joined by two of the Great Barrington authors as plaintiffs, accused the administration of inappropriately coercing the major social media platforms into accepting what constitutes reliable health education versus disinformation. A finding on the part of the plaintiffs, according to UC Law San Francisco professor and vaccine advocate Dorit Reiss, would have had "a chilling effect on the government . . . especially for the CDC."[86] It was reassuring that even the conservative majority US Supreme Court saw no merit in the case, issuing a six-to-three sharp rebuke of the plaintiffs.[87]

The dangers of antisocial media became ever clearer over the course of the pandemic. The first concerted wave of health disinformation began with hydroxychloroquine (either with or without the antibiotic azithromycin) before switching over to ivermectin. Both ivermectin and hydroxychloroquine are established drugs to treat parasitic infections, and because they are produced in bulk for mass treatment of populations, they could ostensibly be repurposed for a COVID-19 pandemic. But there's one problem: neither drug works for COVID-19. In fact, evidence refuting the effectiveness of these antiparasitic drugs has been published in the most prestigious journals in biomedicine, including the *New England Journal of Medicine*, *JAMA*, and the *Annals of Internal Medicine*.[88] In the case of hydroxychloroquine, the medicine in high doses was toxic enough to cause death, according to one group of investigators.[89] Despite this reality, an entire health-disinformation machine was manufactured to push hydroxychloroquine as the pandemic unfolded, with the plutocrats, pros, propagandists, and press serving as intermeshed cogs. The motivations of the different players were varied. For some of the bit players, it was about simple, immediate profit, by selling the drug either online or through virtual visits to a doctor. For others, like fossil-fuel interests and those promoting them, it was about a larger agenda. Taking hydroxychloroquine to treat or prevent COVID-19 would, after all, obviate the need for enacting social-distancing measures or stay-at-home orders, thus keeping the fossil fuels flowing and the revenue coming. For Trump and his supporters, it was viewed as a way to keep the economy humming through the election.

The Trump White House was very much on board with this agenda during the first few months of the pandemic. In fact, Trump—leader of the American petrostate and disseminator of climate denialism—became one of the biggest promoters of hydroxychloroquine and chloroquine misinformation. The West Wing successfully tag-teamed with pros and propagandists to convince the American people of its benefits. According to one study, in March and April 2020 at the start of the pandemic in America, Trump "made 11 tweets about unproven therapies and mentioned these therapies 65 times in White House briefings, especially touting hydroxychloroquine and chloroquine." His claims were amplified by the right-leaning media such as Fox News. Google searches and purchases of these treatments accelerated following his tweets and his March 19, 2020, press conference where he extolled the virtues of these dubious treatments.[90] Peter Navarro, Trump's deputy assistant, further added to a White House disinformation campaign about hydroxychloroquine, taking to Twitter to denounce its opponents: "At the White House, I had a million tablets of hydroxy that could have saved thousands of lives but @cnn crusaded against it to beat @realDonaldTrump. Negligent homicide at a minimum. @fda was also implicated in hydroxy suppression."[91] Besides the fact that hydroxychloroquine did nothing beneficial as a COVID-19 treatment or preventative,[92] as Holden Thorp, the editor of *Science*, pointed out, touting miracle cures like hydroxychloroquine also served to undermine science and the scientists themselves: "Political overhyping of such approaches is extremely dangerous—it risks creating false expectations and depleting drugs needed to treat diseases for which they are approved. And it sets science up to overpromise and underdeliver. . . . And I worry about lasting damage if science overpromises."[93]

Elon Musk, the owner of Tesla and the world's richest man, added to the hype by becoming a highly visible hydroxychloroquine adherent, not to mention the most followed on Twitter. Commenting on hydroxychloroquine's supposed benefits, Musk, on March 16, 2020, tweeted, "Maybe worth considering chloroquine for C19," including a link to a document titled "An Effective Treatment for Coronavirus (COVID-19)."[94] One of the authors of the document went on Fox News to promote the faux cure.[95] Musk also invested in a COVID Early Treatment Fund established by Steve Kirsch, another tech entrepreneur.[96] Early on in the pandemic before vaccines or new antiviral drugs were available, creating such an early-treatment fund made a lot of sense—especially if a repurposed drug showed promise. However, a hydroxychloroquine randomized trial funded by CETF and led by University of Minnesota infectious-disease researcher Dr. David Boulware revealed that the drug was not effective.[97] Additional trials backed up these findings.[98]

The reaction by the right to Boulware's findings was alarming. Because his studies failed to show effectiveness of either hydroxychloroquine or ivermectin, he came under immediate attack by them.[99] Kiera Butler of *Mother Jones* wrote an article about Boulware with the chilling headline "He Was Just Trying to Study COVID Treatments. Ivermectin Zealots Sent Hate Mail Calling Him a Nazi."[100] Soon thereafter, the US FDA rescinded its emergency-use authorization for hydroxychloroquine (and chloroquine) based on its finding that these medications "are unlikely to be effective in treating COVID-19" and because of the toxicity of these medications for the heart and other serious side effects.[101] Later, the FDA also issued a consumer advisory warning against ivermectin for COVID-19, also due to both a lack of efficacy and potential toxicities.[102] In time, many of the same activist individuals or groups who promoted either hydroxychloroquine or ivermectin also worked to discredit COVID-19 vaccines. These two elements—pushing repurposed drugs and railing against vaccines— eventually became dual components of a more comprehensive anti-biomedicine agenda. Several members of this alliance coalesced at an antivaccine rally in Washington, DC, in the winter of 2022 where Steve Kirsch joined forces with antivaccine activist pros.[103] Kirsch announced that "vaccines are killing 15 people for every person it might save" and that "we will kill 100 kids for every child we might save"; both statements are unsupported by any scientific evidence.[104] In the meantime, Elon Musk's Twitter support for hydroxychloroquine morphed into public opposition to vaccines or vaccine mandates, leading to Canada's Freedom Convoy trucker-rally protest. Musk even compared Canadian prime minister Justin Trudeau to Hitler simply for blocking Musk's efforts to fund the protests with cryptocurrency.[105] At the end of 2022, Musk took over Twitter, renamed it X (we continue to refer to it as Twitter), and reinstated or reinvited the return of three prominent antivaccine activists. He also hosted a special Twitter event, giving RFK Jr. an extensive platform to spread antivax disinformation.[106] Then he took to Twitter to declare: "My pronouns are Prosecute/Fauci."[107]

Despite the increasing toll in human lives, we saw the emergence of even more outlandish conspiracy theories, including the claim that it was the vaccines that actually killed Americans rather than COVID-19 or even that the virologists may have invented the virus in the first place. These absurd notions fed into an alternative universe of antiscientific COVID disinformation. Heading into 2023, the pandemic's fourth year, we witnessed an expanded alliance of malevolent billionaires, tech bros, and high–net worth individuals— plutocrats, prominent podcasters, and far-right extremists, including Steve Bannon and the "Proud Boys"—marching at antivaccine rallies and joining forces with the more established antivaccine activists.[108] Much of this played out on Twitter and several high-profile podcasts. Peter became the next target of this new antivaccine and antiscience juggernaut in what journalist Karam Bales termed the "Twitter/X dog-pile."[109]

DEBATE ME, BRO

An extraordinary and almost unprecedented pile-on by antiscience propagandists unfolded in June 2023. On June 12, 2023, Grace Chong, who self-identified on Twitter as the chief financial officer and chief operating officer of Steve Bannon's *War Room* podcast, tweeted a screenshot of Steve Bannon's posting from his Truth Social (an alternative social media platform popular with the Far Right) stating that "Hotez is a criminal . . . ," along with a picture of Peter from his home—a familiar pandemic broadcast-interview sight from early in the pandemic.[110]

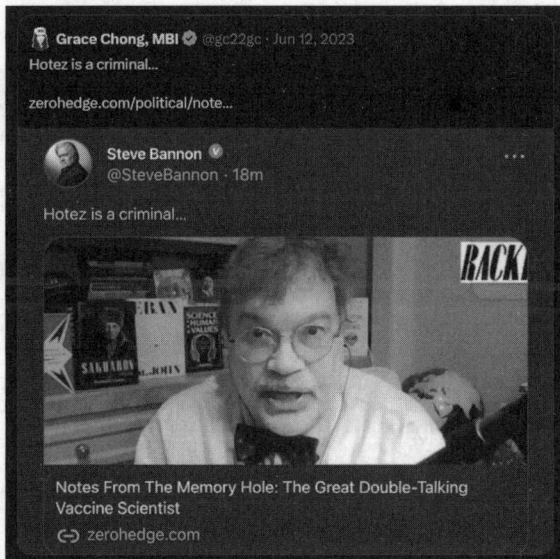

Figure 12. "Hotez is a criminal . . ." social media posting

Three days later, on June 15, 2023, RFK Jr. was invited onto *The Joe Rogan Experience* Spotify podcast. Kennedy brought back long-discredited theories about vaccines causing autism and big pharma's role in pushing vaccines, the substance of his (since-retracted) 2005 article in *Rolling Stone* and *Salon*.[111] Both Rogan and Kennedy complained that big pharma was blocking information showing that ivermectin may actually work against COVID (it doesn't), in addition to other health fallacies about vaccines.

Promoting antivaccine misinformation was not new for Joe Rogan—during the latter half of 2021 as COVID deaths were soaring among the unvaccinated during the delta wave, he touted ivermectin for the treatment of COVID and invited prominent antivaccine activists to appear on his podcast.[112] In an effort to quell the damage, Peter reached out to Rogan via email and Twitter to request going on the show to explain the importance of COVID vaccines. There was no response. At least initially. But then came an intervention on a massive scale. At the beginning of 2022 more than a thousand doctors, scientists, and health professionals wrote an open letter to Spotify demanding that they cease and desist in promoting COVID-19 misinformation.[113] Several high-profile music artists including Neil Young and Joni Mitchell made public protests of having their work associated with Spotify. Mitchell wrote, "Irresponsible people are spreading lies that are costing people their lives."[114]

Rogan addressed the protests in a ten-minute Instagram posting, stating, "I do not know if they're right." Rogan said, "I'm not a doctor. I'm not a scientist. I'm just a person who sits down and talks to people and has conversations with them. Do I get things wrong? Absolutely. I get things wrong. But I try to correct them. . . . I'm interested in telling the truth. I'm interested in finding out what the truth is."[115] The June 2023 RFK Jr. episode occurred despite Joe's attempted reassurances. Peter retweeted a summary of it by Anna Merlan for *Vice News* and commented that "Spotify has stopped even sort of trying to stem Joe Rogan's vaccine disinformation." What followed was Peter's Father's Day weekend from hell and weeks of a harassment campaign against him led by Rogan, Musk, RFK Jr., and their followers. Anna Merlan, again writing in *Vice*, summarized the affair:

> In a series of events no more explicable if you happened to have a front-row seat to them, Joe Rogan and Elon Musk spent the holiday weekend fomenting a crusade against Dr. Peter Hotez, the pediatrician and vaccine expert. Hotez's crime was tweeting approvingly about a story I wrote on Friday concerning Rogan's recent, fawning interview with anti-vaccine activist and presidential candidate Robert F. Kennedy Jr. In response, Rogan, Musk, Kennedy himself, a host of Twitter's biggest anti-vaccine and right-wing personalities, rich investors, and people who paid for Twitter Blue spent the next day demanding that Hotez debate Kennedy on Rogan's show. The brigading ultimately inspired a publicity-hungry YouTuber who's dubbed himself a "predator poacher" to harass the 65-year-old Hotez at his Houston home.[116]

For years, science deniers have tried to goad scientists into debating them. It's a shopworn tactic. There's even a term—*Gish gallop*—for the primary tactic they attempt to exploit when they're able to lure a scientist onto the debate stage. Wikipedia defines it as "a rhetorical technique in which a person in a debate attempts to overwhelm their opponent by providing an excessive number of arguments, with no regard for the accuracy or strength of those arguments."[117] The term originated in the mid-1990s with Mike's friend Eugenie Scott, the founding director of the National Center for Science Education. It's named after the American young-Earth creationist Duane Gish, who honed this technique in his efforts to discredit mainstream evolutionary biologists. The basic idea is to throw so much mud on the wall that the other side (the scientists) spend all their time trying to scrape it off. A losing battle.

The tactic has of course been used in the climate-change arena. Mike has often been asked to debate climate deniers. In February 2022, just months before Peter's ordeal with Rogan, Mike had been asked to go on Rogan's show to debate a climate denier named Steven Koonin. He declined to do so as a matter of principle, tweeting, "Sorry @JoeRogan: I don't debate deniers & disinformers."[118] As Mike has put it, couched in somewhat technical statistical jargon, "It's a simple matter of Bayesian statistics. You get up on the stage with a charlatan, you signal to the audience a prior of 50/50. You have to spend the whole debate trying to work back to the true weight of evidence (99.999 to 0.001)." Translation: Simply getting up on the same stage with a science denier signals to the audience that the denialist viewpoint is a credible one. You've lost the debate before it started. Peter was goaded in precisely this way. What began with Steve Bannon and RFK Jr.'s Joe Rogan podcast quickly exploded. Rogan tweeted, "Peter, if you claim what RFK Jr is saying is 'misinformation' I am offering you $100,000.00 to the charity of your choice if you're willing to debate him on my show with no time limit."[119] Rogan followed this up by insisting that Peter had some type of moral obligation to appear on Rogan's podcast and debate RFK Jr.

A parade of trolls joined in, as Elon Musk tweeted to his 160 million followers: "He's afraid of a public debate, because he knows he's wrong."[120] Exactly what Hotez was wrong about was never made clear—that vaccines cause autism, or vaccines modify your DNA, or vaccines cause cancer? Over the next few days, a range of tech bros, contrarians, and political pundits such as Steve Kirsch, Jordan Peterson, and Andrew Tate pushed Hotez to debate, in some cases offering to match or exceed Joe Rogan's initial $100,000.[121] Not surprisingly, the antivaccine activists who by now had been reinstated on Twitter by Musk had the time of their lives, declaring that Peter was somehow hiding pharma assets or filthy lucre or that he was a professional "pharma shill." Of course, none of this was true, since as Peter often points out he coheads a group of dedicated scientists who develop low-cost, often patent-free vaccines for global health. If anything, the big-pharma companies view him as a threat to their profits (Peter was even nominated for the Nobel Peace Prize in 2022 for his efforts). After refusing to debate RFK Jr., despite the financial incentives, Elon Musk tweeted the absurd claim, "Maybe @PeterHotez just hates charity."[122]

Peter offered valid reasons for not engaging RFK Jr. in a debate. On his MSNBC appearance with Mehdi Hasan, he explained, "I offered to go on Joe Rogan but not to turn it into the Jerry Springer show with having RFK Jr. on."[123] But it was much more than that—in fact, there were three important reasons for Peter refusing to accept the financial inducements or serve as the paid entertainment for billionaires. First, debating RFK Jr. would be a waste of time, based on his experiences in 2017 ahead of RFK Jr.'s potential appointment in the Trump administration. After Peter's multiple discussions with RFK Jr., it became apparent that RFK Jr. had no interest in learning anything new about vaccines or accepting the detailed evidence that vaccines do not cause autism or other conditions. Instead, he stuck to his outdated and discredited talking points and big-pharma conspiracies. Second, Peter felt no obligation to help RFK Jr. By this time, RFK Jr. had declared his intention to run for the US presidency, and Peter didn't want to be exploited as a vehicle for legitimizing him as a serious candidate.

Perhaps the most important reason for not debating RFK Jr. was that doing so would send the wrong message about how science is done and practiced. Professional scientists do not typically "debate" science. Instead, they write scientific papers that are submitted to serious scientific journals for external peer review by experts in the field. Sometimes it can take six months or up to a year or more for papers to be adequately reviewed and comments sent back to the author, and then the author makes revisions.[124] Even then, further revisions are often required. The same can be true of scientific-grant applications. In addition, scientists are expected to present their findings in front of the students and faculty colleagues at their university, or through guest seminars at other universities, and it is especially important for them to present at specialized scientific meetings of their peers. At these venues a scientist can have their findings discussed, considered, and sometimes criticized. It is a process that has gone on for more than three centuries, since the founding of the Royal Society in London, the first major scientific society.

The scientific process has had unparalleled success in revealing the secrets of our world and universe. It has established and confirmed discoveries in biomedicine and climate science, not to mention physics, chemistry, and other areas of science and engineering.[125] But we can't really think of any instances in which scientific discovery was achieved through public debate. While Albert Einstein and Niels Bohr participated in celebrated public colloquies about relativity and quantum physics,[126] they were honest and open discussions between respected and valued colleagues, both trying hard to understand the natural world. Such engagements are a far cry from the sort of spectacle that arises from a debate with unqualified individuals like RFK Jr. who would exploit any such opportunity to spread falsehoods about vaccines and biomedicine.

Peter received significant support for his principled stance not just from members of the scientific community but from opinion leaders and celebrities alike (Mark Hamill, George Takei, Mark Cuban, and CNN's Erin Burnett come to mind). *New York Times* opinion columnist Farhad Manjoo weighed in with support: "So far, Hotez has courageously refused to take the bait. . . . And that's not scientists' method, anyway. They have established ways of assessing empirical questions—you know, things like lab experiments and clinical trials—and none of them involve owning an interlocutor on a popular podcast."[127] In the fall of 2023, Peter received the inaugural Anthony Fauci Courage in Leadership Award from the Infectious Diseases Society of America. He was profiled in the *Lancet*, which called him a "physician-scientist-warrior combating anti-science."[128] He has received numerous other awards from academic and scientific institutions for the stand that he's taken in defense of science and the scientific method.

Unfortunately, many outside the scientific community showed decidedly less appreciation. He was relentlessly hounded on social media by followers of Rogan, Musk, and RFK Jr., with an aggregate reach of more than two hundred million people. Like-minded podcasters such as Russell Brand[129] and Jimmy Dore[130] piled on, as did Alex Jones and *InfoWars*, proclaiming on Twitter,[131] "Over a billion people awaken to the lies of big pharma: Globalist gremlin Peter Hotez has sunk the NWO ship."

Later came physical confrontations, as Peter was stalked and confronted at his home or at lecture venues where he was accosted by YouTubers who shouted provocative questions or offered their opinion about why he refused to debate RFK Jr. before posting these exchanges online. Peter required security to accompany him to venues. For a time, a Houston Police Department or Harris County sheriff patrol car remained parked in front of his home to provide full-time surveillance and protection. Defying the propagandists came at a heavy personal cost. Even then New York Jets quarterback Aaron Rodgers, who was widely criticized for his outspoken antivaccine positions during the pandemic, weighed in, claiming, "@robertfken nedyjr would mop this bum."[132] Conservative *New York Times* columnist Ross Douthat wrote an opinion piece with the title "Go Ahead. Debate Robert F. Kennedy Jr.,"[133] although he was one of the only writers who expressed this sentiment in an ostensibly mainstream publication and perhaps the last one after RFK Jr. was captured on video the following month in July 2023 making deeply inflammatory statements about Jews and the origins of the COVID-19 virus.[134]

Helping to enable the assault on Peter and other scientists was the 2022 post-Musk-takeover decision by Twitter to halt enforcement of its prior COVID misinformation policies.[135] This resulted in a wave of incendiary attacks as the accounts of the leading antivaccine activists (and climate deniers) were restored and bot armies and troll farms were allowed to run amok. The same bot message, claiming that "Hotez is just another satanic agent of the Luciferian globalists seeking to populate the world. Like all liars he will end up in 'the lake which burns with fire & brimstone, which is the second death,'" was spread, for example, by multiple accounts.[136]

Unfortunately, such attacks may come to seem quaint as a whole new level of threat now emerges: artificial intelligence. Generative AI can be used now to manufacture not just fake images or audio recordings but also inauthentic video content that is increasingly difficult to distinguish from reality. Add chatbots that generate text that is indistinguishable from human beings and the ability of the internet to massively spread any and all of this content, and we have the threat of what Stanford's Renée DiResta describes as "infinite" disinformation.[137] A 2023 article in *JAMA Internal Medicine* documented the disturbing speed with which the widely used "ChatGPT" generative AI platform can be used to generate antivaccine disinformation, producing more than a hundred antiscience blog articles, each targeted to particular groups, in roughly one hour.[138] In 2024, AI gained notoriety for generating explicit but fake images of Taylor Swift with findings they may have originated from 4Chan, an anonymous imageboard and fringe social media platform.[139] During the pandemic 4Chan generated mRNA antivaccine disinformation, possibly including state-sponsored antivaccine messages from Russia.[140]

DEPLATFORMING THE PROPAGANDISTS

The propagandists now benefit from the extraordinary amplifying power of social media, including a Musk-weaponized Twitter, disinformative podcasts with huge reaches, and now, most recently, AI. Slowing or stopping them is a complicated endeavor that requires disrupting their weapon of choice, *antisocial* media and other high-visibility disinformation-promoting vehicles.

In the public health domain, some possible strategies were outlined by the US surgeon general in his 2021 advisory.[141] His prescription included better enforcement of social media policies and data sharing with researchers as well as government communication channels to counter misinformation. The problem with this approach is that it relies on the buy-in by platforms in fighting antiscience disinformation. Yet the largest and most aggressive disinformation platform, Twitter, remains defiant of such efforts, and even its leader promotes antiscience dishonest information. It is clear that voluntary compliance has failed, and greater governmental regulation of media companies is critical. The European Union has provided a model for the sort of actions that can be taken by governments, threatening Twitter with an all-out ban of the platform if it fails to engage in acceptable content moderation.[142]

Nina Jankowicz, the former director of the US Department of Homeland Security's Disinformation Governance Board, wrote in February 2024 about how the Biden White House had to pause efforts to counter disinformation due to lawsuits from the GOP, similar to the *Murthy v. Missouri* case.[143] Even efforts to counter disinformation are being challenged by bad actors. Then there is the problem of AI. One glimmer of hope was a recent finding in *JAMA Internal Medicine* that two of the major AI platforms, from Microsoft and Google, contained guardrails to slow or halt efforts to generate health disinformation, leaving ChatGPT as a potential offender.[144] Working with AI platforms to quell science disinformation will certainly be important. But it would be foolhardy to expect voluntary compliance. Commonsense regulations will be critical. Once again, the European Union is showing leadership on this front, having implemented a legal regulatory framework governing the use of AI (the EU Artificial Intelligence Act) in 2024.[145]

Mike has advocated against engaging directly with disinformation-spewing trolls and bots and instead blocking and reporting inauthentic behavior; it lowers, rather than raises, the accounts' influence, given the prevailing algorithms. That doesn't mean that it isn't important to rebut misinformation. It is, particularly when it appears to be gaining traction in mainstream circles. But there are several rules Mike advocates following, which are informed by careful public opinion research. First, you must "inoculate" the potential victims of misinformation by first explaining how it is they've been misinformed by science deniers. There is research showing this works with both climate[146] and COVID-19[147] misinformation. Second, you must replace "sticky" myths with even stickier—where possible—truths. This is possible to do with powerful analogies and metaphors, clever use of storytelling and imagery, and, where appropriate, emotion—exploiting the wiring of the human brain for good rather than bad.

Still another challenging aspect of the online antiscience attacks is the manner in which antiscience has been incorporated into the larger conservative culture wars. Bigotry, racism, misogyny, nativism, and other of the basest human frailties are exploited to advance an antiscience agenda. As we highlighted in Chapter 2, "The Plutocrats," antisemitism is a key part of it as well. While the two circles of the Venn diagram—antiscience and antisemitism—do not completely overlap, the fact that many scientists traditionally have been of Jewish descent creates an opportunity for bad actors to hijack the reptilian brain. Both of us are attacked not only as scientists but also as Jewish scientists, accused of being part of a secret cabal or of somehow profiting from the pandemic or the climate crisis. The Anti-Defamation League (ADL) has investigated some of these attacks, noting how dark elements portray both climate change and the COVID-19 pandemic as hoaxes or conspiracies designed by well-placed Jews (think MTG's "Jewish space lasers") or in some cases blame the Jews for making the SARS-2 virus or the vaccines (note, for example, the outrageous comments of RFK Jr. highlighted earlier) or for supposedly advising brokerage houses on manipulating stocks by altering weather patterns.[148] Accordingly, some of the hate emails we receive or threats on antisocial media emphasize the fact that we are Jewish, sometimes linking us to George Soros or members of the Rothschild banking family. Alternatively, we receive ominous emails or comments on antisocial media that compare us to Dr. Josef Mengele, the infamous Nazi concentration-camp doctor, or make claims that one day there will be a day of reckoning when the Jewish scientists will be brought before a Nuremberg Nazi war criminal tribunal, sometimes named "Nuremberg2," and eventually executed by hanging, just like Nazi war criminals.[149]

Such grisly portrayals represent an excuse to show or invoke Nazi imagery and have been condemned by the ADL and other Jewish organizations as a form of Holocaust trivialization or denialism. There is a sorrowful history of targeting Jewish scientists that must be acknowledged. Some of the worst attacks against us occurred in Weimar Germany during the 1920s and 1930s, when the theories of relativity of Albert Einstein were considered a form of Jewish fraud, as were some of the psychoanalytic constructs of Sigmund Freud.[150] Josef Stalin also targeted Jewish scientists and doctors in Communist Russia, including the infamous "doctors' plot" (in which a supposed cabal of prominent Jewish medical professionals was alleged to be plotting to murder leading government officials). Even as far back as the mid-1300s when the bubonic plague, also known as the "black death," spread across Europe, the Jews were accused of igniting the epidemic, in some cases through purported poisoning of the wells. This launched terrible pogroms in Germany and elsewhere, which ultimately forced the Jewish people to immigrate to Poland and eastern Europe.[151] It is tragic that centuries later, the Jewish people are still being singled out in this manner. The threats against us as Jewish scientists now occur on a regular basis often through online venues. While one of us (Peter) had swastikas sent to his home on two occasions, the other (Mike)—as we learned earlier—had his photo, labeled "Jew," posted on a notorious neo-Nazi website. When such events occur, we are grateful for the protection of law enforcement and the advice of nongovernmental organizations that fight intolerance and bigotry. The ADL has noted a steep increase in reported antisemitic incidents since the beginning of the COVID-19 pandemic, and we expect that as antiscience aggression increases, so too will antisemitic sentiment.

This lived reality represents yet another terrible element of modern antiscience—its coalescence with many of the darkest elements of humanity. The attacks continue to worsen on Twitter and other antisocial media platforms. The private sector can play a role in opposing this trend. Commenting on Musk's weaponization of Twitter, for example, Quina Jurecic of the Brookings Institution writes, "His desire to turn Twitter into a playground for the worst parts of human nature has been mitigated only by the squeamishness of advertisers who don't want their products displayed alongside posts by neo-Nazis."[152] It's good that some advertisers are pushing back, but we need to see far more of this collective resistance.

As we've alluded to in this chapter, the pros and propagandists get a fair amount of assistance from another of the *p*'s, the *press*, or, more generally, the media. In some cases—for example, Fox News and the

Wall Street Journal—they are willing accomplices. However, mainstream media and even media outlets on the progressive side of the spectrum can be unhelpful either through their almost reflexive neutrality in the war on science and, in some cases as we've seen, through unwittingly playing into the tactics of the forces of antiscience. We explore this further in the next chapter.

6

THE PRESS

There is evidence that individuals and organized groups particularly in the USA, who have historically been climate deniers or policy sceptics, have turned their attention to the COVID-19 crisis and found a voice in the (right-leaning) media.

—James Painter, Reuters Institute for the
Study of Journalism, June 2020

The press, as we have already seen, has engaged in widespread attacks on both science and scientists. Beyond the usual suspects—Fox News, the rest of the Murdoch media empire, and other conservative media—even mainstream outlets like the *New York Times* have in recent years miscommunicated the science behind climate change and COVID. Onetime conservative turned progressive critic John Cole runs a very popular satirical Twitter account known as the "New York Times Pitchbot." He uses the account to satirize the often risible commentaries that appear these days in the *New York Times*, mocking the "Gray Lady" with op-ed titles that he imagines could, despite their absurdity and audacity, nonetheless find their way onto today's *New York Times* editorial pages.[1] As we were working on this very chapter, he offered up one of his characteristic sardonic takes:

"There are many topics on which people can reasonably disagree: climate change, vaccination, and January 6, for example. But every decent person should admit that the college protests, like the budget deficit and Bill Clinton's extramarital affairs, are objectively bad."[2] This ridiculing of the *New York Times* for their "both-sides" treatment of climate and COVID-19 captures the basic premise of this chapter.

This problem is hardly limited to the *New York Times*. False equivalences or "bothsidesism," now recognized as their own form of media bias, have become pervasive in our mainstream media. Even *ProPublica*, often considered a paragon of objective, fact-based reporting, has pushed the revisionist history that scientists themselves might have created the COVID-19 virus. That's despite the overwhelming scientific evidence for COVID-19's natural origins through the transmission of a new virus from wild animals to humans, sometimes referred to "zoonotic spillover." The misguided victim-blaming narrative that the scientists are to blame for public misunderstanding has become widespread not just in the far-right press but also with mainstream outlets like the *New York Times* and *Washington Post*. Perhaps that's to be expected, since it conveniently deflects from their own failures in communicating the dire warnings from scientists.

Mike founded a center at the University of Pennsylvania that is devoted to addressing these challenges. Both of us spend much of our time speaking to journalists and participating in television news programs in an effort to communicate climate, pandemic, and related threats to the public. At times it can feel like a losing battle. But we still believe there is a way forward.

FOX IN THE HENHOUSE; OR, DEN OF DENIAL

The right-wing media—which includes outlets such as *Breitbart News*, the *Daily Caller*, the *National Review*, and the Sinclair news syndicate in the United States; the *Daily Mail* in the United Kingdom; and *Die Welt* in Germany—have built an echo chamber of climate denialism over the years. But no media entity has done more to spread falsehoods and lies about climate than Fox News. Created by Rupert Murdoch, an Australian who acquired newspapers first in Australia and then in the United States and United Kingdom to form News, Ltd., and eventually News Corp., Fox News was fashioned as the conservative cable news rival to CNN.[3] Together with Roger Ailes, his founding chief executive officer, Murdoch built Fox News into a juggernaut that now reaches almost one hundred countries, and he created a global media and publishing empire, News Corp., which is a powerful disseminator of disinformation. It includes the *Wall Street Journal* and the *New York Post* in the United States and other outlets around the world, such as the *Times* and the *Sun* in the United Kingdom and the *Australian*, the *Herald Sun*, and SkyNews television network in Australia.

Headquartered in Midtown Manhattan, and today run by Rupert Murdoch's son Lachlan Murdoch, Fox News is almost unrivaled for its ability to generate massive volumes of ideologically motivated antiscience propaganda. From the Dominion Voting Systems lawsuit against the network, we have learned about how it spread lies about the 2020 US presidential election.[4] Documents presented during the trial confirmed a troubling business model designed not to inform but instead to feed viewers the conspiratorial content the network knew would keep them glued to their television screens, while weaponizing them as spreaders of right-wing propaganda.[5] For those of us who have noticed just how closely Fox News' political rhetoric resembles that of Putin's propaganda network, Russia Today, it was troubling but sadly thoroughly unsurprising to learn that Russia media uses Fox News to promote its agenda at home, often replaying Fox News prime-time segments replete with conspiracy theories and questioning of the goals or agenda of NATO, the US government, and its Western allies.[6] Such examples highlight a growing convergence of disinformative framing promoted by the Russian government and Fox News, particularly when it comes to antiscientific disinformation about climate and COVID. Though branding itself "Fair and Balanced," Fox News is neither. The network—which is actually considered entertainment rather than news[7]—has worked tirelessly to spread climate-change denial for years. It has subjected its viewers to an alternate reality where climate change is depicted as a hoax and the scientists as charlatans and fraudsters.

The Chinese have a saying that "the fish rots from the head down." Which brings us back to Rupert Murdoch. He has often revealed himself to be a climate-change denier. He has tweeted dismissive tidbits like "Just flying over N Atlantic 300 miles of ice. Global warming!"[8] and "Climate change has been going on as long as the planet is here, and there will always be a little bit of it." and "If the sea level rises 6 inches . . . we can't mitigate that, we can't stop it. We've just got to stop building vast houses on seashores . . ."[9]

Climate change has been going on as long as the planet is here. There will always be a little bit of it. We can't stop it.

—Rupert Murdoch

Figure 13. Tom Toles / *Washington Post*

Far more consequential than Rupert Murdoch himself and what he believes, however, is the damaging role his media empire has played in its endless promotion of climate-denial propaganda. In 2019, the Boston PBS affiliate WGBH characterized Fox News as "helping to destroy the planet," offering this advice: "Want to fight climate change?

Tell your elderly relatives to turn off Fox News." In 2020, Rupert Murdoch's other son, James Murdoch, resigned from the News Corp. board in protest of the father's use of his media empire to "legitimize disinformation."[10] It all came to a head in the wake of the devastating 2020 Australian wildfires.[11] The News Corp. disinformation machine in Australia, including the *Herald Sun, Sky News*, and the *Australian*, worked overtime to sell the myth that the unprecedented Australian bushfires occurred due to "arson" or "back-burning" fire-management strategies, deflecting attention from the overwhelming evidence that climate change was leading to more widespread wildfires on the continent. An official governmental review in 2011 conducted by economist Ross Garnaut had concluded that more damaging and deadly wildfires from climate change–induced increases in heat and drought "should be directly observable by 2020."[12]

Then there's the role the Murdoch media empire played in the fake Climategate scandal. Murdoch, as we know, at the time had close ties to the Saudi royal family, who attempted to use Climategate to undermine the proceedings at the 2009 United Nations Climate Change Conference in Copenhagen. Fox News, meanwhile, featured almost continuous coverage of the pseudoscandal. Murdoch's *Wall Street Journal* ran numerous Climategate opinion pieces in the lead-up to the summit. Murdoch papers in the United Kingdom and Australia also helped promote false allegations and untruths about climate scientists and climate science.

The larger Murdoch media empire has in fact served as a megaphone of climate denial for decades now, plastering our television screens and flooding the internet with untruths about climate science and smears and innuendo about climate scientists. Fox News has provided a platform for many of the climate deniers we encountered in Chapter 4, "The Pros," such as Steve Milloy and Marc Morano, as well as Christopher Horner and Myron Ebell of the fossil fuel–funded Competitive Enterprise Institute. All have been seen on Fox News denying climate change and attacking climate scientists. Bjorn Lomborg frequently writes opinion pieces for the *Wall Street Journal*—as we saw in Chapter 4—downplaying the impacts of climate change.

Another modus operandi of the Murdoch media is to attack and undermine prominent climate advocates. Al Gore, known for his climate policy advocacy, has been under assault by them for years. They have falsely accused him of exaggerating the climate crisis[13] (a favorite tactic of polluters is to accuse environmental impacts of being "Chicken Littles") and of attempting to "control" people's behavior[14] (exploiting the conspiratorial framing discussed in Chapter 5, "The Propagandists"), and he has been portrayed as a hypocrite. They have disparaged everything from his weight to his home to his electrical bills.[15]

Leonardo DiCaprio has similarly been tarred by the Murdoch media, simply because he is a highly visible climate advocate: "Hollywood Hypocrite's Global Warming Sermon" read one headline in the Murdoch-owned *Herald Sun* of Australia.[16] "Leo DiCaprio Isn't the Only Climate Change Hypocrite" read a headline in the Murdoch-owned *New York Post* (the article also attacked President Barack Obama and even Pope Francis).[17] Youth climate activist Greta Thunberg has been the target of dozens of Fox News hit pieces.[18] After Greta famously crossed the Atlantic in a zero-emissions boat to attend the UN Climate Action Summit in New York City in 2019, Fox News depicted her as a hypocrite for traveling on a "yacht."[19]

Alleged "hypocrisy" has continually been wielded as a weapon against high-profile climate advocates by the Murdoch media. Bill McKibben, founder of the international organization 350.org, is a favorite target. He has been attacked for his use of air travel by Murdoch's *Herald Sun*.[20] Green New Deal advocate and Fox bête noire Congresswoman Alexandria Ocasio Cortez (D-NY) has been maligned by Murdoch's *New York Post* for driving an automobile (the horror!).[21] The longest-serving mayor of Sydney, Australia, Clover Moore, is a strong advocate for climate action. She was shamed by the Murdoch press for flying. She wasted no time, however, in pointing out that this was in fact a textbook example of the tactic of "deflection" discussed in Chapter 5, "The Propagandists": "While those against action on climate change used to flatly deny climate science, their tactics have matured. Now they don't deny; they deflect."[22]

Fox News doesn't just seek to undermine public faith in the scientific evidence for human-caused climate science. They actively work to thwart climate progress by casting doubt on the viability of clean energy. An accidentally well-titled April 2024 Fox News commentary by Robert Bryce of the Koch-funded Manhattan Institute betrays the modus operandi: "Backlash Against Wind and Solar Projects Is Real, It's Global and It's Growing."[23] To the extent the statement may be true, it's ironically because of nonstop propaganda efforts for years now by the fossil-fuel industry[24] and their mouthpiece, Fox News. Along with other right-wing media such as the Koch brothers–funded *Daily Caller*,[25] Fox News used the failure of the solar-energy company Solyndra to discredit former president Barack Obama's clean-energy policies. What they failed to tell their viewers was that Solyndra had not benefited one cent from the president's budget. The president's budget in fact didn't even increase funding for the extremely successful loan-guarantee program that had helped fund Solyndra.[26]

Fox News has worked vigorously to sour their conservative base on climate action. This is evident in their war on AOC and the Green New Deal. They've run stories such as "AOC Accused of Soviet-Style Propaganda with Green New Deal 'Art Series'" and "Stealth AOC 'Green New Deal' Now the Law in New Mexico, Voters Be Damned."[27] "I guess government-forced veganism is in order," insisted Fox News ultraconservative firebrand Sean Hannity.[28] "They want to take your pickup truck, they want to rebuild your home, they want to take away your hamburgers. . . . This is what Stalin dreamt about but never achieved," said Fox News contributor Seb Gorka at a conservative conference.[29] It takes real chutzpah for a network known for constantly parroting Kremlin talking points to criticize climate advocates of "Soviet-style propaganda" and compare them to Stalin.

Studies show that the sustained assault was successful. For conservative Republicans, there was a significant decline in support for the Green New Deal—a drop from 57 percent to 32 percent—over the course of just a few months, which can be tied directly to Fox News' adverse messaging.[30] However, the Murdoch media has also promoted myths and distortions that appear aimed at creating a false dilemma for environmentalists. That includes numerous stories on how deployment of solar energy will bring environmental peril. Fox News has run execrable stories such as "Solar Energy Plants in Tortoises' Desert Habitat Pit Green Against Green," "Environmental Concerns Threaten Solar Power Expansion in California Desert," "Massive East Coast Solar Project Generates Fury from Neighbors," and our favorite, "World's Largest Solar Plant Scorching Birds in Nevada Desert."[31]

Then there are the dangers of *wind*! Robert Bryce, mentioned earlier, has promoted the myth that wind turbines pose a significant threat to birds on the editorial pages of the *Wall Street Journal* (as well as in other ultra-right-wing outlets like the *National Review*).[32] And please shed a crocodile tear for the poor whales, who—in the alternative universe of Fox News—are under grave threat from wind turbines. Fox News has played a major role in promoting this myth. But as the *Guardian* points out, it has been amplified and spread by the usual suspects and "can be traced back to anti-wind activism in the US linked to the energy industry and rightwing thinktanks with connections to fossil fuel industries."[33] More birds are killed every year by house cats. Yet Rupert Murdoch, curiously enough, has failed to use his massive platform to condemn the feline fiends. Climate change is an *actual* threat to whales worldwide.[34] Where do Murdoch, Fox News, and News Corp. stand on that? Oh, never mind.

When all else fails, bad actors resort to scare tactics. In "Fox News' Wind Power Hypochondria," media watchdog group Media Matters for America detailed how Fox News used a tall tale of wind turbines purportedly causing "devastating" health problems on Cape Cod. As Media Matters notes, multiple studies have found no evidence for the claim.[35] Wind turbines also became a vehicle to attack AOC and the Green New Deal. According to Fox News's Jesse Watters: "They have this new green deal or whatever. Ok, where they want to eliminate all oil and gas in 10 years. If you're in the polar vortex, how are you going to stay warm with solar panels?"[36] On one occasion—and we kid you not—a Fox News host literally blamed a plane skidding off the runway on climate policy: "Democrats want this. Because if you're afraid to fly, you're going to have less carbon pollution. This is about the green agenda." His bemused cohost asked him, "You *really* believe that?"[37]

And then there's always the old standby "It's just not going to work." In a 2013 Fox News segment attacking President Obama's renewable energy–incentives program, host Gretchen Carlson asked Shibani Joshi, their business reporter, about solar power and why it has fared better in Germany than in the United States. "What was Germany doing correct?" Carlson asked. "Are they just a smaller country, and that made it more feasible?" "They're a smaller country," Joshi said, "and they've got lots of sun. Right? They've got a lot more sun than we do. . . . [H]ere on the East Coast, it's just not going to work."[38] As Media Matters pointed out in their response to the segment, estimates from the US Department of Energy's National Renewable Energy Laboratory show that "virtually the entirety of the continental United States gets more sun than even the sunniest part of Germany." In fact, NREL senior scientist Sarah Kurtz said via email, "Germany's solar resource is akin to Alaska's," the US state with by far the lowest annual average of direct solar energy."[39] The logical conclusion of this carefully cultivated right-wing disinformation campaign was Trump's executive order in February 2025 aimed at banning wind power in the United States altogether.[40]

THE COVID NEWS CORPSE

Joseph Azam, a former senior vice president at News Corp., blamed Fox News for "single-handedly" politicizing public health in America and for accelerating vaccine hesitancy in much of the US population.[41] Due to its wide viewership, especially for its nighttime opinion programming, Fox News became the go-to source for what ultimately turned into lethal health disinformation.[42]

The ontogenesis of the disinformation campaign is worth close examination because, while greatly accelerated in terms of the time frame on which it unfolded, it bears a truly uncanny resemblance to the longer-term climate-disinformation campaign. Health disinformation around COVID first took off when the pandemic reached the United States in March and April 2020 and the evening Fox News anchors heavily promoted the use of hydroxychloroquine. According to the *New York Times*,[43] Fox News nighttime anchor Laura Ingraham referred to it as a "game changer" as she and her fellow anchors Tucker Carlson and Sean Hannity—which together represented the most widely viewed nighttime news broadcasts with roughly two to three million viewers each[44]—took their cues from an aggressive hydroxychloroquine-promotion campaign by the Trump White House. Peter said to NBC News that the White House was unwilling to do the hard work needed to contain the pandemic in the United States, "so, as a substitute, they veer towards magical solutions like hydroxychloroquine."[45] Not surprisingly, this coincided with the end of his invitations to appear on Fox News at least during their highly viewed evening broadcasts (Peter still appeared occasionally on their afternoon shows until 2021 when the evening anchors began to attack him directly).

The Daily Show, in April 2020, compiled a "best of" reel they called the "Heroes of the Pandumic,"[46] featuring assorted right-wing pundits and Fox News talking heads downplaying the pandemic or engaging in pandemic denial. We see Fox News's Sean Hannity complaining that the "media mob" wants you to think it's "an apocalypse" and Rush Limbaugh dismissing it as "hype" or saying that "the Coronavirus is the common cold, folks." (Limbaugh—a smoker for decades and a longtime denier of the health threat of smoking— died soon thereafter from lung cancer.)[47] Fox Business's Lou Dobbs warned that "the national left-wing media [is] playing up fears of the Coronavirus." Fox News commentator Tomi Lahren mocked those raising concern as shouting, "The sky is falling because we have a few dozen cases." She added that she was "far more concerned with stepping on a used heroin needle." Then there are Fox News personalities Jeanine Pirro and Geraldo Rivera, each of whom was dismissive of coronavirus, as being tantamount to the flu. It was clearly a carefully coordinated Fox News chorus. Other Fox News guests or talking heads insisted they were not "afraid" of the virus, that it was "very difficult to contract," and that it was "milder than we thought." Members of a Fox News panel audaciously insisted, "It's actually the safest time to fly." In the *Wall Street Journal* opinion section, professional climate deniers Benny Peiser and Andrew Montford trespassed as supposed coronavirus experts, insisting that "scary" virus forecasts were built on "bad data" and that preventative actions would constitute "draconian measures" that would damage the economy.[48] It is noteworthy how almost identical antiscience rhetoric to that used to dismiss climate-change projections was being appropriated here to dismiss coronavirus projections.

Unsurprisingly, Fox News and other right-wing media tag-teamed to launch a barrage of attacks against Dr. Anthony Fauci because he supported evidence-based decision-making and was unwilling to reinforce Trump's COVID denialism. Fauci also refused to go along with White House policies that contradicted the consensus advice and wisdom of the American biomedical science community. Despite more than three decades of public service as director of the NIAID, he was relentlessly attacked and unfairly blamed for the negative economic impact of the pandemic in the United States and was blasted for undermining Trump.[49] Eventually, members of the House Freedom Caucus and senators with extremist far-right views even accused him of providing financial support to Chinese scientists to create the COVID-19 virus, despite the overwhelming evidence for natural origins.

One could be forgiven for mistaking "News Corp." for "News Corpse" as the denialism-induced body count continued to rise. Fox News participated in a coordinated effort to convince the elderly that they should just "take one for the team"—literally. Texas lieutenant governor Dan Patrick insisted that grandparents should even be prepared to die to rescue the economy for their grandchildren in one Fox News segment.[50] Fox News's Brit Hume "sane washed" these remarks, suggesting that's an "entirely reasonable viewpoint" and basically that it was fine for older Americans to take one for the stock market.[51] This dangerously misguided framing quickly spread through the conservative media ecosystem. One right-wing talk show sought to assure viewers by explaining that "while death is sad for the living left behind, for the dying, it is merely a passage out of this physical body."[52]

The evolving narrative, once again, bore an uncanny resemblance to how the longer-term climate-disinformation campaign has unfolded, wherein the central thesis evolved from "It's not real" to "It's actually good for us" over several decades. With COVID-19, however, that all happened in just a few months.[53] As one observer noted,

"The right wing's instantaneous flip from 'it's a hoax' to 'let millions die in service to the "market"' is the same script they play with climate change, to a tee. They want you to do nothing."[54] Dan Rather, the great former CBS News anchor, aptly appraised the situation: "After years when we should have learned of the dangers of 'false equivalence' it baffles me that we are seeing a framing that pits the health of our citizens against some vague notion of getting back to work."[55]

The most disturbing gambit from the Fox News anchors was yet to come, however. It came in the spring of 2021 when they sought to denigrate mRNA COVID vaccines just as they were becoming widely available.[56] As summer approached, the uptake of COVID-19 vaccines in the United States had reached a plateau as many refused to be immunized despite the proven effectiveness and safety of vaccines. Tragically for the nation, a new delta variant of the SARS-2 coronavirus was sweeping across the conservative southern region of the United States throughout the summer and into the fall.[57] Even though COVID-19 vaccines were 90 percent protective against symptomatic illness or death, many Americans needlessly succumbed to the virus because they shunned vaccinations in the last half of 2021. The deaths occurred overwhelmingly in Republican-majority, or "red," states as a predictable consequence of the predatory disinformation amplified by Fox News and other conservative media, which contributed at least two hundred thousand and perhaps as many as four hundred thousand American deaths, including at least forty thousand in Texas where Peter lives and works.[58]

Media Matters has carefully documented the contribution of Fox News to the needless deaths during America's delta wave, including two periods from June 28 through July 11, 2021, and June 28 to August 8, 2021, when they did an in-depth analysis of Fox News broadcasts and transcripts. By this time, delta had become the dominant COVID variant in the United States. Their findings were stark and chilling, revealing that almost 60 percent of 628 coronavirus segments aired on Fox News included claims "undermining or down-playing vaccinations" or "undercutting immunization," while 47 percent of Fox News vaccine segments said immunization efforts were "coercive" or "violated personal freedom or choice," and 33 percent of Fox News vaccine segments reported that vaccinations were either "unnecessary or dangerous." The overall finding was that the Fox anchors or their guests voiced almost nine hundred claims deriding COVID immunizations, with more than 90 percent of the evening opinion news "undermining vaccination." Media Matters listed Fox News evening news-opinion broadcasts, *The Ingraham Angle, Hannity*, and *Tucker Carlson Tonight*, as among the worst offenders, noting how antivaccine statements made by Tucker Carlson were endorsed at the highest levels of Fox News leadership, including Lachlan Murdoch.[59]

MURDOCH MEDIA CULTIVATES COVID CONTRARIANS

Several Murdoch outlets contributed to the flood of COVID-19 disinformation, especially the *Wall Street Journal* opinion section, which ran articles that falsely claimed the United States would achieve herd immunity just before the deadly delta variant struck and downplayed the severity of the COVID pandemic or unnecessarily frightened Americans about vaccine side effects.[60] In his 2023 book, *We Want Them Infected: How the Failed Quest for Herd Immunity Led Doctors to Embrace the Anti-vaccine Movement and Blinded Americans to the Threat of COVID*, Dr. Jonathan Howard, a neurologist and psychiatrist at NYU Langone Medical Center, detailed how a new group of COVID pros—including contrarian physicians from prestigious academic health centers, such as Stanford and Johns Hopkins medical schools—began writing opinion pieces in the *Wall Street Journal* and appearing as regular talking heads on Fox News. Several were connected to the Great Barrington Declaration.[61] Many of them professed strong antivaccination viewpoints, even though their expertise was in other fields. New links were thus emerging between conservative media—especially the Murdoch media—and this new generation of pros. An example is Dr. Martin Makary, a surgery professor at the Johns Hopkins medical school who wrote a *Wall Street Journal* op-ed making the preposterous assertion that the United States could achieve herd immunity in a matter of a few months.[62] He was identified by Media Matters for his numerous "claims undercutting vaccines," second only in this regard to Fox News's Tucker Carlson and Brian Kilmeade.[63] Media Matters reported how Makary, in his more than 140 Fox News appearances, repeatedly downplayed the severity of COVID-19, including the threats of new variants and the need for vaccinations.[64] While he lacked expertise in the fields of infectious diseases, virology, and vaccinology, the chyron running along the bottom of the screen during his Fox New appearances could impress viewers with the fact that he was a professor at the world-renowned Johns Hopkins medical school. Donald Trump appointed Makary FDA director in early 2025.

Others cited by Media Matters include two Stanford medical school professors, Drs. Jay Battacharya and Scott Atlas,[65] who were also frequent Fox News guests and wrote for the *Wall Street Journal*. Bhattacharya, who would go on to lead the NIH in Trump's second term, is a physician, health economist, and professor of health policy. He was also one of three lead authors of the Great Barrington Declaration,[66] which, as readers will recall, proposes the misguided notion of cultivating herd immunity in lieu of more aggressive and far more effective public health interventions (such as immunization, vaccination mandates, and stay-at-home orders to limit hospital emergency-room and intensive care–unit surges). To be specific, it proposes the following: "As immunity builds in the population, the risk of infection to all—including the vulnerable—falls. We know that all populations will eventually reach herd immunity—i.e. the point at which the rate of new infections is stable—and that this can be assisted by (but is not dependent upon) a vaccine. Our goal should therefore be to minimize mortality and social harm until we reach herd immunity." Most mainstream public health agencies consider reliance on herd immunity a deeply flawed public health policy, with organizations such as the World Health Organization countering that "the death toll from a herd immunity approach would be intolerable and overwhelm healthcare systems."[67]

During his time in the first Trump administration serving on the White House Coronavirus Task Force, Scott Atlas was a leading proponent of herd immunity, which was widely condemned by the public health community and led the *Washington Post* Editorial Board to publish the June 2022 commentary "How One Doctor Wrecked the Pandemic Response."[68] Months later, Atlas responded with his own commentary, in the *Wall Street Journal* of course, titled, with no sense of irony, "When Will Academia Account for Its Covid Failures?"[69]

And then there is Tucker Carlson. Carlson, having honed his COVID disinformation messaging at Fox News, continued with it on his streaming platform and Twitter after he left the network.[70] The Murdoch media had nurtured and spun off a major COVID antiscience propagandist. In January 2024 at a public speaking event in Alberta, Canada, he pushed vaccine denialism, saying vaccines "didn't work."[71] On his podcast, Carlson invited another podcaster and former biology professor, Bret Weinstein, to amplify false claims that mRNA vaccines had caused seventeen million deaths globally and that the World Health Organization through its pandemic preparedness was laying plans to strip away "personal and national sovereignty."[72] The truth is that in the United States alone, COVID mRNA vaccines have saved between two and three million lives[73] and globally millions more.[74] In the meantime, Georgetown University health-law professor Larry Gostin rightly points out that no accord or agreement with the World Health Organization would override the US Constitution.[75] Weinstein went on Joe Rogan's podcast where he promoted a discredited theory about HIV not being the cause of AIDS.[76]

As the pandemic went on, the Murdoch media dusted off their now more than a decade old Climategate playbook. But this time, they deployed it against COVID-19 scientists rather than climate researchers. Peter was a target. On Fox News he was called a charlatan by Tucker Carlson (ironically, on the very same day he had been nominated for the 2022 Nobel Peace Prize for his work on a low-cost COVID vaccine). He was mocked by Laura Ingraham and Florida governor Ron DeSantis for his (accurate) prediction of the deadly impact of the delta virus.[77] But subsequently, Fox News set their sights on a *different* Peter, Peter Daszak. Daszak heads up an NGO known as the EcoHealth Alliance that engages in global health research to unravel how viruses and other pathogens emerge among animals and eventually transfer or spill over to humans to cause epidemics or even pandemics. The organization is at the cutting edge of the "one-health" concept that looks holistically at disease through three pillars of humans, animals, and the environment.

During 2023–2024, Daszak found himself at the center of attacks by antiscience activists—including congressional Republicans—over alleged collaborative gain-of-function research EcoHealth had done with the Chinese working with coronaviruses in bats. GoF involves the insertion or manipulation of genes in a virus (or other microorganisms) to increase their transmissibility or virulence, that is, severity of illness. Even though GoF research played no role in COVID-19 origins, the GoF rumors fed the conspiracy theory that the SARS-2 virus, the causative agent of COVID-19, was made or augmented by humans or that it was leaked from a laboratory, mostly likely the Wuhan Institute of Virology, which is one of many coronavirus research centers that were established in China after the emergence of SARS in 2002. Antivaccine activists and other COVID conspiracy-mongers insisted that the COVID-19 virus originated from research conducted in Wuhan that was supported by US government grants funneled through the EcoHealth Alliance, and Daszak must naturally be in on the conspiracy.

As described in Chapter 1, "The 1-2-3 Punch," SARS and related SARS-like viruses originate in bats before they are transmitted to humans either directly or through another animal such as a civet or a raccoon dog, a process known as zoonotic spillover. Despite at least a half-dozen peer-reviewed scientific articles detailing the natural origins of COVID-19 due to spillover events, none on either GoF or lab leak, both Daszak and his organization have been hounded relentlessly by right-wing front groups with vexatious FOIA demands, and their private correspondences have been seized and weaponized against them in a massive disinformation campaign by bad actors. As with Climategate, the *Wall Street Journal* once again served as a megaphone amplifying the pseudoscandal, promoting the notion that EcoHealth and Daszak were involved in a "lab-leak" cover-up and cherry-picking words and phrases from private correspondence to smear them. *Wall Street Journal* Editorial Board member Allysia Finley—naturally also a climate-change denier[78]—wrote a commentary in the paper promoting this narrative and dragging both Peter (Hotez) and of course Anthony Fauci into the matter for good measure.[79] Fool me once!

MAINSTREAM OF MISINFORMATION

Beyond the flagrantly antiscience Murdoch media empire and other ultraconservative media outlets, like the *Daily Mail* in the United Kingdom, we have seen a steady erosion in the resolve and commitment of the MSM to defend science and scientists in the face of ideologically motivated attacks. This often manifests as a tendency to "present both sides" in any purported (often manufactured) scientific controversy. *False balance*—the notion that mainstream science and antiscience propaganda deserve equal prominence in our public discourse—and *performative neutrality* are terms that have been used to describe this veneer of objectivity.[80] Historically, mainstream journalists have generally worked hard to get it right, and both of us have spent countless hours working with them to explain the state of scientific understanding, whether it's climate or COVID. But this commitment to truth and accuracy has now started to wane with many mainstream outlets.

False balance about the *reality* of human-caused warming has largely dissipated in mainstream media coverage over the past decade, as the evidence has become literally undeniable. But while "hard" denial has largely disappeared, in its place instead is "soft" denial—the downplaying of the seriousness of the problem. Along with other tactics, such as division, deflection, delay, and another *d* word we will encounter shortly, they are multiple fronts in what Mike has termed "the new climate war." And all of them are evident in abundance today in our press.

Look no further than the so-called paper of record, the *New York Times*. In both their news coverage and their opinion pieces, they have often emphasized individual responsibility over the needed systemic change and have downplayed the role of regulatory policies to rein in polluters.[81] We expect that in Murdoch's *Wall Street Journal*. We shouldn't expect that in the *New York Times*. They could be forgiven, perhaps, for publishing an occasional op-ed by a climate-contrarian pro like "skeptical environmentalist" Bjorn Lomborg, the very titles of which (for example, "The Poor Need Cheap Fossil Fuels")[82] typically drip with mendacity. But they cannot be forgiven for providing a regular column to Bret Stephens, a former opinion writer for the *Wall Street Journal* who has for years been dismissive about the climate crisis. The result has been predictable. As *Slate* headlined a piece about his new stint at the *Times* in 2017, "Bret Stephens' First Column for the *New York Times* Is Classic Climate Change Denialism."[83] Of course, people change their views. And accordingly, Stephens purportedly no longer denies climate change. At least, that's the implication of the title of his November 2022 column "Yes, Greenland Is Melting but . . ."[84] You might suspect that the word *but* is doing a lot of work in that sentence. And indeed it is. Writing for *Gizmodo*, Molly Taft had this to say: "Stephens claims that a visit to Greenland changed his mind on climate change, but the uniquely infuriating essay is still full of Stephens-style bullshit . . . and bad faith arguments on climate. Not much has changed, it seems."[85] It is perhaps worth noting that the *Times* tops the list of fossil fuel–industry media enablers, according to the environmental watchdog outlet DeSmog. They even allow polluters like BP to do podcasts with their branding, quite literally selling the imprimatur of the paper to the highest fossil fuel–industry bidder.[86]

And now consider arguably the most important of the *d* words today: *doomism*. Ironically, it may constitute the greatest threat of all to climate action these days, because it targets not climate deniers or delayers, but those most likely to support action. Consider David Wallace-Wells, author of the notably doomist *New York Magazine* article (and book of the same name) *Uninhabitable Earth*, which argues that we've already triggered irreversible, civilization-ending climate change. In February 2016, the *New York Times* published a commentary by Wallace-Wells titled "Time to Panic," reinforcing that narrative.[87] It's important to emphasize the urgency of climate action. But "panic" messaging can also be counterproductive. We saw earlier in Chapter 5, "The Propagandists," for example, how doomist framing has led to support for potentially dangerous geoengineering schemes as desperation measures.

When it comes to the promotion of doom and despair, it is ironically some of the most progressive media outlets that are most to blame. Let's look at a case in point. Mike and his colleague Katharine Hayhoe, a leading climate scientist and climate communicator, wrote an op-ed about the growing threat of doomism for the *Financial Times* in May 2024, wherein we singled out the *Guardian*—which has in general done excellent reporting and commentary on climate—for promoting unhelpful doomist framing in an article they titled "Hopeless and Broken: Why the World's Top Climate Scientists Are in Despair."[88] The article was based on a flawed "survey" of scientists performed by one of their reporters, which fed a social media frenzy of doomism. It had implied that there was now a scientific consensus that we will blow past "dangerous" warming limits like 1.5°C or 2.0°C, heading instead toward 2.5°C warming. That would be devastating were it true. But that's not what the science indicates. Current pledges and targets would limit warming to around 2°C, and ambitious action could still limit warming below 1.5°C. We are largely in control of our own destiny. So, while there is clearly *urgency*, there is also *agency*. In the war over climate misinformation, we expect to do battle with the *Wall Street Journal*. We don't expect to have to do battle with the *Guardian*.

So far, we've talked about examples of problematic coverage of the climate crisis by the MSM. The situation with COVID is every bit as bad. Reporters for the *New York Times* and *Washington Post* have generally maintained a commitment to science-driven reporting on the SARS-2 virus, its comparison to other coronaviruses, and the COVID-19 vaccines, frequently seeking out experts like Peter in the process. But as the pandemic enters its sixth year, some of those reporters moved on to other venues and topics. And both papers have openly begun to promote either discredited deliberate GoF or lab-leak COVID-origin theories and de-emphasize the strong body of scientific literature on the natural origins of COVID-19. In so doing, they neglect the consensus of the virology-science communities and mislead the public.

Apropos of all of this is an episode involving Dr. Anthony Fauci. Dr. Alina Chan is a molecular biologist and a postdoctoral researcher and adviser at the Stanley Center for Psychiatric Research at Harvard-MIT's Broad Institute, specializing in "gene therapy and cell engineering."[89] Dr. Chan has regularly coauthored op-eds in right-wing media such as the *Telegraph* and *Wall Street Journal*, collaborating with Matt Ridley, a writer who also happens to be a noted coal baron and promoter of both climate-change and COVID-19 conspiracy theories.[90] Together, Chan and Ridley even wrote a book attacking the mainstream science of COVID-19 and promoting lab-leak conspiracy theories.

In early June 2024, on the eve of Fauci's testimony to a hostile, conspiracy theory–driven Republican House of Representatives, the *New York Times* published a misleading op-ed by Chan with the head-scratching title "Why the Pandemic Probably Started in a Lab, in 5 Key Points" that further promoted lab-leak conspiracy theories, replete in the online edition of the *Times* with sophisticated animated maps and graphs that seemed designed to add the patina of scientific rigor and authority that were actually lacking in the scientific claims themselves. The *New York Times* promoted it above all other stories with a commentary by their graphics director in their newsletter to subscribers.[91] Commenting on the piece, leading virologist Dr. Angela Rasmussen wrote, "Key points aren't actually very key when they are factually incorrect. Good of the [*New York Times* opinion section] to help the mob sharpen their pitchforks for Fauci's Select Subcommittee hearing by enshrining the lies that he will be attacked with as truth in the paper of record."[92] This was hardly a one-off. In May 2025, for example, none other than David Wallace-Wells, the *Times* climate columnist we encountered earlier in the chapter, trespassed into public health territory to pen a similarly themed opinion piece, "Why Are So Many

People Sure Covid Leaked from a Lab?," in which he characterized the science behind zoonotic spillover as "ambiguous and unsettled."[93]

Rasmussen was spot-on in her criticism of the Chan piece. It helped feed the Fauci pile-on. Of course, Fox News couldn't resist entering the fray, with Laura Ingraham exclaiming, "We're gonna be paying for the Fauci effect for a generation. Perhaps more," while blaming Fauci for parents dying, divorces, suicides, overdoses, and pretty much all that is bad in the world.[94] But ABC News wasn't much better, emphasizing the theater of the moment over the substanceless nature of the accusations.[95] Even NPR could offer up only the anemic exemplar of false balance: "As Republicans probe COVID's origins, some see an attack on science; others say it's long overdue."[96] Democratic congressman Jamie Raskin publicly apologized to Fauci for the shameful state of affairs: "Some of our colleagues in the United States House of Representatives seem to want to drag your name through the mud. They're treating you, Dr. Fauci, like a convicted felon."[97] But he could have just as easily substituted "the mainstream media" for "House of Representatives."

While it was appropriate to emphasize scientific uncertainty early in the pandemic when the first humans became infected with the SARS-2 coronavirus, by 2022 multiple peer-reviewed articles in top journals such as *Science* had shown how the virus arose through zoonotic spillover. Specifically, the evidence supported a chain beginning with bats and going through a second intermediate wild-animal host—such as raccoon dogs or civets—in much the same way that SARS originated in China in 2002.[98] The analyses even found that there were two distinct lineages that arose, designated A and B, with B arising around November 18, 2019, and A separately at a later period, in other words through multiple zoonotic events.[99] A subsequent analysis of the Huanan Wet Market samples led by Sorbonne professor Dr. Flo Débarre, published in the prestigious journal *Cell* in 2024, revealed the presence of virus RNA colocalized to suspected wild-animal hosts for the virus such as raccoon dogs and civets. The bottom line is that all roads pointed to an origin involving zoonotic spillover in 2019 much like the original SARS back in 2002[100] and as vividly illustrated by the hypothetical pandemic in the 2011 film *Contagion* (discussed in Chapter 4). Such a conclusion is consistent with the major findings of a summary report by the Office of the Director of National Intelligence that determined that COVID-19 "probably emerged and infected humans through an initial small-scale exposure that occurred no later than November 2019" and that "the virus was not developed as a biological weapon" and "probably not genetically engineered."[101]

In addition, the DNI found that "China's officials did not have foreknowledge of the virus before the initial outbreak of COVID-19 emerged."[102] While the intelligence report does not entirely rule out the possibility that the virus was leaked from a laboratory, this is highly unlikely given that there is no record of this virus in any Chinese laboratory inventory despite extensive cooperation between US and Chinese scientists. To this day, there are no published scientific studies that provide any evidence the virus was leaked, and now we have consistent and coherent evidence for zoonotic origins, as well as multiple scientific studies, including those finding related SARS- and SARS-2-like coronaviruses across the face of China and Southeast Asia.[103]

The nuanced language used by scientists is often misunderstood or misinterpreted in public discourse, and that was clearly a problem here. In a *Los Angeles Times* column titled "U.S. Government Debunks COVID Lab-Leak Conspiracy Theory, Enraging Conspiracy Theorists," Michael Hiltzik wrote: "The intelligence report says that both hypotheses 'remain plausible.' But anyone with the slightest capability for critical thinking will see from the text that it comes down against the lab-leak hypothesis." He then went on to condemn a 2023 *ProPublica* article elevating the probability of a lab leak, which turned out was factually incorrect due to a misreading of Mandarin Chinese. The article should have been retracted by the editors. As Hiltzik pointed out: "The lab-leak conspiracy gang has smeared scientists and misled the public into believing a theory that has no factual support whatsoever. They should be ashamed."[104]

In the meantime, the *Washington Post* Editorial Board has published multiple opinion pieces suffering from the same fallacy. They include pieces with the following titles: "China Pressured Experts Away from a Lab-Leak Investigation. What Is It Hiding?" and "Two Possible Theories of the Pandemic's Origins Remain Viable. The World Needs to Know."[105] Both pieces were published early on during the pandemic in 2021, but even in 2023 when considerable scientific evidence had mounted against the lab-leak hypothesis, the *Washington Post* Editorial Board continued with their false equivalency, asking: "What is China hiding and why?" and stating, "Two broad hypotheses exist about the origins of covid-19." They went so far as to offer a "third, plausible explanation" that "might lie between these two" when again there is no evidence at all to support this suggestion.[106]

For those of us experienced in the battle against antiscience groups, we are used to their "just asking questions" gambit but surprised to see something akin to this in MSM news outlets. Another *Post* Editorial Board opinion piece followed a month later with the provocative title "There's New Light—and Lingering Questions—in the Mystery of Wuhan."[107] Around that same time, the *Times* once again weighed in, publishing an article by seasoned reporters carrying the headline "Lab Leak or Not?" insisting, "A lab leak was once dismissed by many as a conspiracy theory. But the idea is gaining traction, even as evidence builds that the virus emerged from a market."[108] Such unhelpful media framing enabled an aggressive, politically motivated GOP-led House Select Subcommittee on the Coronavirus Pandemic, which Michael Hiltzik of the *Los Angeles Times* summarized as "House Republicans give a crash course in how to conduct a conspiracy theory about COVID's origin."[109]

With both COVID and climate, the mainstream media have helped keep alive a pernicious but misguided narrative that—to the extent there is distrust of the science by some members of the public—it's the *scientists'* fault. The erosion of trust, they say, is because we did not communicate the climate and COVID threats effectively to the public (despite aggressive efforts for years to do just that). In a June 2023 article in the *Hill*, Saul Elbein insisted that "scientists failed for decades to communicate" the seriousness of the climate crisis.[110] The scientific study upon which Elbein was commenting in fact said no such thing.[111] It was instead providing a nuanced discussion of the possibility that global sea-level rise could exceed standard climate-model projections. But it is instructive to see how mainstream media coverage framed it within a "scientists aren't telling you how bad it really is" conspiracy-theory narrative.

It's convenient, of course, for MSM outlets to promote this sort of framing, for it deflects attention from their own failures in communicating the dire warnings that scientists have indeed been offering for years. And it ignores the mountain of antiscience aggression we have detailed in this book and the concerted effort to impede scientists in their efforts to communicate the science and its implications to the public and policymakers. And while throwing scientists under the bus may be a convenient ruse for media outlets that have failed to report for duty, it provides dangerous red meat for MAGA conservatives looking for their next target in the science wars. Not only does it—quite literally—endanger scientists, but the denigration of scientists also endangers us all by hindering their efforts to warn the public of the growing threats we face.

WAIT, THERE'S MORE

There are also systemic changes in our media environment that add to the challenge of media misinformation and false equivalencies. Among them is the diminished space for science-related content on news pages and reduced time for it in the television news cycle. Meanwhile, science journalists are being laid off in unprecedented numbers. Both *National Geographic* and *Wired* announced major layoffs in 2023, including the former's entire reporting staff, while *Popular Science* halted its magazine publication after being in existence for more than 150 years.[112] Chelsey Coombs, the former social media editor of *Popular Science*, wistfully opined, "There's not many places left to do science journalism." A general consequence of this decline in science reporting is that the defining crises of our time— climate and pandemics—take second fiddle to the latest Trump scandal or human-interest story. And it couldn't come at a worse time. As Sabrina Imbler of *Defector* put it: "Science journalism has arguably never been more important, as the harsh impacts of climate change are hitting the planet faster than many scientists expected and the biodiversity crisis threatens all corners of life on Earth."[113]

Compounding this situation further is a dearth of working scientists committed to engaging in the public discourse. There is more than enough blame to go around here. As we have already witnessed, there are numerous bad actors who have worked to intimidate and silence scientists in an effort to blunt their efforts to engage with the public and policymakers. Mike and Peter can attest to this firsthand. But there are also scientists who have established a niche for themselves in criticizing scientists for anything they consider to constitute advocacy. In Chapter 4, we learned about how some scientists profited, literally, by attacking Carl Sagan and undermining his advocacy for nuclear de-escalation. That's a defining example. But there are numerous more recent instances of the phenomenon.

Consider Roger Pielke Jr., a political scientist at the University of Colorado who frequently admonishes scientists (particularly, climate scientists) engaged in what he considers to be policy advocacy. What's ironic here is that this itself is a sort of policy advocacy—advocacy for disengagement and "neutrality." Pielke fashions himself an "honest broker" (in fact, that's the title of his book). But as University of Melbourne scholar Darrin Durant points out in his article "Are Honest

Brokers Good for Democracy?"[114] positioning yourself as a putative neutral arbiter of what scientists should and shouldn't say or do is fraught on many levels, leading to "ambiguity about which and whose consensus ought to guide scientists," "an implicit tendency to insulate politics from science," and "a possible replication of the anti-pluralism of political populism." Durant notes the internal inherent contradiction in this case: "Pielke says issue advocacy is 'fundamental to a healthy democracy and is a noble calling,' except when issue advocates don't follow his lead, then they're scurrilous dictators?"[115]

Consider too, in this vein, a recent commentary by Ulf Büntgen published by the journal *Nature*'s online *npj Climate Action* journal. In his commentary, Büntgen insisted, among other things, that "climate science and climate activism should be separated," and "the latter should not be confused with science communication and public engagement"—a curiously bizarre and one-sided policing of scientific discourse.

Büntgen insisted that "scholars should not have a priori interests in the outcome of their studies." The statement seems to exploit the ambiguity of "outcome"; certainly, one's interests shouldn't have any bearing on the outcome of a calculation. But the broader implication seems to be that scientists shouldn't care about the possible applications or implications of their scientific work. It is difficult to imagine anyone with even a cursory familiarity with the history of the Manhattan Project—recently thrust back into the spotlight by the movie *Oppenheimer*—believing that to be true. Such unilateral disarmament in the battle against antiscience is curious indeed. One might liken it to a policy of scientific appeasement. As one of us (Mike) argued a decade ago in a *New York Times* op-ed, "If You See Something, Say Something," "There is a great cost to society if scientists fail to participate in the larger conversation—if we do not do all we can to ensure that the policy debate is informed by an honest assessment of the risks. In fact, it would be an abrogation of our responsibility to society if we remained quiet in the face of such a grave threat."[116]

The forces of appeasement are indeed now gaining an upper hand. We've seen a vast diminishment of the influence of scientists in our public discourse. Our colleagues at Research!America, a policy think tank based in Washington, DC, find that 72 percent of Americans cannot name a living scientist (while more than one-half cannot name a research institution).[117] And the few Americans who can identify a living scientist tend to name celebrities rather than working scientists who regularly write scientific papers, apply for scientific grants, and present their findings at meetings. Scientists who are in the mix and actively doing science are largely invisible to the American public. Tragically for the nation, these three forces—false equivalency by the MSM, the evaporation of the science-journalist workforce, and the invisible working scientist—reinforce each other in ways that greatly obstruct efforts to counteract the misinformation and disinformation of the conservative press.

THE RISE OF CLIENT JOURNALISM
AND FOURTH ESTATE COLLAPSE

In the ancien régime of pre-Revolutionary France, the three estates of the realm were composed of the clergy, the nobility, and the other subjects, mostly the bourgeoisie and the peasants. In "The Fourth Estate as the Final Check," Delbert Tran, deputy attorney general of California, argues that the "fourth estate"—our press—constitutes a critical restoring force in the American system of government. Writing just after Donald Trump was elected in 2016, when one party—the GOP—gained control of all three branches of the government (the presidency, Congress, and a conservative-stacked judiciary), Tran consoled readers, "Beyond the three traditional branches of government, there is another that has often been described as a fourth branch: the free press." Tran emphasized the critical role the press has played throughout our nation's history. He noted how even the Continental army had a printer so citizens would have access to newspapers during the American Revolution. He underscored how the press helped us defeat McCarthyism and exposed the executive branch's malfeasance during Watergate.[118]

This critical check on our system is now mortally threatened. For decades, our judicial branch served as a critical restoring force, yielding progressive policies supporting civil rights and environmental protection in the face of opposition from other branches of our government. Those days are over. As Senator Sheldon Whitehouse of Rhode Island argues in his book *The Scheme*,[119] plutocrats have essentially now purchased our court system, yielding a series of Supreme Court decisions that overturn decades-old precedent, including a woman's right to choose. Plutocrats are now purchasing our media too, effectively removing this final check on the actions of the rich and powerful.

As a result, we have seen the rise of the "client journalism" and "access journalism" that elevates, rather than challenges, the wealthy and powerful, sacrificing principles of objectivity and integrity that are the hallmarks of responsible journalism. The *New York Times*, as we have seen, exemplifies this cancer on the body journalistic. "Why is *New York Times* campaign coverage so bad?" asks Dan Froomkin of *Press Watch*. "Because that's what the publisher wants." Froomkin cites a speech by *Times* publisher A. G. Sulzberger, which he boiled down to two rules that Sulzberger enforces: "One: You will earn my displeasure if you warn people too forcefully about the possible end to democracy at the hands of a deranged insurrectionist [Donald Trump]. And two: You prove your value to me by trolling our liberal readers." Froomkin concludes that Sulzberger's troubling journalistic philosophy "explains a lot of the Times's aberrant behavior." Though the speech was supposed to reflect Sulzberger's response to the numerous critics of the *New York Times* behavior of recent, Froomkin laments that Sulzberger "never honestly engaged with their most serious and urgent concerns. Instead, [he] built straw men and blew them down, from start to finish." Longtime *New York Times* opinion writer (and Nobel Prize winner in economics) Paul Krugman, who had penned commentary for the *Times* for two and a half decades, often on topics such as climate change (about which he and Mike have interacted on several occasions), resigned in January 2025. The *Columbia Journalism Review* quoted him as to why: ""I've always been very, very lightly edited on the column. . . . And that stopped being the case. The editing became extremely intrusive. It was very much toning down of my voice, toning down of the feel, and a lot of pressure for what I considered false equivalence [and, increasingly, attempts] to dictate the subject."[120]

The Jeff Bezos–owned *Washington Post* has an arguably as if not more troubled recent history.[121] On the very same date, June 3, 2024, that the *New York Times* launched their aforementioned attack on Anthony Fauci, the *Washington Post* replaced their first female executive editor, Sally Buzbee, with a Murdoch acolyte, Matt Murray. The *Post* reported on the development. In the "Style" section:

Sally Buzbee, the executive editor of The Washington Post since 2021, has stepped down, publisher and CEO William Lewis announced late Sunday. She will be replaced by Matt Murray, the former editor in chief of the Wall Street Journal, Lewis said. After the presidential election in November, Robert Winnett, most recently the deputy editor of Telegraph Media Group, will take over in a newly created role of editor. . . . The abrupt shake-up at the top of The Post—which Lewis announced alongside ambitious plans for a new division of the newsroom—is the biggest move by far from the British-born journalist since he took over as CEO in January.[122]

There has been a mass exodus of journalists, opinion writers, and editors following Bezos's decision to slant their editorial framing sharply to the right, with a focus on what he calls "personal liberties and the free market." Longtime *Washington Post* editor Marty Baron, who retired in 2021, recently stated of Bezos's recent appeasement to Trump and the MAGA Right that he had "handled his ownership admirably for more than a decade. But his courage failed him when he needed it the most."[123]

If this all sounds like the collapse of the fourth estate, that's because that's precisely what it is. Those of us who value objectivity over "balance" and truth over "controversy" might have to recognize the possibility that our corporatized media can no longer guarantee that. Mike's friend Will Bunch, columnist for the *Philadelphia Chronicle* (which is owned by a nonprofit, rather than a billionaire plutocrat), was blunt in the wake of the 2024 presidential election, drafting a column titled "2024's Other Big Loser? The Mainstream Media . . . ," describing an "electorate besieged by disinformation" that "turned to Trump," noting that "trust in traditional media has imploded" and asking, "What comes next?"[124] The answer, of course, is that nobody knows. We're in uncharted territory.

It's doubtful that the founding fathers could have envisioned the modern corporatocracy that largely now owns and runs our media. Our system of governance wasn't designed to deal with a concerted plutocratic takeover of our entire communication infrastructure. The only true long-term solution is fundamental political change, rebuilding the regulatory environment that has been steadily eroded by Republicans. The dismantling began in earnest with the Federal Communications Commission's elimination of the "Fairness Doctrine"—which required that mandated broadcast networks offer differing viewpoints on topics of public significance—by Ronald Reagan in 1987, including his veto of a Democratic-sponsored bill to reinstate it. Some claim that the Fairness Doctrine wouldn't have prevented the emergence of ideological media networks like Fox News, since it governed the rules for broadcast and not cable programming.[125] That argument is tendentious at best. Congress could have tried to pass similar legislation that would have built upon the Fairness Doctrine, extending it to cable. But how do you build upon something that has been dismantled? Only political change, including massive turnout to support politicians who favor people over plutocrats, can ultimately solve this larger systemic problem. We didn't see that happen in the 2024 presidential election that ceded both the presidency and Congress to an antidemocratic, antiscience-embracing Republican Party. Perhaps we will in the next election. In the meantime, we must nonetheless do what we can to address the other more immediate factors, namely, the "triple threat" of invisible scientists, missing science journalism, and performative neutrality. We'll have more to say about how we might accomplish all of this in the next and final chapter.

7

THE PATH FORWARD

Courage will now be your best defence against the storm
that is at hand—that and such hope as I bring.

—J. R. R. Tolkien, *The Lord of the Rings:*
The Return of the King

Central to J. R. R. Tolkien's *Lord of the Rings* trilogy is an existential quest to achieve victory over evil. We indeed face a loom-
ing antiscience threat of Tolkienesque proportions today—a battle,
in this case not for "middle earth," but for Earth itself. The forces
mounted against us—the plutocrats, petrostates, pros, propagandists,
and the press—are numerous and daunting. To quote elfin Lord
Elrond, "Our list of allies grows thin."

At the risk of extending this metaphor too far (we will stretch
it even further later in the chapter), let us continue on: The ring of
power? It would be the fossil fuels that empower authoritarians,
petrostates, and plutocrats. The dark lord Sauron we are fighting is
the polluter-petrostate juggernaut, which poses a threat to the planet.
Just as the ring had to be destroyed in the fires of Mount Doom, the
coal, natural gas, and oil that imperil us today must be kept in the
ground, together with deadly infectious and zoonotic diseases. And
just as with Saruman, the old "friend" turned foe, our press (and

media)—the once hallowed fourth estate—has been increasingly unhelpful or outright malicious and antagonistic in the narratives they promote. The Uruk-hai and Orcs are the politicians, pros, and the propagandists down in the trenches of social media parroting and spreading antiscience talking points and disinformation.

We wish this epic struggle was no more than a medieval fantasy. We wish that humanity had followed a more enlightened path decades ago when the climate crisis had clearly emerged, or back in 2020 when we were given a golden opportunity to implement pandemic policies—guided by the best available science—to ensure the health of both our species and our planet. Or in 2024 when both fact-based discourse and democracy themselves were on the ballot. But that was not to be. And the reality, now, is that human civilization is in actual grave danger. In the absence of a concerted worldwide effort to address the twin crises of climate change and pandemic threats—and the antiscience that currently obstructs our efforts to take mitigative actions—it could ultimately collapse. In an August 1, 2024, commentary in the *Journal of Virology* titled "The Harms of Promoting the Lab Leak Hypothesis for SARS-CoV-2 Origins Without Evidence," several leading public health scientists issued this stark admonition:

> Science is humanity's best insurance against threats from nature, but it is a fragile enterprise that must be nourished and protected. What is now happening to virology is a stark demonstration of what is happening to all of science. It will come to affect every aspect of science in a negative and possibly a dangerous way, as has already happened with climate science. It is the responsibility of scientists, research institutions, and scientific organizations to push back against the anti-virology attacks, because what we are seeing now may be the tip of the proverbial iceberg.[1]

We couldn't agree more with the entire sentiment. The antiscience superstorm, as we have seen, packs a one-two-three punch, and each of the threats must be addressed in concert with the other. What is required is mobilization unlike any we have yet witnessed. We must treat the rising tide of scientific disinformation with the same urgency as the rise of Nazi Germany and global fascism nearly a century ago or Vladimir Putin's invasion of Ukraine and Middle East conflict today. Antiscience has become a lethal force and a major global security issue. The US government *should* be spearheading steps to halt the spread of antiscience, including a "stop disinformation" campaign, taking actions that disincentivize the plutocrats, pros, propagandists and press from advancing a dangerous and deadly antiscience agenda. But for the time being, bad actors have all three branches of government working with them rather than against them. President Donald Trump has even indicated that he will "ban federal money being used to label domestic speech as 'mis- or dis-information.'"[2]

The UN General Assembly, the UN Security Council, NATO, and future G7 and G20 summits could nonetheless prioritize efforts to slow and halt antiscience disinformation and provide basic protections for the scientists. Scientists shouldn't have to endure death threats online or be attacked by publicity-seeking demagogues or politicians at frivolous congressional hearings. However, we expect this to worsen, rather than get better, in the years ahead.

Ultimately, it comes down to us, as individuals, working toward the needed change. It is all too easy to become disillusioned—and far too many are now throwing up their hands in defeat. But this is what the forces of inaction want. They recognize that both denial and despair can lead to the same destination—disengagement and inaction. They win. We lose. So we must remain focused on pushing back against the tide of antiscience and on advancing the cause of evidence-based science and science-based policy.

Across human history we have learned how social transitions tend to happen through "tipping points" in collective consciousness. A 2018 study found that the "opinion of the majority could be tipped to that of the minority" if it reaches a "critical mass," estimated by some to be roughly 25 percent of the public.[3] For instance, this may explain how we achieved a tipping point in public support for marriage equality in the United States.[4] To paraphrase *Guardian* columnist George Monbiot, social change seems *impossible* . . . until it becomes *inevitable*.[5] A quote often attributed to Mahatma Gandhi, "First they ignore you, then they laugh at you, then they fight you, then you win," may in fact be the paraphrased comments of early-twentieth-century trade-union activist Nicholas Klein: "First they ignore you. Then they ridicule you. And then they attack you and want to burn you. And then they build monuments to you."[6]

But the point is clear—we must push forward, confident in the knowledge that this benevolent tipping point in public consciousness could be near, while mindful of the fact that it must occur before we experience malevolent tipping points in public and planetary health. In this final chapter, we offer a way forward in combating the deadly forces of antiscience. We begin with a discussion of what it will take to win the war on science ("Winning the War"). We then assess the formidable current political landscape and what it will take to change it ("Democratic Values"). We end by presenting our "Battle Plan."

WINNING THE WAR

The good news is that there is still time to oppose the antiscience siege, but in the meantime, it gains power and reach with each passing day. The antiscience disinformation enterprise has expanded, professionalized, and organized into a vast and well-financed entity that exerts influence at the highest levels of the US government; it spreads disinformation and propaganda through the news media, podcasts, books, websites, and antisocial media; it blocks the legislation needed to slow our advance toward climate disaster, while encouraging people to take ineffective or harmful medications in lieu of lifesaving vaccines. Antiscience has already caused serious illness and mass casualties in the near term. Unmitigated, it will in the long term take millions more lives, produce misguided national policies, and have long-lasting catastrophic consequences, including, potentially, the destabilization of our civilization. We have seen this cancer begin to metastasize across the Western Hemisphere, spreading from the United States and Canada into Europe, Australia, and beyond. It now pervades low- and middle-income countries on the African continent and in Asia.

Both the science and the scientists continue to be targeted. Elected members of Congress, state legislators, and their financial backers—fossil-fueled plutocrats and tech bros—have joined forces and work through the conservative media, social media, and the internet to harass and discredit prominent scientists, portraying us as enemies of the people. Tragically, they are succeeding, and our scientific infrastructure is beginning to crumble. Because America is a nation built on science, this assault threatens our place in the world. And without American leadership, there is little hope for addressing the global crises we face. In the absence of a concerted worldwide effort to mitigate the climate crisis and further pandemics, our future is at risk. Both the planet and human civilization are imperiled. The challenge we face is truly daunting.

So, what can we actually do to oppose the juggernaut of civilization-threatening antiscience? There's actually a *lot* we can do. Although it is easy to become overwhelmed, we find it helpful to characterize the needed actions in terms of three overlapping circles of a Venn diagram (Fig. 14).

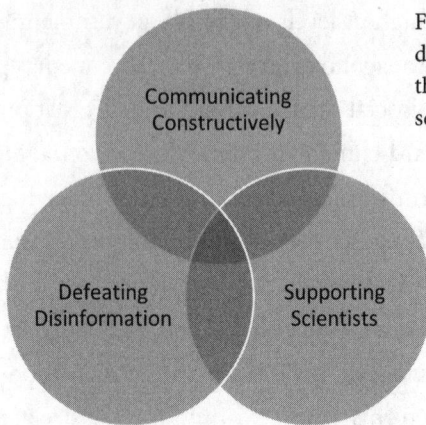

Figure 14. The Venn diagram for winning the war against antiscience

One circle describes ways to expand the visibility of scientists, while providing the tools for scientists to better engage with the public. Another characterizes efforts to protect scientists. And the remaining circle emphasizes the battle against the intensifying flow of antiscience disinformation. We propose a framework for accomplishing this tripartite mission.

Communicating Constructively

The forces of antiscience are extremely well funded and well organized, and message discipline is their name and game. The scientific community has been greatly outmatched. We have let the forces of antiscience define and frame the public debate. So yes, we must battle against disinformation and antiscience if there is to be any prospect for prevailing in this battle for people and the planet, and we'll have more to say about that shortly. But that alone is not enough. As the best defense is indeed a good offense, we need to do a better job communicating science and its implications to the public. We must proactively and constructively address the "triple threat" of invisible scientists, missing science journalism, and performative journalistic neutrality.

Scientists are vulnerable to bad-faith attacks, in part, because in many instances the public does not have a deep understanding of what it is that we actually do as "working scientists." They do not understand how we struggle over revisions of scientific papers and grants, prepare to present our findings at scientific meetings, and mentor our students and postdoctoral researchers. They don't understand the process of scientific-grant applications, the competition for funds, or the reviews by independent scientists. They're unaware, for example, that grants go to the institution, not the individual, and fund our research rather than going to our pockets. During the pandemic, one of the talking heads on Fox News—now Trump NIH director Dr. Jay Bhattacharya—accused Peter of being "funded by Fauci's group"[7] because his lab receives support from the NIAID of the NIH: in fact, the funds go to the Baylor College of Medicine and Texas Children's Hospital, not Peter, and Fauci had no role in the grant decision-making process, which was scored and ranked by an independent study section of outside scientists. Instead, Fox News viewers were given the impression that funding to Peter's laboratory represented some type of unsavory backroom deal. Therefore, part of science education relies on explaining the *processes* of the scientific endeavor.

The scientific community must provide a more open and easily recognizable public face so we are not seen as white-coated oddballs lurking in dark corners. Paul Sutter at the Institute for Advanced Computational Science at Stony Brook University argues: "The public needs to see scientists as people, rather than simply sources of information."[8] While measuring shifts in science literacy is not straightforward, Ethan Siegel, a cosmologist and science communicator, suggests that focusing on asking people to master scientific and quantitative skills is less important than making all citizens better understand "what the enterprise of science actually is," while "fostering an appreciation for how applying the best-known science to our societal problems positively impacts all of us."[9] We couldn't agree more.[10]

Increasing the visibility of scientists and providing the public with a better understanding of who scientists are and what they actually do would help to demystify our words and actions. Unfortunately, there are just not enough working scientists today who are willing and able to operate in the public square. Several recent polls find that trust in science is diminishing, and undoubtedly the far-right press and the false equivalencies of the MSM are to blame for much of this. But the general invisibility of scientists, together with collective lack of preparation "to deal with the media," as Sutter argues, has helped to enable this situation.[11] A recent Pew Research Center poll supports this thesis, demonstrating declines in the number of Americans who believe that science positively impacts society or that scientists seek to act in the public interest. The Pew findings divide along partisan lines, as does acceptance of the science of vaccines and climate change.[12] We must ultimately address this larger structural problem in our politics, but in the meantime, Siegel's and Sutter's goal of educating Americans about the processes and values of science and scientists seems achievable.

Both of us learned science communication at the school of hard knocks. This is not the way it should be. We must create a new cadre of early-career scientists who are both skilled in engaging the public and schooled in the tactics used by the forces of antiscience to disseminate disinformation and propaganda and familiar with best practices in opposing them.[13] Among other things, we need a better incentive structure that encourages PhD-trained scientists to enter into STEM (science, technology, engineering, and mathematics) education at the high school and middle school levels.[14] A well-funded program for top-flight scientists to work in the public education system—a Peace Corps for science, if you will—would mean that kids get to know, and hopefully be inspired by, actual scientists early on in their educational experience. Our scientific institutions—colleges and universities, scientific societies, and academies—need to provide media-training opportunities for young scientists with a proclivity and interest in communication. Such opportunities certainly exist today, thanks in large part to increasing demand among young scientists. But they don't exist at the scale that is necessary.

Creating a cadre of scientists committed to public engagement requires, in turn, a dramatic shift in the incentive structure in academia, science laboratories, and professional societies that properly reward scientists who spend substantial time and effort on public outreach. Part of the remedy is helping university administrators understand the value of public science communication. Since such activities do not generally translate into grant funding for the college, institute, or university and are not incorporated into most of the traditional academic metrics of scientific productivity, they tend to be undervalued or even discouraged. Sadly, both of us hear regularly from scientists who feel disinclined to engage the public because their university or academic health-center offices of communication warn them of placing their institutions at risk. Too often, it goes something like this: "Well, you are an academic and free to speak out, of course. But . . . don't screw things up and place the institution at risk. Maybe it's better if you just keep your head down and focus on your research papers and grant applications." We consider such sentiments to be the academic equivalent of the "sword of Damocles," given that scientists are disinclined to speak out without assurances that they have strong institutional backing.

The opportunities to incorporate science communication into our doctoral and postdoctoral training programs are significant, and many young scientists are especially eager for them.[15] We need to build this into the DNA of the science profession. And our institutions need to join forces in this endeavor. The newly formed "Coalition for Trust in Health & Science," which reflects a partnership between many of the leading health-science organizations in the United States to create a line of communication from the scientific community to the public, provides one possible model.[16]

What about the challenges of *missing science journalism* and the *performative neutrality* of our press? Both phenomena are undoubtedly a consequence of larger structural and systemic changes in our media environment, namely, the massive numbers of science and environmental journalists who have been laid off as media companies have downsized and eliminated science desks. The absence of journalists with the relevant training and background means that science-related stories are more likely to get covered by journalists on the policy or politics beat. They are poorly equipped to litigate the contentious, often technical, debates about the science. That in turn makes them more susceptible to "bothsidesism," the default journalistic tendency to give equal weight to "both sides" in a supposed "controversy" (this has become a huge problem with our MSM as detailed in Chapter 6, "The Press"). Or, conversely, because they are less equipped with historical knowledge of the science and an epistemological understanding of science, they are more likely to fall victim to the "single-study fallacy," treating a brand new study (especially if touted by an uncritical institutional press release overstating its significance) as if it overthrows the entire existing body of scientific understanding. This phenomenon is all too familiar these days in the climate discourse, where a single questionable article is seized upon to justify clickbait climate "doom porn" promising imminent and unavoidable human extinction.

Performative neutrality is even more pernicious, as it happens at the editorial and managerial levels. Even the best science journalist can see the framing of their article sabotaged by a misleading headline and selective editing by editorial staff. There is lot of room to operate when editors, publishers, and media outlets are looking to skew their coverage for political or ideological reasons or simply to appease conservative critics, who will never be satisfied anyway.

Addressing these problems in any fundamental way will require major changes in journalistic standards and policies as well as political change. We will return to that later. But there are still opportunities for making near-term progress. Attempts to expand the pool of doctoral-trained scientists who might be inclined toward journalism would be helpful here. The two of us are actively engaged in such efforts ourselves. Mike is currently working with colleagues at Penn's Annenberg School for Communication to help cultivate a cadre of science-savvy communicators. There are numerous other journalism and communication programs in the United States and abroad that are working to educate a new generation of science and environmental journalists, equipping them to handle the challenges of the current journalism environment.

We need to support the journalists who are working hard to communicate science objectively and responsibly in a challenging media environment while being unafraid to call out those who are not. Social media is today just one means of weighing in. We can also speak with our pocketbook, subscribing to news outlets that promote responsible coverage and canceling subscriptions to and tuning out those that don't. We can encourage others to do the same.

Those of us in the world of science communication must work toward establishing a collaborative relationship with members of the journalistic community. In April 2024, Mike's center at Penn hosted the annual meeting of the Society of Environmental Journalists, the largest member organization of journalists who correspond on the environment and sustainability. It provided an opportunity for scientists, academics, and leading journalists across the nation to spend a week together getting to know each other in plenaries, field trips, and informal events as well as forming relationships: relationships that will benefit all involved, while enabling more effective science and environmental communication.

In instances where the journalistic world is failing, scientists and science communicators themselves need to step up to the plate. They can be proactive about writing op-eds and commentaries and submitting them to local, national, and international publications. They can offer themselves up for media interviews and work with others to amplify their message.

The other side is actively doing all of this. They have cultivated a stable of well-trained, articulate, charismatic individuals with seemingly impressive credentials. They get them on cable television and talk shows and place their commentaries in leading news outlets. They work with an array of conservative celebrities and influencers to amplify their messaging on social media. And mainstream media organizations compromised by performative neutrality happily serve as enablers of it all. Look no further than the *New York Times's* recent embrace of COVID antiscience on their opinion pages in the form of commentaries by nonexpert but authoritative-sounding Ivy League–credentialed individuals. We need our own stable of (actual) experts who are willing and able to write for and speak to popular audiences. We furthermore need buy-in from science-forward organizations and institutions willing to provide resources to train such a cadre and help them get their message out. And we need help from influencers and opinion leaders who can amplify the message using their formidable platforms.

Let us, finally, not neglect conservatives altogether, even if the returns on our investment might seem diminishing. Uncoupling antiscience from the bedrock of conservative thinking is critical to winning over the hearts and minds of the more than one-third of the country that self-identifies today as "conservative." Everyone is entitled to their political views but not their own facts, to quote former New York senator Daniel P. Moynihan. As Jonathan Chait notes in "Donald Trump Has Finally Killed the Pro-science Wing of the Republican Party," the thorough alignment of the Republican Party with antiscience is a relatively recent development.[17] We must convince the libertarian think tanks, conservative colleges, and other right-leaning institutions that by adopting antiscience as a shibboleth today, they are undermining American strength and values and harming our country. They must be brought to understand why this is not their fight and why it makes no sense for them to be in it.

Such outreach is ideally done by trusted messengers within the conservative sphere. That includes members of the religious community, the business world, and national-security leaders. It's worth reminding conservatives that the Republican Party was once a party of environmental stewardship. Think of Nixon's founding of the EPA or Reagan's support of the Montreal Protocol. It once championed science and technology as a driver of progress and prosperity. There are conservative figures who are well positioned to carry this message. A great example is former US congressman Bob Inglis, a "Reagan Republican" House member from South Carolina who lost his congressional seat because of opposition from the Koch brothers after he voiced concerns about global warming and advocated for addressing climate action. Now Inglis travels the country advocating market-driven climate solutions to conservative audiences.

But it's important to recognize that a majority of the people are already on board. Polls show that most Americans do recognize the threats posed by the climate crisis and pandemics and support meaningful policy interventions. In a system of democratic governance, that should be enough. But it isn't, when plutocrats, polluters, and petrostates use the considerable wealth and influence they have to place their thumbs firmly on the scale, in their favor and against ours. That is a reminder, once again, of the importance of political change—we'll get to that soon.

Supporting Scientists

One reason many scientists choose not to engage with the public is the fear that they will find themselves at the center of ideologically and politically motivated attacks aimed at discrediting and intimidating them. Indeed, the intent of these attacks is to serve notice to others who might think of speaking up and speaking out. In *The Hockey Stick and the Climate Wars*, Mike coined a term for the phenomenon, the *Serengeti Strategy*, or the strategy of trying to pick off vulnerable scientists and make an example of them for the rest of the community.

While Mike focused on the intimidation campaign against climate scientists, the principle holds in public health science and any area of science where vested interests perceive themselves to be threatened. Look no further than the more recent attacks on Peter and Tony Fauci by members of the US Congress. They are unlike anything we've seen since the McCarthy era of the 1950s. They are intended, once again, to dissuade other scientists from speaking out.

That is why individual scientists must stand up to the attacks. It sends an important message to others that we, as a community, will not take these attacks lying down. While there are broad US constitutional protections for free speech, false and defamatory statements receive no such protection. Mike speaks from personal experience here. Back in 2012, he was subject to false allegations of fraud by two right-wing writers (Mark Steyn, in the *National Review*, and Rand Simberg, for the Competitive Enterprise Institute) who, adding insult to injury, drew parallels between Mike, a Penn State professor at the time, and Jerry Sandusky, the former assistant football coach of Penn State who was convicted of child molestation. Mike demanded a retraction and apology. When neither individual was willing to do so, he took them to court. As a public figure, there's a high bar for winning a defamation suit. The plaintiff must demonstrate what's known as "actual malice," that is, that not only were the defendants' statements false, but they either knew they were false or (citing the famous *Sullivan v. New York Times* standard) or showed "reckless disregard for the truth."

While it took twelve long years to play out, Mike prevailed, with a Washington, DC, jury unanimously finding in his favor in early

February 2024. Mike received countless congratulatory and celebratory phone calls, text messages, and emails from fellow scientists, policymakers, and heads of major scientific institutions thanking him for persevering. They understood that this wasn't just a victory for a scientist. It was a victory for science and fact-based discourse. At a time when scientists are being harried by conservative politicians and receiving death threats from unhinged individuals who have been weaponized by antiscience disinformation, this was a small but significant victory.

There is a bit of a postscript to this episode that deserves mention. Steyn, later that same year, was slapped down in the UK courts for his wanton and dangerous promotion of antiscience, this time about COVID rather than climate. A right-wing "shock jock," Steyn used to guest host for Rush Limbaugh and has appeared numerous times on Fox News. More recently, he hosted a show in the United Kingdom on the right-wing network GB News. On more than one occasion, he subjected his audiences to COVID-19 misinformation. In April 2022, he falsely asserted that official UK health data demonstrate vaccines caused higher infection, hospitalization, and death rates. Then in October 2022, he had conspiracy theorist Naomi Wolf—whom you may recall from Chapter 5, "The Propagandists"—come on his show and insist to viewers that COVID vaccines were part of an effort "to destroy British civil society" and that this constituted "mass murder," akin to "doctors in pre-Nazi Germany."[18] In response to numerous complaints about the two episodes, the British media watchdog commission OfCom ruled in March 2023 that GB News had violated British media codes of conduct, finding that Steyn had given a "materially misleading interpretation" of COVID data "without sufficient challenge or counterweight," causing potential "harm to viewers." They determined that Wolf had promoted "a serious conspiracy theory," with GB News failing to take "adequate steps to protect viewers" from "potentially harmful content." Steyn insisted that these actions "killed" his career and sued OfCom. The high court of the United Kingdom rejected the suit, ruling that OfCom was "entitled to conclude" that Steyn had violated its rules and that their deliberations had been "detailed and comprehensive."[19] Steyn was ordered to pay OfCom substantial legal costs.[20]

Such legal victories—important as they are—are nonetheless the exception to the rule. Scientists typically depend on the backing of their employers, that is, universities or government science agencies, for legal protections. In some cases, however, this does not happen, and the scientists must arrange their own legal defense, often at considerable expense. Some of these same scientists are often abandoned by their employers after receiving baseless attacks online or, in many cases now, actionable threats of physical harm. These threats are typically encouraged through the sort of stochastic terrorism we've discussed elsewhere, wherein bad actors, be they polluters, propagandists, or politicians, stoke the fires of irrationality and hate, often today through social media campaigns.

We must consider expanding protections for scientists before the attacks get worse. A possible model is the Climate Science Legal Defense Fund that Mike played a role in establishing more than a decade ago.[21] The CSLDF supports climate scientists who are threatened with legal action over their scientific work or subject to frivolous and vexatious open-records and FOIA demands, the only intentions of which are to harass and humiliate them and find grist for the right-wing smear mill. Such protections need to be extended to biomedical scientists and scientists in other fields who face bad-faith, ideologically motivated attacks aimed at discrediting and intimidating them. Peter has suggested creating a clearinghouse of individuals and organizations generating antiscience disinformation and providing legal advice and access to pro bono legal representation for scientists under attack.[22] In the United States, we could also create federal protections for scientists along the lines that Canada now has had in place for two years in the form of laws to protect health-care providers from threats and bullying.[23] In the meantime, the Texas-based Cynthia and George Mitchell Foundation is exploring with Peter the prospects of creating a CSLDF-like structure, but for biomedical scientists.

Governmental agencies and nonprofits need to step up to the plate. Normally, we would look to the White House Office of Science and Technology Policy and the President's Council of Advisors on Science and Technology, but these institutions are compromised now by Trump administration antiscience priorities. However, independent scientific bodies such as the National Academies of Science, Engineering, and Medicine are still in a position to take action. These institutions must prioritize protecting science in America, for if they fail to do so, they will have no purpose in existing anyway. On April 28, 2024, Peter and Mike participated in a plenary panel at the annual meeting of the National Academy of Sciences titled "Scientists Under Fire" along with Anthony Fauci and Yale Medical School immunobiologist Akiko Iwasaki. It was moderated by academy president Marcia McNutt and jointly sponsored by the National Academy of Sciences, the National Academy of Engineering, and the National Academy of Medicine. It included a discussion of the panel members with the audience of academy members, many of whom expressed strong support for the national academies taking a more proactive stance in supporting scientists who find themselves subject to attack. Based on both the audience response and the comments by McNutt following the panel discussion, we believe there are reasons for optimism that we may see a more proactive stance on the part of the academies in the years to come. We also need expanded federal protections for scientists, potentially including protections such as those afforded to health-care professionals in Canada who face bullying and threats,[24] or something akin to the CSLDF but for all professional scientists.[25] At the international level, the other national academies, including the Royal Society of the United Kingdom, must step up as well. The recent call to action by UNESCO advocating for the "promotion of scientific freedom and the safety of scientists" provides a model for the sort of action that is needed at the international level.[26]

Defeating Disinformation

Disinformation is a plague that now threatens us all; it was identified in a recent report of the WEF as the number-one threat our civilization faces in the years ahead.[27] But unlike coronavirus, it cannot be cured with a vaccine—at least not in a literal sense. On the other hand, knowledge and facts and clear, simple explanations that have authority behind them are the metaphorical vaccine for the disinformation virus. We can all contribute to the cure by rebutting antiscientific claims and talking points, blocking and reporting trolls and bots online, correcting the misinformation they spew with links to authoritative sources, and working together to form an army of science communicators and myth busters. On Twitter, one such group—the "#ClimateNinjas"—regularly refutes climate-denial propaganda and reports malignant trolls and bots.

We must fight back against the disinformation that is infecting our schools and misleading our children. Mike serves on the board of directors of an organization—the National Center for Science Education—that is fighting back against efforts by ideologically motivated groups that are targeting the teaching of environmental science and evolution in schools and seeking to replace it with antiscientific propaganda. The NCSE offers a list of actions you can take yourself. They include "harness the enthusiasm of youth activists." If you have young ones, make sure they feel empowered to speak up and speak out. The NCSE also encourages us to "emphasize the injustice of not providing climate change education, remember that politics is the art of the possible." Parents and those who care about the next generation can request that teachers incorporate climate into their teaching. Last but not least, expect political partisanship to be a barrier, communicate with your legislators, and be persistent.[28] In addition, you can, of course, also contribute to groups like the NCSE that are leading the charge.

In their report "The Russian 'Firehose of Falsehood' Propaganda Model," Christopher Paul and Miriam Matthews of the RAND Corporation, suggest ways of countering the Russian disinformation campaign.[29] Drawing from psychology research, they suggest a three-pronged strategy to increase the effectiveness of refutations, which involves warning those who have been exposed to misinformation as early as possible, repetition of the refutation to reinforce it, and providing an alternative, compelling "sticky" counternarrative. However, as they note, forewarning is more effective than refuting falsehoods that have already taken hold. In the science communication literature, this is often referred to as "inoculation." They note that this offsets the advantage propagandists gain through "first impression." Finally, they note, "It may be more productive to highlight the *ways* in which Russian propagandists attempt to manipulate audiences, rather than fighting the specific manipulations."

Most important of all, however, we must defuse disinformation at its source. We must oppose the plutocrats and petrostates who are behind it. Among other things, that means shining a light on "dark money"—the mechanism by which they are now able to funnel unlimited amounts of secretive money to their preferred political candidates. We have the Republican-stacked Supreme Court of the United States (SCOTUS) to thank for that, courtesy of the infamous *Citizens United* decision. Senator Sheldon Whitehouse (D-RI) has led the effort to fight back. He has spearheaded legislation (the DISCLOSE Act)[30] requiring that "dark-money" groups disclose donors giving more than $10,000 or more during any given election cycle. The bill also requires groups funding advertisements supporting or opposing judicial nominees to disclose their donors. Any such efforts are now on hold until at least next year's midterm elections, when Democrats have an opportunity to try to take back the Senate.

Of course, this is only scratching the surface. Plutocrats such as the Kochs, Leonard Leo, and the organizations they fund—including the Heritage Foundation and the Federalist Society—have now fundamentally undermined our democracy and corrupted our politics. In his book *The Scheme*,[31] Whitehouse documents how right-wing plutocrats and corporate interests have successfully implemented a decades-long campaign to purchase the US Congress and the SCOTUS. As a result, we've seen longtime norms disregarded, as Republican presidents and conspiring Republicans in the Senate have blocked Democratic nominees while green-lighting thoroughly inappropriate and unqualified nominees who have been prescreened by the Federalist Society for their support of its antiregulatory agenda. And as a result, we have seen the radical overthrowing of established court precedent by the Trump appointee–loaded Supreme Court, be it in the public health (the 2022 reversal of *Roe vs. Wade*,[32] which codified a woman's right to choose) or planetary health (the 2024 reversal of the Chevron doctrine,[33] which confirmed the federal government's authority to regulate the harmful actions of polluters) domain (Fig. 15).

It isn't just dark-money and right-wing organizations that are the problem. Sometimes the money is on display in broad daylight and comes from ostensibly nonpartisan sources. When it comes to pandemics, for example, Walker Bragman, Alex Kotch, and other investigative journalists have revealed how brokerage firms such as Fidelity, Vanguard, and J.P. Morgan are funneling money to dangerous groups that tout unproven or dangerous cures or harmful disinformation. Exposing this practice and seeking mechanisms to halt the money flow is paramount. The same holds true for the PACs that shuttle funds to antivaccine activist groups that oppose climate action and the renewable energy transition, and lobbying firms that promote an antiscience agenda. In her book *Demagoguery and Democracy*, Patricia Roberts-Miller emphasizes the importance of reducing "the profitability of demagoguery" by reducing its consumption.[34] So far, this has proven to be an aspirational goal. But in the meantime, we must seek ways to cut the flow of funds to antiscience groups that contributed to the hundreds of thousands of American deaths and long-COVID suffering during the pandemic.

We can also use the power of our pocketbook to fight back. Some nonprofits have proposed engaging advertisers to shun Fox News, just as they have Twitter and other promoters of dangerous disinformation,[35] but for the most part these efforts have been underfunded and fragmented. We can encourage our elected leaders to levy pressure. As an example, a group of Democrats who served as their state's election officials wrote to Mark Zuckerberg asking him to cease posting

Figure 15. "Supreme Court Announces New Robes to Reflect Ownership." Benjamin Slyngstad on X

Facebook ads claiming the 2020 election was stolen, underscoring that such falsehoods threaten violence, lead to an erosion of trust in our electoral system, and cause election workers to leave their jobs in fear.[36] Ultimately, however, slowing the tide of societally threatening disinformation will, among other things, require legislation and executive actions.

We might take guidance from other Western nations. In Australia there is a "media diversity" movement afoot to deal with the aftermath of deregulatory policies that have led to media consolidation and self-interested, socially irresponsible, and unaccountable media conglomerates like the Murdoch media empire.[37] In Canada, a recent "anti-greenwash" bill was passed requiring companies to provide evidence to support any environmental claims they make. No sooner had the bill passed than did Pathways Alliance, representing one of the largest Canadian oil-sands companies, scrub their website and social media accounts of offending claims out of fear of legal liability.[38]

In the United States, the First Amendment provides broad "free-speech" protections, and the conservative-stacked Supreme Court, as we know, has been overly sympathetic to corporate polluters. However, there have been some wins on the disinformation front. In June 2024, the Supreme Court actually sided with the Biden administration in a dispute over the executive branch's rights to regulate disinformation.[39] In a six-to-three decision, the Court rejected a lower-court ruling that the administration had overstepped its authority in asking social media companies to curtail dangerous online content threatening the safety or security of the public, such as antisemitic and anti-Muslim posts or posts involving national security, public health, and election integrity. Of course, the legal battles go on.

Meanwhile, there is actually some small degree of bipartisanship in the United States—exceedingly rare in today's charged political environment—for regulation of social media companies. Citizens across the political spectrum generally support regulation of the activities of the technology giants, and especially social media platforms. Bipartisan concern about China's ownership of TikTok—now the most widely used social media platform by youths in America—seems to be a primary impetus for bipartisan support of tech giants, as parents increasingly worry about what their kids are being exposed to online.[40] In May 2024, President Joe Biden's surgeon general, Vivek Murthy, proposed adding a warning label to social media platforms, given the clear scientific evidence that it poses a severe hazard to children in the form of anxiety and clinical depression and bodily dysphoria.[41]

The current Trump administration is unlikely to support those efforts. But several bipartisan bills to regulate social media companies have been introduced in the US Congress in recent years, even if contentiousness remains when it comes to precisely how that should be done and what the government's role should be.[42] Any such efforts today face an uphill battle, as these companies are using their immense wealth and influence to fight back against even the most commonsense regulation. Indeed, the methods they are using—including the creation of front groups to do their dirty work—have a lot in common with past propaganda campaigns funded by big tobacco and the fossil-fuel industry.

Let's look at Mark Zuckerberg and Facebook, which have been widely criticized for platforming extreme right-wing rhetoric and pro-Trump fake news articles during recent elections. With its reputation in tatters, Facebook has not only changed its name (to "Meta") but also created front groups to do its dirty work. A group with the patriotic and noble-sounding name "American Edge Project" is actually a front group founded[43] by Zuckerberg-led Facebook/Meta. It has been running ads attempting to scare Americans into believing that any regulation of social media will have dire consequences for American economic and national security. The ads are clearly intended to intimidate politicians who might support regulation. As the

Washington Post notes, "American Edge ads bash policymakers seeking to regulate the Internet as having a 'misguided agenda,' or wanting to 'take away the technology we use every day.' That tone stands in contrast to public relations campaigns that run under Meta's name, which position the company as an eager partner aiming to update outdated Internet regulations and features Facebook employees calling for 'better guidance' from policymakers on thorny issues like content moderation."[44]

The same group has placed op-eds around the country from purportedly independent and authoritative individuals that promote similar messaging. As one anonymous Facebook insider conceded, "Facebook can't be the messenger. . . . If we are out there saying it, people won't believe it as much, so the conversation is how can you set up a proxy."[45] *If it looks like a propaganda campaign*, swims like a propaganda campaign, and quacks like a propaganda campaign, it's probably a propaganda campaign. One might hope a name-and-shame approach could convince Facebook to cease and desist engaging in these dubious activities. But if not, you have the power of your pocketbook. You might switch to more socially conscious social media platforms like LinkedIn and encourage friends and family to do the same.

Perhaps the most significant question is what to do about new artificial intelligence technology. The US State Department has classified AI as an "extinction-level" threat.[46] But it's not for the reasons you might think. The real threat isn't the Terminator's "Skynet" or *The Matrix*'s "machine world" or any of the scenarios depicted in the technodystopian futures of Hollywood. The threat is how AI can readily be weaponized by the bad actors we've encountered in this book. AI has the capacity to make disinformation—and antiscientific lies—go infinite. It would be nice if big tech—Twitter, Google, Microsoft, ChatGPT, and other companies that generate AI products—would willingly support actions that would mitigate the adverse use of their technology. And there's at least some recognition of this need on their part. Facebook/Meta, for example, has required political advertisers on their platform to disclose any use of AI.[47] Of course, that's a rather minimal action on their part.

Past experience tells us that voluntary compliance of rogue social media companies is unlikely. Consider Twitter's pivot to antiscience, courtesy of Trump-friendly tech plutocrat Elon Musk and the petrostate actors—Russia and Saudi Arabia—that leveraged his buyout of the platform.[48] For years, Twitter was an important medium for the communication of climate information. During the pandemic, it became a valuable tool for the rapid exchange of meaningful scientific ideas and concepts. However, since the Musk takeover in late October 2022, there has been a massive science brain drain from the platform as scientists find it simply too toxic, rife not just with antiscience (much of it quite clearly from inauthentic accounts, troll farms, and bot armies) but with fascism, homophobia, misogyny, racism, and pretty much every negative attribute of the human species. Some sort of functioning public square is essential in today's world if we are to facilitate the sort of real-time interactions between scientists, journalists, policymakers, and members of the public that is so critical to tackling the climate crisis or pandemic threats. But while there has been some movement toward the alternative platform Bluesky,[49] it is uncertain whether another social media platform or combination of social media platforms can fill the void left in the wake of Twitter's collapse in the near term. As long as plutocrats and petrostates are allowed to control these platforms, there's little to protect us from the malicious weaponization of social media platforms by bad actors. Once again, political change remains paramount. The European Union, as noted back in Chapter 5, "The Propagandists," has provided a model for how governments can implement reasonable regulatory constraints on social media platforms.

We must, meanwhile, play defense against the assault on our existing infrastructure for combating disinformation by the Right.

Look no further than what transpired in June 2024 with the Stanford Internet Observatory and their research director, Renée DiResta. (Renée, whom we encountered earlier in Chapter 5, is, we should note, a friend and colleague of ours. She published a must-read book, *Invisible Rulers: The People Who Turn Lies into Reality*, about her experiences on the front lines of the disinformation wars in 2024.) The SIO was the first group to report on Russia's efforts to weaponize social media and fake news in support of Donald Trump's 2016 presidential bid. It developed a curriculum for teaching college students how to navigate increasingly fraught, weaponized social media platforms. It had continued to expose the "bad actors—spammers, scammers, hostile foreign governments, networks of terrible people targeting children, and yes hyper-partisans actively seeking to manipulate the public."[50]

Such efforts made the SIO and other individuals and groups fighting disinformation a threat to Putin and MAGA Republicans who are curiously supportive of him and his agenda. Foremost among them is the 2020 election denier and Republican House member Jim Jordan (R-OH). Investigative journalist Allison Neitzel of *Who, What, Why* explains:

> Ahead of Trump's third run and with consequences for at least a few of his lies and acts finally beginning to catch up to him, the GOP is launching an all-out war against online disinformation research—the collective endeavors of scientists, journalists, and fact-checkers to monitor the online spew of disinformation and hate—with the twin goal of both protecting its own lie machines and distracting from a politically ruinous focus on those consequences.
>
> Central to this effort is Rep. Jim Jordan (R-OH), who, ironically in light of his involvement in the effort to overturn the 2020 election for Trump, now chairs the House Select Subcommittee on the Weaponization of the Federal

Government. He has been, according to a June report in *The Washington Post*, abusing his position of political power to harass and silence disinformation researchers with "a flurry of records requests, subpoenas and lawsuits."[51]

Jordan and like-minded Republicans who, as Colgate University professor John Naughton points out in the *Guardian*,[52] feel threatened by the scrutiny placed on them and their allies by work like DiResta's have engaged in a full-on assault against the SIO and DiResta, using all of the weapons available to them. That includes vexatious congressional subpoenas, expensive (for Stanford and the SIO's staff) lawsuits, constant assault by the right-wing echo chamber of Fox News and Elon Musk's weaponized Twitter, and of course massive Russian troll armies online. They seek to "work the refs" by depicting DiResta and the SIO's efforts at fact-checking and exposing falsehoods as "censorship." And unfortunately, working the refs actually works.

The attacks on DiResta and Stanford have been remarkably successful.[53] In the wake of the unrelenting and expensive assault, Stanford has caved. It has defunded the observatory despite its five years of high-impact work. The founder and director, Alex Stamos, is gone. Stanford refused to renew DiResta's contract and told the other staff to leave. Stanford denies that it intentionally dismantled the observatory. It insists it is committed to supporting independent research and academic freedom. But its actions say otherwise. As Naughton notes, the university has incurred massive legal bills defending itself and "may have decided that enough is enough."

This was hardly an isolated recent instance of appeasement on Stanford's part. In October 2024, one month before the 2024 presidential election, they hosted a high-profile summit of COVID contrarians. Despite widespread condemnation by the scientific community—many of whom were victims of scurrilous personal and ad hominem attacks by the conference speakers—Stanford University president Jonathan Levin chose to kick off the conference himself. It was both shocking and demoralizing to watch one of our ostensibly great universities be complicit in such a display, seemingly oblivious to the ethics of platforming discredited theories that contributed to thousands of American deaths and enabling scurrilous ad hominem attacks against leading health scientists.[54]

If the right-wing disinformation war were World War II, you might say that Stanford's leadership is Neville Chamberlain. Their acts of appeasement to the antiscience Right have been warned against by Yale historian Timothy D. Snyder: "Do not obey in advance. Most of the power of authoritarianism is freely given. . . . A citizen who adapts in this way is teaching power what it can do."[55] Especially during this rise of authoritarianism in America with the reelection of Donald Trump, it is critical that academia—and other civic and governmental institutions in our society—fight back, tooth and nail, against the authoritarian assault on democracy and fact-based discourse. As an example, a group of climate scientists and climate-oriented organizations—including Mike—signed an open letter shortly after the election and prior to the November 13 start of the twenty-ninth UN Climate Change Conference (COP29), demanding that governments take measures to oppose the spread of climate disinformation. Quoting from the letter:

> We, the undersigned educational, climate and information integrity organisations, including the members of the Climate Action Against Disinformation (CAAD) coalition and climate experts, call on *governments worldwide* to take immediate and decisive action to address this crisis. With the COP negotiations setting the stage for global climate action and the G20 Summit offering a crucial platform for international cooperation, it is imperative that governments *recognize the threat of climate disinformation* and take concrete steps to ensure information integrity, paving the way for meaningful

climate action. Governments also need to encourage *social media companies, advertising technology providers, and broadcast and publishing companies to be accountable and stop acting as enablers to planetary destruction.*[56]

DEMOCRATIC VALUES

Dark-money groups and plutocrats have rigged the rules in their favor. Petrostates—Russia in particular—have flooded the zone with disinformation, and fascist elements within the United States and other Western nations are openly collaborating with them. Meanwhile, professional propagandists are spreading disinformation far and wide, overwhelming the defenses of our information infrastructure and saturating our limited attention economy. As if that weren't enough, our press and media have lost their way, showing little appetite for separating fact from fiction.

We now find ourselves on a uniquely challenging battlefield for fighting the rising tide of antiscience that threatens our civilization today. The good news is that the fundamental obstacles aren't physical, or biological, or technological. They are political. And political obstacles—even in today's fraught geopolitical environment—can be overcome.

Science—as an enterprise—strives to be nonpartisan. Historically, this has served its interest, leading to steady support and funding even as the balance of power in the US government has shifted back and forth between Democratic and Republic regimes. But such neutrality is no longer possible at a time when one of the two major parties in the United States—the Republican Party—has an agenda that is so clearly committed to antiscience, including the denial of basic environmental and health science simply because of its policy implications. There is a solution to that problem: vote them out of power. This applies not just to members of Congress but also to state-wide offices, local races, and, of course, as we discussed earlier, the presidency.

We must, in short, take back our politics. We need to restore the rightful role of science in our political and societal discourse if we are to maintain the capacity to address the major challenges we face, including the climate crisis and worsening pandemics. We must get rid of the obstructions that remain to implementing science-based policymaking.

It starts in the United States, at the top, with the executive branch. Among the regulatory agencies directly under the president's control is the EPA, which enforces the laws governing corporate polluters, including the fossil-fuel companies that generate the lion's share of the carbon pollution behind human-caused warming. The FDA and CDC are responsible for regulation of treatments and protocols for dealing with public health threats like COVID-19.

We witnessed an across-the-government approach to climate action during the Biden era, with multiple agencies, including the EPA, Department of Energy, Commerce Department, State Department, and Treasury, each playing a role in executive policies to reduce carbon emissions. We saw the presidency work with Congress—during the Biden presidency—to pass comprehensive climate legislation in the form of the Inflation Reduction Act, whose provisions, if implemented, will lead to a roughly 40 percent reduction in carbon emissions over the next decade. We also saw them implement a "seven-point plan" to deal with the COVID-19 threat. The first bullet point? "Listen to Science."[57] This philosophy stands in stark contrast to what we witnessed—and are now witnessing again—under the prior and now current president, Donald Trump. While Joe Biden, and Barack Obama before him, substantially advanced climate US climate policy, their efforts still fell short of the goal—50 percent reduction in carbon emissions by 2030—required to avert catastrophic warming of more than 1.5°C/3°F. The United States must strive for even more aggressive reductions.

In the case of COVID-19 and pandemics, the executive branch is responsible for regulating harmful medications and bolstering professional credentialing. Hydroxychloroquine, ivermectin, and other unproven or harmful treatments have no value to society, and even worse, they are killing Americans. It is important that the FDA and their legal counsel work to curtail efforts to profiteer from such modern snake oil, while continuing to work with state medical and licensing boards to limit the actions of health-care practitioners who hide behind health-freedom propaganda to extract money from patients or cause them harm.

Unfortunately, it is Trump that we are once again stuck with for another three-plus years. His agenda is codified by Project 2025—a plan that, as noted in previous chapters, involves the thorough dismantling of climate policy in the United States through eliminating agencies overseeing policies related to energy or the environment and repeal of the Inflation Reduction Act and other congressionally enacted climate measures. As if to emphasize its equal-opportunity adoption of antiscience, among the Project 2025 provisions are a substantial downsizing and restructuring of the NIH, FDA, and other science-driven agencies, replacing career scientists and civil servants with agenda-driven political appointees,[58] and the reinstatement of all service members discharged for not getting vaccinated against COVID-19.[59] With Republicans now in full control of the US Congress, at least for one more year, Project 2025 and its antiscience policies on climate and pandemics are in the process of being fully implemented by both the executive and the legislative branches of our government.

That brings us to the remaining branch of government, the judicial branch. It is the president who nominates federal judges and Supreme Court justices. And it is the Senate that confirms them.

Only a Democratic president working with a Democratic Senate can, in the prevailing political climate, undo the tremendous damage that a rogue Trump-packed Court has done to American democracy and all that depends upon it.

In June 2024 the six conservative SCOTUS jurists overturned the long-standing so-called Chevron Deference Doctrine. The doctrine stated that federal courts must defer to expert opinions of federal agencies regarding matters of government regulation. This decision has direct implications for climate action in the United States. The original case in question, after all, was about the right of the EPA to set emissions standards for polluters like Chevron. The SCOTUS decision hampers the executive branch's ability to address carbon pollution through executive actions, such as EPA pollution standards. Such executive actions are the primary means, until now, by which the United States has been able to pursue its obligations to the rest of the world in reducing carbon emissions.

The implications of the Supreme Court's recent actions, however, go far beyond the climate crisis. In writing the Chevron decision, SCOTUS chief justice John Roberts included the assertion (emphasis added): "Perhaps most fundamentally, Chevron's presumption is misguided because agencies have no special competence in resolving statutory ambiguities. *Courts do*."[60] The highlighted statement is profoundly disturbing, as it stands in contradiction to the separation of powers encoded in the US Constitution. "Ambiguities" can be defined so broadly as to include the key details of almost any government statute, so Roberts has in essence now replaced scientifically trained heads of government agencies with politically appointed judges and justices when it comes to the administration of science-related government policy. That includes enforcement of the regulatory statutes of the EPA, the FDA, and CDC, hamstringing the federal government's ability to take action to mitigate the two great crises—climate and pandemics—that we face today.

It seems remarkable to have to state it, but it's simply, objectively true. The SCOTUS now constitutes an existential threat to American democracy. It has permitted hyperpartisan gerrymandering, it has dismantled the Voting Rights Act, and it has allowed unlimited dark money to flood our politics. But worst of all, it has now usurped the role of the executive and legislative branches in legislating and enforcing policy governing human and planetary health. We find ourselves in extraordinary circumstances. And extraordinary circumstances require extraordinary actions.

The SCOTUS has been packed by the plutocrat-funded Federalist Society—with Trump, and before him George W. Bush, as their conduit. They effectively stole four SCOTUS seats over the past six years, leading to a lopsided six-to-three conservative advantage. That action cannot be undone directly. SCOTUS appointments are for life. However, it is possible to restore rightful balance to the Court. And once again, there is no way to avoid being nonpartisan here, because the antidemocratic assault on the Court by the Right has been wholly partisan.

Democrats, if they win back the Senate and presidency in 2028, must consider doing away with the filibuster, an antidemocratic tool originally put in place by proslavery southern whites that requires a supermajority of sixty votes to pass legislation or approve Supreme Court justices. A simple majority of fifty-one senators could then restore the balance of the Court by eliminating the filibuster. An organization of law and public policy experts called "Take Back the Court," whose advisory board includes notable constitutional scholars such as Harvard's Lawrence Tribe, argues convincingly that the Supreme Court should be expanded to thirteen rather than nine justices (equal to the number of circuit courts that now exist in the United States). In this scenario, a Democratic president and US Senate could restore the balance by appointing four more justices.[61] That is a must in any successful effort to turn back the authoritarian tide, but it requires that Democrats run the table in the 2026 midterm election and the 2028 presidential election. That will be difficult to accomplish given the skewing of our political system for Republican partisan gain that is underway today.

THE BATTLE PLAN

The measures outlined above for preserving our scientific infrastructure—enhancing science communication, defending the scientists, and countering antiscience disinformation, while promoting democratic values—are necessary but not sufficient measures. We must ultimately take back our *entire* government from the antidemocratic, plutocratic forces that have temporarily seized control of it. With three more years of Trump (and at least a year more of a Republican Congress), this will be a challenge in the near term. But here's our four-part battle plan in the meantime.

Penalize the Pros, Propagandists, and Petrostates

First, we must create far greater disincentives for bad actors. They have access to weapons that can destroy our civilization. We're not talking about nuclear arms—that threat cannot be written off, of course. But a far more imminent threat has now emerged: cyberwar. Bad actors like Russia have the ability to cripple critical infrastructure; we've already seen them probing vulnerabilities in our energy grid.[62] But even more significantly, we've seen them engage in an assault on our system of democracy through online cyberwarfare.

Russia's tampering in the 2016 US presidential election in favor of electing Donald Trump was simply a taste of things to come.[63] We saw a ratcheting up in the 2020 election where Russia exploited existing racial tensions, creating fake social media accounts aimed at depressing turnout by Black voters.[64] They also played a key role in promoting the trending hashtag #DefundThePolice on social media, seeking to tie the Democratic Party to caricatures of radical Black activism in an effort to scare white voters away from the party.[65]

With the prospect of weaponized AI and "infinite disinformation," we find ourselves in even more dire straits. Whether we like it or not, the technological horse is now out of the barn. It is out there for those looking to abuse it. During the 2024 presidential campaign, we saw the Republican National Committee create an AI-generated video depicting the ostensible dystopian future in which we would find ourselves if Biden were reelected. This included fake AI-generated video of Biden and Vice President Kamala Harris celebrating their 2024 election victory amidst the fabricated scenes of global and national crises that we're supposed to believe would surely follow.[66] This is a decidedly ironic notion given the internal chaos and global upheaval we witnessed in just the first few months of Trump's second presidency. In July 2024, the BBC reported on Russian websites masquerading as US news outlets that were generating phony stories in hopes of influencing the 2024 presidential election—the BBC report indicated that this was part of a larger "AI-powered operation."[67] Russia attacked our electoral process itself during the 2024 election, using bomb threats, for example, to target major Democratic population centers. Their collective efforts were determinative in returning Donald Trump and Republicans to power.[68]

A large segment of the American electorate has now been effectively weaponized by massive disinformation. We need the most punitive possible actions the law allows—to punish state actors like Russia who, in apparent collaboration with prominent members of the GOP, are engaged in a cyberwar against the United States and other Western nations. We need powerful-enough disincentives to dissuade them from doing so. That will require international agreements and protocols regarding cyberwarfare and governments willing to act and work together to sanction bad actors who violate these agreements. We need to treat weapons of mass disinformation as weapons of mass destruction and consider all possible remedies for those who abuse them. Petrostates like Russia and Saudi Arabia are of particular concern and should be served notice of repercussions, including at the very minimum severe sanctions.

With the election of Donald Trump—and a Republican Congress that is in lockstep with his authoritarian agenda—the United States is now itself, for the time being, a petrostate and a bad actor in this space.[69] It is therefore incumbent upon other Democratic nations such as the European Union, United Kingdom, Australia, and Japan to band together (perhaps even joined by the authoritarian nation of China were it to recognize it as being in their long-term interest) to take whatever punitive actions are necessary against bad state actors—the United States sadly now included—to rein them in. A glimmer of hope emerged at the COP29 meeting in Baku in November 2024, as China appeared to step up as a new global climate leader, filling the void created by the intransigent incoming Trump administration.[70] We have now seen the European Union issue stark warnings over what they now perceive as a hostile United States.[71]

Pressure the Plutocrats

Second, we must greatly reduce the influence of plutocrats. They have a stranglehold on our politics, our press, and all who are subject to their disinformation. Only fundamental reforms, including not just disclosure but strict limits on dark money, have any hope of stemming the tide. With a restored SCOTUS, a Democratic president, and a Democratic Congress, it would eventually be possible to finally pass the needed reforms on campaign financing and roll back the absurd *Citizens United* decision that opened the floodgates of dark money that has compromised our politics. We must also restore common-sense regulation and oversight of the media. Reinstating the Fairness Doctrine, eliminated by Ronald Reagan in 1987, and updating it to reflect the realities of the twenty-first-century media environment (that is, extending to apply to cable and streaming) would go some way toward limiting the disinformative propaganda promoted by blatantly partisan plutocrat-owned outlets like Fox News. Stricter regulation of social media companies is also paramount—and that can be accomplished through a combination of executive and legislative action. Unfortunately, once again, these are mostly all pipe dreams for now. Any such efforts are effectively on hold for more than three years while Trump remains president.

In the meantime, we can and must use our voices, organize, speak out, pressure our elected representatives, call out and ridicule the bad actors, be brave, speak truth to power, and back up others willing to do the same. In early March 2025, Mike spoke in Washington, DC, at the "Stand Up for Science" rally held at the Lincoln Memorial, along with other notable science figures such as Bill Nye, Francis Collins, and proscience (former) Republican congressman Fred Upton (R-MI)—and many others—in an effort to do just that.[72] Midterm elections—which are just a year away—are an opportunity to potentially win back at least part of our government to the side of science, reason, and responsibility. This ship won't be turned around on a dime. It will take sustained effort.

Mend the Media

The months leading up to the 2024 election, when our very democracy was on the line, were a test for our media. Many outlets—we've named them and shamed them repeatedly in this book—failed us, but some did not. Mike's hometown newspaper, the *Philadelphia Inquirer*, rose to the challenge, warning readers on a daily basis of the threat of a Trump presidency. The *Guardian*, the *Boston Globe*, and *ProPublica* performed admirably. Unsubscribe from the outlets that failed the test, and subscribe, instead, to those that passed it. Similarly, tune out from the cable news channels that failed us, and watch independent and public television news programming instead. Your pocketbook, once again, is a powerful tool for change.

We will need to use the courts here as well. We have seen in recent years that they are a potent remedy for pushing back against defamatory lies by partisan media outlets. Consider the nearly billion-dollar ($965 million) judgment in 2022 against Alex Jones for promoting fake conspiracies about the Sandy Hook school massacre and the nearly billion-dollar ($787 million) settlement in 2023 by Fox News in the libel suit Dominion Voting Machines filed against them over their lies about the 2020 presidential election.[73] And then there's the judgment in February 2024 that one of us (Mike) obtained against the authors of two defamatory articles making false allegations about him and his research.[74] That leads us to the third and perhaps most important battle-plan item of all.

Be the Change

As the great Mahatma Gandhi reportedly counseled, let us *be the change we wish to see in the world*. This applies to us scientists. Fighting back in the legal realm, as Mike has done, is one way to do that, of course. But there are far more ways that everyday scientists can effect change. As we argued earlier, we can join with our fellow scientists and organize and pressure academic and scientific institutions to take a more proactive stance against antiscientific disinformation and to provide support and defense for scientists subject to concerted right-wing attacks. They must stand up for academic freedom and not bow to the pressure of corrupt, agenda-driven politicians. If we don't put equal pressure on these institutions to do the right thing, they will assuredly cave into the bad-faith demands of polemicists, propagandists, and pressure groups. Look no further than Stanford's pitiful capitulation to right-wing critics in dissolving their Internet Observatory for the study of disinformation because it came under attack by . . . disinformation promoters like Putin-loving Ohio congressman Jim Jordan.

As Mike put it in a recent BBC interview, "Scientists have to step up to the plate and consider playing a more public role in the science policy decisions that will shape our future."[75] We've seen some progress here over the past decade. Back in 2012, Mike's friend and colleague Andrew Weaver, a leading climate scientist from the University of Victoria in British Columbia, ran for higher office. He was elected as the first Green Party member of British Columbia's legislative assembly in 2013 and went on to become the leader of the Green Party of British Columbia in 2015. He used this platform to push for clean energy and oppose the expansion of liquefied natural gas.[76] Climate policy scholar Claudia Sheinbaum, however, took it to a new level in June 2024, running for and being elected president of Mexico. It remains to be seen just what she will do with this platform.[77]

Of course, you hardly need to be an expert to play an important role. Ordinary citizens are the key actors in a democratic system. All of us can work toward increased support for science education and objective and comprehensive science standards in schools. It often comes down to voting, and not just at the presidential level, but at the state and local levels, where school boards are elected and other key policy decisions are made. Even the 2024 election, which handed full power of our federal government to a Republican Party opposed to science-based policies, offered at least one silver lining in the climate domain. Climate initiatives did well across the country. Voters in Washington rejected a ballot measure that attempted to repeal the state's cap-and-trade system for emissions reductions, while voters in California and Hawaii overwhelmingly passed measures to invest in climate resilience. Voters in the fossil-fuel stronghold of Louisiana approved new incentives for clean energy.[78]

We must use not just our votes but our voices as well. Youth climate protesters have helped change the entire conversation, refocusing the discussion on basic matters of ethics and justice. It is incumbent upon all of us to support and join their efforts. At a time when multilateral climate negotiations have become compromised by petrostates and fossil-fuel lobbyists, understandably losing the trust of climate advocates worldwide,[79] we must—as youth climate activists have done—speak truth to power and put pressure on our elected representatives to work toward global climate agreements that truly meet the moment.

We can call upon our celebrities and opinion leaders to use their voices and their platforms. We would love to see major performing artists like Taylor Swift reach out to their immense fan base and mobilize them to support proscience politicians. Any lasting solution to the antiscience crisis will require limiting the ability of vested interests and plutocrats to seize control of our media environment, increasing support for public media, enforcing basic rules of journalistic integrity, and getting special-interest money out of our politics. That comes down to each of us—voting, speaking out, engaging others in conversation. In other words, *being the change*.

Now, as we end, we will revisit the *Lord of the Rings* (*LOTR*) metaphorical framing with which we began the chapter. We will start with a discussion of the dual threats to Middle Earth: the dark lord Sauron and the once-benevolent Wizard Saruman, who has fallen under Sauron's influence and now does his bidding. The two leading threats we face today in the real world are instead the climate crisis and the prospect of increasingly deadly pandemics. In place of Sauron and Saruman and the armies they wield are the plutocrats and petrostates and the armies *they* wield.

In *The Two Towers*—the second book in the *LOTR* trilogy—the two hobbits Merry and Pippin find themselves among Treebeard and his army of tree-like beings known as Ents, the protectors of the forests, as they mobilize against Saruman, who is destroying the forests for his military operation (yes, *LOTR* is replete with environmental themes). The diminutive Pippin questions the purpose of such a small creature as he in this great war. He concedes to Merry that at least "we've got the Shire" and that "maybe we should go home." Merry admonishes Pippin, explaining that the war will spread, "and all that was once green and good in this world will be gone." He chillingly warns, "There won't *be* a Shire, Pippin."

It might be tempting to see the battle against antiscience as too removed from your everyday life. Perhaps you're convinced by this book that it's a worthy fight, but it's not *your* battle; it's the scientists' battle. Yes, you say, it's important that *they* push back against the forces of antiscience, but you've got other priorities to attend to—your own livelihood, your family, your home. In response, we offer this stark warning: if humanity fails to combat the great global crises we face today, there won't *be* an Earth—at least not one that we'd recognize. Yes, there will still be a large spherical planet rotating around the sun. There will be life. But we will lose the welcoming planetary home we know today, with its rich forests and oceans and ecosystems teaming with diverse, interconnected life forms. Yes, there might still be human beings, but nothing that resembles the thriving civilization we have today.

That's stark. You might even say "doomist." So let's talk about doom and despair. It is true that in the absence of a functioning American democracy and cooperation between the nations of the world, we will likely fall victim to our own two towering threats: the climate crisis and ever more widespread and deadly pandemics. But the choice is ours. As we have explained in this book, the obstacles are not physical or technological. They are political. We can do this—*if* we can garner the political will. We still have agency. Those who wrongly insist it's too late—for example, to avert catastrophic planetary warming—are wrongly implying that we are consigned to a dystopian future of runaway planetary warming and widespread extinction no matter what we do. *That* is doomism. And that is wrong.

Today, harm is being done by the spread of such despair and defeatism, some of it—as we have seen—weaponized by bad actors like Russia to create division and disengagement. We are in fact far from defeat. The United States remains close to being on track to meet its commitment to cutting carbon emissions by 50 percent by 2030 and reaching zero by 2050, despite the opposition by Trump, the Republicans, and polluters and petrostates. A path to limiting warming below 1.5°C still exists, though it is becoming increasingly narrow. Yes, we may miss the 1.5°C target. But keeping warming below 2.0°C would still avoid much harm and suffering. It's never too late to make a difference.

Let us consider once again the lament by Elf Lord Elrond, that "our list of allies grows thin." This is true of us today: we are opposed, in our efforts to address both climate and public health threats, by politicians, polluters, and petrostates, propagandists, and, alas, the press too. Yet while Lady Galadriel of the Woodland Realm warns the company of men, elves, and dwarves (the "Fellowship of the Ring") that if they "stray but a little," their quest will fail "to the ruin of all," she provides reassurance that "hope remains while all the Company is true." We have company too. Not a coalition of men, elves, and dwarves but one of scientists, activists, enlightened policymakers, thought leaders, and ordinary citizens who care about our world and our future. For lack of anything better, we'll call it the "Fellowship for the Planet." And so we must remain allied and focused on the battle ahead.

The diminutive hobbit Frodo was saddled with the crucial task of destroying the ring of power and with it the dark lord Sauron whose life force is inextricably tied to it. He is accompanied on this venture by his fellow hobbit and friend Sam. Frodo grows to doubt his resolve. His task begins to feel hopeless. The film adaptation of *The Two Towers* contains this exchange between the two:

> **Frodo**: I can't do this Sam.
>
> **Sam**: I know. It's all wrong. By rights we shouldn't even be here. But we are. It's like in the great stories, Mr. Frodo. The ones that really mattered. Full of darkness and danger, they were. And sometimes you didn't want to know the end. Because how could the end be happy? How could the world go back to the way it was when so much bad had happened? But in the end, it's only a passing thing, this shadow. Even darkness must pass. A new day will come. And when the sun shines it will shine out the clearer. Those were the stories that stayed with you. That meant something, even if you were too small to understand why. But I think, Mr. Frodo, I do understand. I know now. Folk in those stories had lots of chances of turning back, only they didn't. They kept going. Because they were holding onto something.
>
> **Frodo**: What are we holding onto, Sam?
>
> **Sam**: That there's some good in this world, Mr. Frodo . . . *and it's worth fighting for.*"[80]

Relevant too is the so-called Battle of the Black Gate, toward the very end of the story. The coalition of men, elves, dwarves, and hobbits prepares for battle against Sauron's forces at the Gate to Mordor. Believing Frodo to be dead and all prospects for victory gone, they nonetheless fight on. Aragorn—the leader of men—delivers, in the film adaptation, a particularly rousing call to battle:

I see in your eyes the same fear that would take the heart of me.

A day may come when the courage of Men fails, when we forsake our friends and break all bonds of fellowship, but it is not this day.

An hour of wolves and shattered shields when the Age of Men comes crashing down, but it is not this day!

This day we fight!

There is something ageless and archetypal about these literary references. They are fictional, but they speak to deep truths and universal themes of good versus evil, heroism, and the time-honored virtue of fighting the good fight, regardless of the expected outcome.

And so we choose to do battle against the forces of darkness, fighting back against a malevolent movement that represents all that is bad in the world—fascism, authoritarianism, racism, misogyny, and bigotry—a movement that uses antiscientific disinformation as its preferred weapon. We do this not because our success is guaranteed. Given the forces mobilized against us, we are clearly the underdog. And no white wizard will come to our rescue. But we have truth and justice on our side. And the stakes simply couldn't be greater. We fight for a livable planet, for us, our children, and future generations. Because it's worth fighting for.[81]

ACKNOWLEDGMENTS

FROM MICHAEL E. MANN

I am grateful to the many individuals who have provided help and support over the years. First and foremost are my family: my wife, Lorraine; daughter, Megan; parents, Larry and Paula; brothers, Jay and Jonathan; and the rest of the Manns, Sonsteins, Finesods, and Santys.

I am indebted to all those who have inspired me, mentored me, and served as a role model to follow, including, but not limited to, Carl Sagan, Stephen Schneider, Jane Lubchenko, John Holdren, Bill Nye, Paul Ehrlich, Donald Kennedy, Warren Washington, and Susan Joy-Hassol. I thank leaders of the Youth Climate Movement, including Greta Thunberg, Alexandria Villaseñor, Jerome Foster, and Jamie Margolin, for the inspiration they have provided.

I offer thanks to my various colleagues and staff at the University of Pennsylvania who have made me feel so welcome in this academic community, including Presidents Larry Jameson, Liz Magill and Amy Gutmann, Deans Steven Fluharty, Mark Trodden and Sarah Banet-Weiser, and Provost John L. Jackson Jr., as well as others too numerous to list here.

I am greatly indebted to the various politicians on both sides of the political spectrum who stood up against powerful interests to support and defend me and other scientists against politically motivated attacks and who have worked to advance the cause of an informed climate-policy discourse. Among them are Sherwood Boehlert, Jerry Brown, Bob Bullard, Bob Casey Jr., Bill Clinton, Hillary Clinton,

Peter Garrett, Al Gore, Mark Herring, Jared Huffman, Bob Inglis, Jay Inslee, Edward Markey, Terry McAuliffe, John McCain, Christine Milne, Jim Moran, Alexandria Ocasio-Cortez, Harry Reid, Bernie Sanders, Arnold Schwarzenegger, Arlen Specter, Malcolm Turnbull, Henry Waxman, and Sheldon Whitehouse and their various staff.

I want to thank my agents, Jodie Solomon and Rachel Vogel, and the PublicAffairs crew, including my editors Colleen Lawrie and Brian Distelberg, production editor Michelle Welsh-Horst, legal editor Elisa Rivlin, and publicists Brooke Parsons, Alcimary Pena, and Miguel Cervantes, for all their hard work and support with this and other projects.

I also wish to thank the various other friends, supporters, and colleagues past and present for their assistance, collaboration, friendship, and inspiration over the years, including John Abraham, Kylie Ahern, Ken Alex, Yoca Arditi-Rocha, Kurt Bardella, Ed Begley Jr., Andre Berger, Lew Blaustein, Max Boykoff, Ray Bradley, Doug Bostrom, Sir Richard Branson, Jonathan Brockopp, Bill Brune, Will Bunch, James Byrne, Mike Cannon-Brookes, Elizabeth Carpino, Nick Carpino, Keya Chaterjee, Noam Chomsky, Shannon Christiansen, Kim Cobb, Ford Cochran, Michel Cochran, Julie Cole, John Collee, Leila Conners, John Cook, Jonathan Koomey, Patrick Coyne, Jason Cronk, Jen Cronk, Michel Crucifix, Heidi Cullen, Hunter Cutting, Greg Dalton, Fred Damon, Kert Davies, Didier de Fontaine, Brendan Demelle, Deirdre Des Jardins, Andrew Dessler, Steve D'Hondt, Henry Diaz, Leonardo DiCaprio, Paulo D'Oderico, Pete Dominick, Andrea Dutton, Bill Easterling, Kerry Emanuel, Matt England, Howie Epstein, Jenni Evans, Morgan Fairchild, Anthony Fauci, David Fenton, Thierry Fichefet, Chris Field, Frances Fisher, Pete Fontaine, Josh Fox, Al Franken, Pierre Friedlingstein, Peter Frumhoff, Jose Fuentes, Andra Garner, Peter Garrett, Peter Gleick, Jeff Goodell, Amy Goodman, Hugues Goosse, Nellie Gorbea, David Graves, David Grinspoon, Alex Hale, David Halpern, Thom Hartmann, David Haslingden,

Susan Joy Hassol, Katharine Hayhoe, Tony Haymet, Megan Herbert, Bill Higgins, Michele Hollis, Rob Honeycutt, Ben Horton, Malcolm Hughes, Amorie Hummel, Kathleen Hall Jamieson, Jan Jarrett, Paul Johansen, Phil Jones, Jim Kasting, Bill Keene, Sheril Kirshenbaum, Barbara Kiser, Johanna Köb, Jonathan Koomey, Miroslava Korenha, Heather Kostick, Kalee Kreider, Paul Krugman, Lauren Kurtz, Greg Laden, Chris Larson, Deb Lawrence, Tony Leiserowitz, Stephan Lewandowsky, Diccon Loxton, Jane Lubchenko, Ed Maibach, Scott Mandia, Joseph Marron, John Mashey, François Massonnet, Roger McConchie, Andrea McGimsey, Bill McKibben, Marcia McNutt, Pete Meyers, Sonya Miller, Chris Mooney, John Morales, Granger Morgan, Ellen Mosely-Thompson, Leilani Munter, Ray Najjar, Giordano Nanni, Jeff Nesbit, Phil Newell, Mary Nguyen, Gerald North, Dana Nuccitelli, Miriam O'Brien, Michael Oppenheimer, Naomi Oreskes, Tim Osborn, Jonathan Overpeck, Lisa Oxboel, Rajendra Pachauri, Blair Palese, David Paradice, Jeffrey Park, Ray Pierrehumbert, Rick Piltz, Phil Plait, John Podesta, James Powell, Kimberly Prather, Stefan Rahmstorf, Cliff Rechtschaffen, Hank Reichman, Ann Reid, Annie Reidl, Catherine Reilly, James Renwick, Andy Revkin, Tom Richard, David Ritter, Eugene Robinson, Alan Robock, Joe Romm, Lyndall Rowley, Mark Ruffalo, Scott Rutherford, Sasha Sagan, Barry Saltzman, Ben Santer, Julie Schmid, Gavin Schmidt, Steve Schneider, Eugenie Scott, Joan Scott, John Schwartz, Marshall Shepherd, Drew Shindell, Randy Showstack, Hank Shugart, David Silbert, Peter Sinclair, Michael Smerconish, Dave Smith, Jodi Solomon, Richard Somerville, Graham Spanier, Amanda Staudt, Eric Steig, Byron Steinman, David Stensrud, Nick Stokes, Sean Sublette, Larry Tanner, Jake Tapper, Lonnie Thompson, Holden Thorp, Kim Tingley, Dave Titley, Lawrence Torcello, Sarah Thompson, Kevin Trenberth, Fred Treyz, Katy Tur, Leah Tyrrell, Ana Unruh-Cohen, Jean-Pascal van Ypersele, Ali Velshi, Dave Verardo, Mikhail Verbitsky,

David Vladeck, Nikki Vo, Bob Ward, Bud Ward, Bill Weir, Ray Weymann, Robert Wilcher, John B. Williams, Barbel Winkler, and Christopher Wright.

FROM PETER J. HOTEZ

I am indebted to my family who have supported me for decades, including my wife, Ann Hotez, and my adult sons and daughters and their spouses (listed in the Dedication) and other family members, including Dr. Lawrence and Linda Hotes, Elizabeth and Warren Kirshenbaum, and Andi Hotes. Ann is someone who has stood by me for forty years and someone I depend on for advice, counsel, and her infinite wisdom, especially in dealing with the complexities of an ever-expanding antiscience ecosystem. I am also indebted to my cousins, the extended Hotez, Goldberg, and Lazowski families; longtime family friends from West Hartford, Connecticut, the Conway family; my brother- and sister-in-law, David Frifield (and his wife, Megan) and Julia Frifield; my mother-in-law, Marcia Frifield, and my late father-in-law, Don Frifield; and the late Rochelle Lurie.

I am indebted to those who have inspired or mentored me during the decades I have spent at universities and academic health centers, some since I was a Yale undergraduate. They include Professors Curtis Patton, Frank Richards, Eugene Shapiro, Warren Andiman, Richard Bucala, Robert Baltimore, Michael Cappello, Marietta Vazquez, George Miller, Alison Galvani, Albert Ko, and Megan Ranney (Yale University); Professors Anthony Cerami and Barry Coller and President Richard Lifton (Rockefeller University); Professors Gerhard Schad, Paul Offit, David Roos, and Stanley Plotkin (Penn); Professors Gary Simon, Jonathan Reiner, Allan Goldstein, David Diemert, Jeff Bethony, John Hawdon, and Paul Brindley (George Washington University); Professor Kevin Tracey (Feinstein Institute); Professors Huda Zoghbi, Bert O'Malley, Richard Gibbs, Cheryl Walker, Anthony

Maresso, Joe Petrosino, Erez Lieberman, Daniel Musher, Sheldon Kaplan, Julie Boom, Lara Shekerdemian, Gordon Schutze, and Mary Estes (Baylor College of Medicine); President and Chief Executive Officer Emeritus Mark Wallace (Texas Children's Hospital); Drs. Rich Roberts and Barton Slatko (New England BioLabs); Drs. Mark Rosenberg and Ruth Berkelman (CDC); Professors Michele Barry, Nathan Lo, and Bonnie Maldonado (Stanford University); Professors Avi Israeli and Howard Chaim Cedar (Hebrew University); Professor Jon Samet (University of Colorado); Drs. Roger Glass and Barbara Stoll (China Medical Board); Professors Anthony Fauci, Ruth Katz, Larry Gostin, and Bruce Gellin (Georgetown University); Professors Anne Moscona, Martin Chalfie, Salim Abdool Karim, and Jordan Orange (Columbia University); Professors Andrew Pollard, Adrian Hill, and Sarah Gilbert (Oxford University); Director General and Professor Yasmine Belkaid (Institut Pasteur); Professors Arthur Caplan, Ruth Ben-Ghiat, and Adam Ratner (New York University); Professors Sallie Permar and Carl Nathan (Weill Cornell Medical College); Professors Carol Greider and William Sullivan (UC Santa Cruz); Professor Art Reingold (UC Berkeley); Professors Ashish Jha, Scott Rivkees, and Jennifer Nuzzo (Brown University); Professors Elijah Paintsil and Gerald Keusch (Boston University); Professor Jermey Farrar and Director General Tedros Adhanom Ghebreyesus (WHO); Professor David Molyneux (Liverpool School of Tropical Medicine); Dr. Richard Horton (*Lancet*); Professors Alimuddin Zumla, Alan Fenwick, Andrew Haines, and Roy Anderson (Imperial College and University College London); Professor Mike Osterholm (University of Minnesota); Professors Anne Rimoin and William Gelbart (UCLA); Drs. Eric Topol and Kristian Anderson (Scripps Institute); Dr. Margaret Hamburg, Dean Maureen Lichtveld, and Dr. Peter Salk (University of Pittsburgh); President Rev. Peter M. Donohue and Provost Patrick Maggitti (Villanova University); Professor Victor

Dzau (National Academy of Medicine); Professor Marcia McNutt (National Academy of Sciences); Dr. Holden Thorp (American Association for the Advancement of Science and the magazine *Science*); Dr. John Nemeth (Sigma Xi); Professors Judd Walson, Ruth Karron, Panagis Galiatsatos, Mathuram Santosham, Lauren Gardner, and Daniel Salmon (Johns Hopkins University); Marc and Jeri Shapiro; Gary and Lee Rosenthal; Professor Robert Langer (MIT); Dr. Wendy S. Levine; and Susan Feigin Harris.

I also want to thank a number of individuals for their advice in navigating the complex antiscience space, including Professor Angie Rasmussen (University of Saskatchewan), Professor Flo Débarre (Sorbonne, Paris); Professor David Gorski (Wayne State University); Professor Dorit Reiss (UC San Francisco Law School); Dr. Jonathan Howard (New York University); Michael Hiltzik (*Los Angeles Times*); Jeff Storobinsky and Dr. Saskia Popescu (RAND); Dr. Allison Netizel, Dr. Andrea Love, Dr. Jess Steier, Dr. Madhu Pai, Dr. David Joffe, and Dr. Neil Stone (University College London); Joshua Cohen (*Forbes*); Terri Burke and Rekha Lakshmanan (the Immunization Partnership); Adrienne Dreiss Ropp (Cynthia and George Mitchell Foundation); Dr. Peter Daszak (EcoHealth Alliance); Anna Merlan (*Mother Jones*); Dr. Jonathan Stea (University of Calgary); Dr. Vince Iannelli, Lee Baird, Dr. Charles Gaba, Dr. Tyler Black, Walker Bragman, Noah Ruderman, IntegralAnswers, Michael Board, Jeff Storobinsky, Art Pronin, HOUmanitarian, Dianne Whitehead, Jason Parker, Deirdre Des Jardins, Morgan Fairchild, Alyssa Milano, Dr. Annette Lee, Dr. Eric Feigl-Ding, Mehdi Hasan, and Dr. Dean Baker (Center for Economic & Policy Research); Professor Madhu Pai (McGill University); Dr. Flint Dibble, Maria de Los Angeles (Maque) Garcia de la Garza, Dr. Brian Goldman, and Professors Bruce Lee, Scott Ratzan, and Ayman El-Mohandes (CUNY); Dr. Phil Markolin and Professor Marion Koopmans (Erasmus University); Mark Thomson (MWPP); Imran Ahmed (CCDH); Professor Tim Callaghan

(Boston University); and Professor Timothy Caulfield (University of Alberta).

Finally, I want to thank the leadership of Baylor College of Medicine, Texas Children's Hospital, Baylor University, Texas A&M University, Rice University, and the Texas Medical Center for their stalwart support in these challenging times. Particular thanks go to Dr. Paul Klotman, president of Baylor College of Medicine and the Baylor Executive Leadership Team; Greg Brenneman (Baylor College of Medicine Board Chair); Dr. Debra Feigin Sukin, the president and chief executive officer of Texas Children's Hospital; Dr. Lara Shekerdemian, the Baylor College of Medicine chair of pediatrics; Dr. Maria Elena Bottazzi, the codirector of the Texas Children's Hospital for Vaccine Development and Pediatric Tropical Medicine Division chief; Robert Corrigan, Baylor College of Medicine senior vice president and general counsel; Patrick Turley and James Banfield, Baylor general counsel; Afsheen Davis, Texas Children's Hospital senior vice president and general counsel; and Crowell & Moring, LLP, president Linda Livingstone, provost Nancy Brickhouse, dean Lee Nordt, and Dr. Richard Sanker (Baylor University); Professors David Satterfield and Scott Solomon, and President Reginald DesRoches (Rice University); Professors John Junkins, Clifford Fry, Gerald Parker, Brian Colwell, Dean John August, and Andrew Natsios (Texas A&M University); and Bill McKeon, the chief executive officer of the Texas Medical Center. I also want to thank the Texas Medical Center and Houston Police Departments, the Harris County Sheriff's Office, the FBI, and Baylor College of Medicine and Texas Children's Hospital Security, as well as the ADL, Southwest Chapter.

I also wish to thank Colleen Lawrie and Brian Distelberg at PublicAffairs Books and Rachel Vogel at Dunow, Carlson & Lerner Literary Agency. I also want to thank my editorial assistant of many years, Nathaniel Wolf, as well as Douglas Soriano and Nathan Harrington.

NOTES

CHAPTER 1: THE 1-2-3 PUNCH

1. US Centers for Disease Control and Prevention (CDC), "COVID Data Tracker: Deaths," https://covid.cdc.gov/covid-data-tracker/#deaths-landing, accessed January 20, 2024; US Environmental Protection Agency, "Climate Change Indicators: Heat-Related Deaths," www.epa.gov/climate-indicators/climate-change-indicators-heat-related-deaths, accessed January 20, 2024; C. J. Carlson, "After Millions of Preventable Deaths, Climate Change Must Be Treated Like a Health Emergency," *Nature Medicine* (2024).

2. P. J. Hotez, *Preventing the Next Pandemic: Vaccine Diplomacy in a Time of Anti-science* (Johns Hopkins University Press, 2023).

3. Michael E. Mann, *The New Climate War: The Fight to Take Back Our Planet* (PublicAffairs, 2021); W. J. Broad, "Putin's Long War Against American Science," *New York Times*, April 13, 2020, www.nytimes.com/2020/04/13/science/putin-russia-disinformation-health-coronavirus.html.

4. Clifford Young, Sarah Feldman, and Bernard Mendez, "The Link Between Media Consumption and Public Opinion," Ipsos, October 18, 2024, www.ipsos.com/en-us/link-between-media-consumption-and-public-opinion.

5. Carl Sagan, *The Demon-Haunted World: Science as a Candle in the Dark* (Random House, 1995). See the review by M. Mann, "Summer Books," *Nature* 548 (2017): 28–30.

6. Kyle Spencer, "Inside the Far Right's Fight for College Campuses," *Rolling Stone*, November 27, 2022, www.rollingstone.com/politics/politics-features/raising-them-right-far-right-fight-college-campus-1234636392/; Naomi Oreskes and Erik M. Conway, "From Anti-government to Anti-science: Why Conservatives Have Turned Against Science," *Daedalus* 151, no. 4 (2022): 98–123, https://doi.org/10.1162/daed_a_01946.

7. P. J. Hotez, "Global Vaccine Access Demands Combating Both Inequity and Hesitancy," *Health Affairs* 42, no. 12 (2023): 1681–1688, https://doi.org/10.1377/hlthaff.2023.00775, PMID: 38048497.

8. P. J. Hotez, *The Deadly Rise of Anti-science* (Johns Hopkins University Press, 2023).

9. A. Lardieri, "America Is on the Edge of a Dangerous 'Vaccine Tipping Point' Says FDA—amid Measles Outbreak and Record High Vaccine Refusers," *Daily Mail*, January 17, 2024, www.dailymail.co.uk/health/article-12975143/united-states-vaccine-rates-tipping-point-says-fda-commissioner.html.

10. https://www.latimes.com/business/story/2024-06-19/this-gop-leaning-polling-firm-has-turned-into-a-purveyor-of-anti-vaccine-propaganda.

11. Julia Métraux, "New Hampshire's Republicans Are Taking a Stand—Against the Polio Vaccine," *Mother Jones*, April 20, 2024, www.motherjones.com/politics /2024/04/new-hampshire-republicans-polio-mmr-measles-vaccine-antivax-bill/.

12. Laura Barrón-López and Jackson Hudgins, "Trump Vows to Defund Schools Requiring Vaccines for Students If He's Reelected," *PBS NewsHour*, June 24, 2024, www.pbs.org/newshour/show/trump-vows-to-defund-schools-requiring -vaccines-for-students-if-hes-reelected.

13. Christina Jewett and Sheryl Gay Stolberg, "RFK Jr.'s Lawyer Has Asked the FDA to Revoke Polio Vaccine," *New York Times*, December 13, 2024, www .nytimes.com/2024/12/13/health/aaron-siri-rfk-jr-vaccines.html.

14. Hotez, *Deadly Rise of Anti-science*.

15. Mann, *New Climate War*.

16. Naomi Oreskes and Erik M. Conway, *Merchants of Doubt: How a Handful of Scientists Obscured the Truth on Issues from Tobacco Smoke to Global Warming* (Bloomsbury Press, 2010).

17. Hotez, *Deadly Rise of Anti-science*.

18. P. J. Hotez, "On Antiscience and Antisemitism," *Perspectives in Biology and Medicine* 66, no. 3 (2023): 420–436.

19. G. Piel, *Science in the Cause of Man* (Alfred A. Knopf, 1961).

20. B. Nogrady, "'I Hope You Die': How the COVID Pandemic Unleashed Attacks on Scientists," *Nature* 598, no. 7880 (2021): 250–253, https://doi.org/10.1038 /d41586-021-02741-x, PMID: 34645996; C. O'Grady, "In the Line of Fire," *Science* 375, no. 6587 (2022): 1338–1343, https://doi.org/10.1126/science.abq1538, PMID: 35324295.

21. Hotez, *Deadly Rise of Anti-science*.

22. Michael E. Mann, *The Hockey Stick and the Climate Wars: Dispatches from the Front Lines* (Columbia University Press, 2014).

23. Hotez, *Deadly Rise of Anti-science*; Oreskes and Conway, "From Anti-government to Anti-science."

24. Oreskes and Conway, "From Anti-government to Anti-science."

25. Mann, *Hockey Stick and the Climate Wars*.

26. Michael E. Mann, "'Widespread and Severe': The Climate Crisis Is Here, but There's Still Time to Limit the Damage," *Time*, August 9, 2021, https://time .com/6088531/ipcc-climate-report-hockey-stick-curve/.

27. Michael E. Mann, "Yes, We Can Still Stop the Worst Effects of Climate Change. Here's Why," *LiveScience*, November 14, 2023, www.livescience.com /planet-earth/climate-change/yes-we-can-still-stop-the-worst-effects-of-climate -change-heres-why.

28. Alister Doyle, "Evidence for Man-Made Global Warming Hits 'Gold Standard': Scientists," Reuters, February 26, 2019, www.reuters.com/article /idUSKCN1QE1ZT/.

29. Michael E. Mann and Susan Joy Hassol, "That Heat Dome? Yeah, It's Climate Change," *New York Times*, June 29, 2021, www.nytimes.com/2021/06/29 /opinion/heat-dome-climate-change.html.

30. Mike Weilbacher, "James Hansen, Whose Senate Testimony Made Waves in 1988, Was Right About Climate Change," *Philadelphia Inquirer*, July 10, 2023, www.inquirer.com/opinion/climate-change-james-hansen-testimony-35-years -20230710.html.

31. Mann, *New Climate War.*

32. Mann, *Hockey Stick and the Climate Wars.*

33. Fred Pearce, "Climate Change Special: State of Denial," *New Scientist,* November 2006.

34. See Mann, *Hockey Stick and the Climate Wars.*

35. Daniel Wolfe and Daniel Dale, "'It's Going to Disappear': A Timeline of Trump's Claims That Covid-19 Will Vanish," CNN, October 31, 2020, https://edition.cnn.com/interactive/2020/10/politics/covid-disappearing-trump-comment-tracker/index.html.

36. Michael E. Mann, "Climate Scientists Feel Your Pain, Dr. Fauci," *Newsweek,* August 11, 2020, www.newsweek.com/climate-scientists-feel-your-pain-dr-fauci-opinion-1524293.

37. Max Roser, "Data Review: How Many People Die from Air Pollution?," Our World in Data, November 25, 2021, https://ourworldindata.org/data-review-air-pollution-deaths.

38. "Extreme Weather Caused Two Million Deaths, Cost $4 Trillion over Last 50 Years," United Nations News, May 22, 2023, https://news.un.org/en/story/2023/05/1136897.

39. Denise Mann, "Workers in U.S. Southwest in Peril as Summer Temperatures Rise," *U.S. News & World Report,* May 18, 2022, www.usnews.com/news/health-news/articles/2022-05-18/workers-in-u-s-southwest-in-peril-as-summer-temperatures-rise; Zachary Hansen, "It's So Hot in Phoenix, They Can't Fly Planes," *AZ Central,* June 20, 2022, www.azcentral.com/story/travel/nation-now/2017/06/19/its-so-hot-phoenix-they-cant-fly-planes/410766001/.

40. "Maricopa's Ozone High Pollution Advisory Extended Through Tuesday, June 20, 2017," www.az-phc.com/2017/06/page/36/.

41. Hotez, "Global Vaccine Access Demands Combating Both Inequity and Hesitancy."

42. Hotez, "Global Vaccine Access Demands Combating Both Inequity and Hesitancy."

43. Hotez, *Preventing the Next Pandemic.*

44. P. J. Hotez and A. D. LaBeaud, "Yellow Jack's Potential Return to the American South," *New England Journal of Medicine* 389, no. 16 (2023): 1445–1447, https://doi.org/10.1056/NEJMp2308420, PMID: 37843124.

45. P. J. Hotez and K. O. Murray, "Dengue, West Nile Virus, Chikungunya, Zika—and Now Mayaro?," *PLOS Neglected Tropical Diseases* 11, no. 8 (2017): e0005462, https://doi.org/10.1371/journal.pntd.0005462, PMID: 28859086, PMCID: PMC5578481.

46. T. Ahmed et al., "Climatic Conditions: Conventional and Nanotechnology-Based Methods for the Control of Mosquito Vectors Causing Human Health Issues," *International Journal of Environmental Research and Public Health* 16, no. 17 (2019): 3165, https://doi.org/10.3390/ijerph16173165, PMID: 31480254, PMCID: PMC6747303; C. Baril et al., "The Influence of Weather on the Population Dynamics of Common Mosquito Vector Species in the Canadian Prairies," *Parasites and Vectors* 16, no. 1 (2023): 153, https://doi.org/10.1186/s13071-023-05760-x, PMID: 37118839, PMCID: PMC10148408.

47. S. J. Ryan et al., "Global Expansion and Redistribution of Aedes-Borne Virus Transmission Risk with Climate Change," *PLOS Neglected Tropical Diseases*

13, no. 3 (2019): e0007213, https://doi.org/10.1371/journal.pntd.0007213, PMID: 30921321, PMCID: PMC6438455.

48. P. J. Hotez, *Blue Marble Health: An Innovative Plan to Fight Diseases of Poverty amid Wealth* (Johns Hopkins University Press, 2016).

49. P. J. Hotez et al., "Hookworm Infection," *New England Journal of Medicine* 351, no. 8 (2004): 799–807, https://doi.org10.1056/NEJMra032492, PMID: 15317893.

50. S. Fuhrimann et al., "Risk of Intestinal Parasitic Infections in People with Different Exposures to Wastewater and Fecal Sludge in Kampala, Uganda: A Cross-sectional Study," *PLOS Neglected Tropical Diseases* 10, no. 3 (2016): e0004469, https://doi.org/10.1371/journal.pntd.0004469, PMID: 26938060, PMCID: PMC4777287.

51. M. L. McKenna et al., "Human Intestinal Parasite Burden and Poor Sanitation in Rural Alabama," *American Journal of Tropical Medicine and Hygiene* 97, no. 5 (2017): 1623–1628, https://doi.org/10.4269/ajtmh.17-0396, errata in *American Journal of Tropical Medicine and Hygiene* 98, no. 3 (2018): 936, PMID: 29016326, PMCID: PMC5817782.

52. K. P. Puchner et al., "Vaccine Value Profile for Hookworm," *Vaccine* 42, no. 19 (2024): S25–S41, https://doi.org/10.1016/j.vaccine.2023.05.013, PMID: 37863671.

53. D. Blackburn et al., "Outbreak of Locally Acquired Mosquito-Transmitted (Autochthonous) Malaria—Florida and Texas, May–July 2023," *MMWR Morbidity and Mortality Weekly Report* 72 (September 8, 2023): 973–978, https://doi.org10.15585/mmwr.mm7236a1, PMID: 37676839, PMCID: PMC10495185.

54. T. Lovey and P. Schlagenhauf, "Augmentation des températures et menace du paludisme en Europe: Un retour indésirable?" [Rising Temperatures and the Threat of Malaria in Europe: An Unwelcome Return?], *Revue Médicale Suisse* 19, no. 825 (2023): 849–852, https://doi.org/10.53738/REVMED.2023.19.825.849, PMID: 37139879.

55. Bat Conservation International, "Bats Feel the Effects of Climate Change," November 1, 2023, www.batcon.org/bats-feel-the-effects-of-climate-change/.

56. M. Gilbert et al., "Presence of SARS-CoV-2-Like Coronaviruses in Bats from East Coast Malaysia," *Tropical Biomedicine* 40, no. 3 (2023): 273–280, https://doi.org/10.47665/tb.40.3.001, PMID: 37897158.

57. C. A. Sánchez et al., "A Strategy to Assess Spillover Risk of Bat SARS-Related Coronaviruses in Southeast Asia," *Nature Communications* 13, no. 1 (2022): 4380, https://doi.org/10.1038/s41467-022-31860-w, PMID: 35945197, PMCID: PMC9363439.

58. M. Worobey et al., "The Huanan Seafood Wholesale Market in Wuhan Was the Early Epicenter of the COVID-19 Pandemic," *Science* 377, no. 6609 (2022): 951–959, https://doi.org/10.1126/science.abp8715, PMID: 35881010, PMCID: PMC9348750; J. E. Pekar et al., "The Molecular Epidemiology of Multiple Zoonotic Origins of SARS-CoV-2," *Science* 377, no. 6609 (2022): 960-966, https://doi.org/10.1126/science.abp8337, errata in *Science* 382, no. 6667 (2023): eadl0585, PMID: 35881005, PMCID: PMC9348752.

59. "It's a Fact, Scientists Are the Most Trusted People in World," Ipsos, September 18, 2019, www.ipsos.com/en/its-fact-scientists-are-most-trusted-people-world; Alison Boshoff, "Eco-warrior or Hypocrite? Leonardo DiCaprio Jets

Around the World Partying . . . While Preaching to Us All on Global Warming," *Daily Mail*, May 23, 2016, www.dailymail.co.uk/tvshowbiz/article-3605779/Eco-warrior-hypocrite-Leonardo-DiCaprio-jets-world-partying-preaching-global-warming-Title-goes-here.html.

60. Robert F. Kennedy Jr., foreword to *An Enemy of the People*, by Henrik Ibsen, translated by R. Farquharson Sharp (Skyhorse, 2021).

61. Katherine Knott, "J. D. Vance Called Universities 'the Enemy.' Now He's Trump's VP Pick," *Inside Higher Ed*, July 16, 2024, www.insidehighered.com/news/government/politics-elections/2024/07/16/trump-taps-jd-vance-sharp-critic-higher-ed-vp.

62. Kathryn Diss and Lucy Sweeney, "How JD Vance Transformed Himself from a Never Trumper to the Republican Pick for Vice-President," ABC News (Australia), July 16, 2024, www.abc.net.au/news/2024-07-17/how-jd-vance-transformed-himself-from-never-trumper-to-vp-pick/104106988.

63. Mike Damiano and Hilary Burns, "Attack the Universities. Trump's VP Pick JD Vance Has Harsh Words for Higher Education," *Boston Globe*, July 16, 2024, www.bostonglobe.com/2024/07/16/metro/vance-trump-rnc-universities/.

64. National Conservatism (@NatConTalk), "@JDVance1 at the National Conservatism Conference: 'The professors are the enemy.' #NatCon," Twitter (now X), November 2, 2021, 8:58 p.m., https://twitter.com/NatConTalk/status/1455700807144415232.

65. Anthony Fauci, *On Call: A Doctor's Journey in Public Service* (Viking, 2024).

66. Yash Roy, "Greene Alleges Fauci Committed 'Crimes Against Humanity' with COVID Response," *Hill*, June 15, 2024, https://thehill.com/homenews/4724215-greene-alleges-fauci-committed-crimes-against-humanity-with-covid-response/; Ellie Quinlan Houghtaling, "Alex Jones Wildly Threatens Fauci in Macabre Rant," *New Republic*, June 12, 2024, https://newrepublic.com/post/182634/alex-jones-gruesome-fantasy-fauci.

67. Alex Shephard, "The Party of Trump Is Trying to Get People Mad at Anthony Fauci Again," *New Republic*, June 8, 2021, https://newrepublic.com/article/162682/anthony-fauci-right-wing-media-attacks.

68. Ramon Antonio Vargas, "Anthony Fauci Says Marjorie Taylor Greene Drove Death Threats Against Him," *Guardian*, June 4, 2024, www.theguardian.com/us-news/article/2024/jun/04/fauci-death-threats-marjorie-taylor-greene-fox.

69. Elon Musk (@elonmusk), "My pronouns are Prosecute/Fauci," Twitter (now X), December 11, 2022, 5:58 a.m., https://twitter.com/elonmusk/status/1601894132573605888.

70. Elon Musk (@elonmusk), "He's afraid of a public debate, because he knows he's wrong," Twitter (now X), June 17, 2023, 7:58 p.m., https://twitter.com/elonmusk/status/1670219488485154816.

71. Alexa Mikhail, "After Elon Musk, Joe Rogan Vaccine Twitter Brawl, Scientists Say 'Vile Rhetoric & Misinformation' Is Forcing Them Off the Platform," *Fortune*, June 20, 2023, https://fortune.com/well/2023/06/20/elon-musk-joe-rogan-peter-hotez-anti-vaccine-twitter-harassment/.

72. S. Wilder-Smith, "TAK-003 Dengue Vaccine as a New Tool to Mitigate Dengue in Countries with a High Disease Burden," *Lancet Global Health* 12, no.

2 (2024): e179–e180, https://doi.org/10.1016/S2214-109X(23)00590-9, PMID: 38245106; E. E. Ooi and S. Kalimuddin, "Insights into Dengue Immunity from Vaccine Trials," *Science Translational Medicine* 15, no. 704 (2023): eadh3067, https://10.1126/scitranslmed.adh3067, errata in *Science Translational Medicine* 15, no. 709 (2023): eadk1254, PMID: 37437017; R. McMahon et al., "A Randomized, Double-Blinded Phase 3 Study to Demonstrate Lot-to-Lot Consistency and to Confirm Immunogenicity and Safety of the Live-Attenuated Chikungunya Virus Vaccine Candidate VLA1553 in Healthy Adults," *Journal of Travel Medicine* (December 13, 2023), https://doi.org/10.1093/jtm/taad156, e-publication ahead of print, PMID: 38091981.

73. N. Côrtes et al., "Integrated Control Strategies for Dengue, Zika, and Chikungunya Virus Infections," *Frontiers in Immunology* 14 (December 2023): 1281667, https://doi.org/10.3389/fimmu.2023.1281667, PMID: 38196945, PMCID: PMC10775689.

74. P. J. Hotez, "A Journey in Science: Molecular Vaccines for Global Child Health in Troubled Times of Anti-science," *Molecular Medicine* 30, no. 1 (2024): 37, https://doi.org/10.1186/s10020-024-00786-y.

75. P. J. Hotez and M. Matshaba, "Promise of New Malaria Vaccines," *BMJ* 379 (2022): o2462, https://doi.org/10.1136/bmj.o2462, PMID: 36241198.

76. S. Cankat, M. U. Demael, and L. Swadling, "In Search of a Pan-Coronavirus Vaccine: Next-Generation Vaccine Design and Immune Mechanisms," *Cellular and Molecular Immunology* (December 2023), https://doi.org/10.1038/s41423-023-01116-8, e-publication ahead of print, PMID: 38148330; P. J. Hotez et al., "From Concept to Delivery: A Yeast-Expressed Recombinant Protein-Based COVID-19 Vaccine Technology Suitable for Global Access," *Expert Review of Vaccines* 22, no. 1 (2023): 495–500, https://doi.org/10.1080/14760584.2023.2217917, PMID: 37252854.

77. S. Y. Tartof et al., "BNT162b2 XBB1.5-Adapted Vaccine and COVID-19 Hospital Admissions and Ambulatory Visits in US Adults," *MedRxiv* (December 2023), https://doi.org/10.1101/2023.12.24.23300512.

78. A. Watanabe et al., "Protective Effect of COVID-19 Vaccination Against Long COVID Syndrome: A Systematic Review and Meta-analysis," *Vaccine* 41, no. 11 (2023): 1783–1790, https://doi.org/10.1016/j.vaccine.2023.02.008, PMID: 36774332, PMCID: PMC9905096.

79. P. Katona, K. Patel, and S. Freeman, "Fatal Attraction: The Seductive Appeal of Irrationality, Anti-science, and Toxic Extremism," *Hill*, December 12, 2023, https://thehill.com/opinion/4356067-fatal-attraction-the-seductive-appeal-of-irrationality-anti-science-and-toxic-extremism/.

80. "Marie Curie: Facts About the Pioneering Chemist," History.com, February 22, 2021, www.history.com/news/marie-curie-facts.

CHAPTER 2: THE PLUTOCRATS

1. R. Talbert, *The Senate of Imperial Rome* (Princeton University Press, 1984).

2. P. Strathern, *The Medici: Godfathers of the Renaissance* (Vintage, 2009).

3. M. Josephson, *The Robber Barons* (Harper Paperbacks, 1962).

4. A. Carnegie, *The Gospel of Wealth* (1889; reprint, Carnegie Corporation of New York, 2017), www.carnegie.org/publications/the-gospel-of-wealth/, accessed December 27, 2023.

5. www.bloomberg.org/?gad_source=1, accessed February 25, 2024.

6. https://tsffoundation.org/about/, accessed February 25, 2024.

7. Damian Carrington, "Climate Denial Ads on Facebook Seen by Millions, Report Finds," October 8, 2020, www.theguardian.com/environment/2020/oct /08/climate-denial-ads-on-facebook-seen-by-millions-report-finds.

8. Sarah Frier and Sarah Kopit, "Facebook Built the Perfect Platform for Covid Vaccine Conspiracies—Mark Zuckerberg Wanted to Make His Social Network a Reliable Source About the Pandemic. Instead He's Helped Spread Misinformation About Vaccines Causing Infertility," *Bloomberg News*, April 1, 2021, www.bloomberg.com/news/features/2021-04-01/covid-vaccine-and-fertility -facebook-s-platform-is-letting-fake-news-go-viral.

9. David Vetter, "How Meta Nuked a Climate Story, and What It Means for Democracy," *Forbes*, April 11, 2024, www.forbes.com/sites/davidrvetter/2024/04 /11/how-meta-nuked-a-climate-story-and-what-it-means-for-democracy /?sh=651c244036fd.

10. Queenie Wong, "Once a Trump Critic, Mark Zuckerberg Pivots Toward the President," *Los Angeles Times*, February 3, 2025, www.latimes.com /business/story/2025-02-03/mark-zuckerberg-once-a-trump-critic-cozies-up -to-the-president.

11. See "Researchers Receive $10.2 Million to Study New Malaria-Prevention Method," www.psu.edu/news/research/story/researchers-receive-102 -million-study-new-malaria-prevention-method/.

12. See J. Blanford et al., "Implications of Temperature Variation for Malaria Parasite Development Across Africa," *Scientific Reports* 3, no. 1300 (2013), https:// doi.org/10.1038/srep01300.

13. See Michael E. Mann, "The Right Path Forward on Climate," *Newsweek*, February 23, 2021, www.newsweek.com/right-path-forward-climate-change -opinion-1571169.

14. G. W. Comer, *A History of the Rockefeller Institute, 1901–1953: Origins and Growth* (Rockefeller Institute Press, 1964).

15. www.rockefellerfoundation.org/, accessed February 25, 2024.

16. Suzanne Goldenberg, "Rockefeller Brothers Fund: It Is Our Moral Duty to Divest from Fossil Fuels," *Guardian*, March 27, 2015, www .theguardian.com/environment/2015/mar/27/rockefeller-fund-chairman-moral -duty-divest-fossil-fuels.

17. P. J. Hotez, "DR Congo and Nigeria: New Neglected Tropical Disease Threats and Solutions for the Bottom 40," *PLOS Neglected Tropical Diseases* 13, no. 8 (2019): e0007145, https://doi.org/10.1371/journal.pntd.0007145, PMID: 31393879, PMCID: PMC6687097.

18. P. J. Hotez et al., "The Great Infection of Mankind," *PLOS Medicine* 2, no. 3 (2005): e67, https://doi.org/10.1371/journal.pmed.0020067, PMID: 15783256, PMCID: PMC1069663.

19. M. L. McKenna et al., "Human Intestinal Parasite Burden and Poor Sanitation in Rural Alabama," *American Journal of Tropical Medicine and Hygiene* 97, no. 5 (2017): 1623–1628, https://doi.org/10.4269/ajtmh.17-0396, errata in *American Journal of Tropical Medicine and Hygiene* 98, no. 3 (2018): 936, PMID: 29016326, PMCID: PMC5817782.

20. C. Keating, *Kenneth Warren and the Great Neglected Diseases of Mankind*

Programme: The Transformation of Geographical Medicine in the US and Beyond (Springer, 2017).

21. J. Ettling, *The Germ of Laziness: Rockefeller Philanthropy and Public Health in the New South* (Harvard University Press, 1981); C. Elman, R. A. McGuire, and B. Wittman, "Extending Public Health: The Rockefeller Sanitary Commission and Hookworm in the American South," *American Journal of Public Health* 104, no. 1 (2014): 47–58, https://doi.org/10.2105/AJPH.2013.301472, PMID: 24228676, PMCID: PMC3910046; J. F. Fox and T. N. Grigoriadis, "Rural Health in the Progressive Era: Revisiting the Hookworm Intervention in the American South," *Journal of Rural Medicine* 17, no. 4 (2022): 236–247, https://doi.org/10.2185 /jrm.2021-061, PMID: 36397794, PMCID: PMC9613373; E. Fee, *Disease & Discovery: A History of the Johns Hopkins School of Hygiene and Public Health, 1916–1939* (Johns Hopkins University Press, 1987).

22. N. R. Stoll, "The Osmosis of Research: Example of the Cort Hookworm Investigations," *Bulletin of the New York Academy of Medicine* 48, no. 10 (1972): 1321–1329, PMID: 4570679, PMCID: PMC1806888.

23. P. J. Hotez, "A Journey in Science: Molecular Vaccines for Global Child Health in Troubled Times of Anti-science," *Molecular Medicine* 30, no. 1 (2024): 37, https://doi.org/10.1186/s10020-024-00786-y.

24. P. J. Hotez et al., "The Human Hookworm Vaccine," *Vaccine* 31, supp. 2 (2013): B227–B232, https://doi.org/10.1016/j.vaccine.2012.11.034, PMID: 23598487, PMCID: PMC3988917.

25. E. Zerhouni, "GAVI, the Vaccine Alliance," *Cell* 179, no. 1 (2019): 13–17, https://doi.org/10.1016/j.cell.2019.08.026, PMID: 31519310; T. Schwab, *The Bill Gates Problem: Reckoning with the Myth of the Good Billionaire* (Metropolitan Books, 2023).

26. Y. Y. Syed, "RTS,S/AS01 Malaria Vaccine (Mosquirix®): A Profile of Its Use," *Drugs and Therapy Perspectives* 38, no. 9 (2022): 373–381, https://doi .org/10.1007/s40267-022-00937-3, PMID: 36093265, PMCID: PMC9449949.

27. Schwab, *Bill Gates Problem*.

28. Michael E. Mann, "The Right Path Forward on Climate Change: Opinion," *Newsweek*, February 23, 2021, www.newsweek.com/right-path-forward -climate-change-opinion-1571169.

29. Emily Kirkpatrick, "Melinda Gates Says Bill Gates's Work with 'Abhorrent' Jeffrey Epstein Led to Divorce," *Vanity Fair*, March 3, 2022, www.vanityfair.com/style/2022/03/melinda-gates-jeffrey-epstein-led-to-bill -gates-divorce-gayle-king-interview.

30. Jenny Rice, "Why Jeffrey Epstein's Death Is the Perfect Bait for Conspiracy Theorists," MSNBC, July 2, 2023, www.msnbc.com/opinion/msnbc-opinion /jeffrey-epstein-death-fuels-many-conspiracy-theories-rcna92112.

31. Courtney Weaver and Valerie Hopkins, "Billionaire Has Become Hate Figure on Right in Both Europe and US," *Financial Times*, November 4, 2018, www .ft.com/content/e2a1ecb0-dc0d-11e8-9f04-38d397e6661c.

32. Armin Langer, "The Eternal George Soros: The Rise of an Antisemitic and Islamophobic Conspiracy Theory," in *Europe: Continent of Conspiracies* (Routledge, 2021).

33. Zack Beauchamp, "Marjorie Taylor Greene's Space Laser and the Age-Old

Problem of Blaming the Jews," *Vox*, January 30, 2021, www.vox.com/22256258/marjorie-taylor-greene-jewish-space-laser-anti-semitism-conspiracy-theories.

34. Ron Dicker, "Rep. Marjorie Taylor Greene's Tweet About NYC Smoke Is a Real Doozy," *Yahoo News (Canada)*, June 8, 2023, https://ca.news.yahoo.com/rep-marjorie-taylor-greenes-tweet-175736707.html.

35. Alex Mahadevan, "Bill Gates and George Soros Are Targets of Another COVID-19 Conspiracy Theory," Poynter Institute, June 1, 2020, www.poynter.org/fact-checking/2020/bill-gates-and-george-soros-the-target-of-another-covid-19-conspiracy-theory/.

36. D. Zuidijk, "Davos' World Economic Forum Is a Favorite Focus of Conspiracies," *Bloomberg*, January 17, 2024, www.bloomberg.com/news/newsletters/2024-01-17/world-economic-forum-popular-target-for-conspiracy-theories.

37. Thomas Grove et al., "Musk's Secret Conversations with Vladimir Putin," *Wall Street Journal*, October 25, 2024, www.wsj.com/world/russia/musk-putin-secret-conversations-37e1c187.

38. P. Krugman, "The Paranoid Style in American Plutocrats," *New York Times*, August 28, 2023, www.nytimes.com/2023/08/28/opinion/columnists/covid-climate-cryptocurrency-plutocrats.html.

39. https://blockchainmagazine.net/what-does-crypto-bros-stand-for-and-how-will-they-help-bitcoin/ (no longer available).

40. J. Hamilton, "Ramaswamy Shares Crypto Plan, Making Him the Only GOP Candidate Who Has One," *CoinDesk*, November 16, 2023, www.coindesk.com/policy/2023/11/16/ramaswamy-shares-crypto-plan-making-him-the-only-gop-candidate-who-has-one/.

41. A. Tabet and K. Koretski, "Vivek Ramaswamy Regrets Taking the Covid Vaccine. His Wife, a Surgeon, Does Not," NBC News, September 18, 2023, www.nbcnews.com/politics/2024-election/vivek-ramaswamy-regrets-taking-covid-vaccine-wife-surgeon-not-rcna105527.

42. N. Corasaniti, "Defending Trump, Ramaswamy Rattles Off Right Wing Conspiracy Theories," *New York Times*, December 6, 2023.

43. Lee Hedgepeth, "Vivek Ramaswamy Called 'the Climate Change Agenda' a Hoax in Alabama's First-Ever Presidential Debate. What Did University of Alabama Students Think?," *Inside Climate News*, December 11, 2023, https://insideclimatenews.org/news/11122023/vivek-ramaswamy-climate-change-agenda-hoax-university-of-alabama-students/.

44. Veronika Melkozerova, "Putin Fears Crypto Mining Could Cause Energy Shortages in Russia," *Politico*, July 17, 2024, www.politico.eu/article/russia-vladimir-putin-crypto-russia-sanctions-west-banking-services-nuclear-power-plant-usa-mining/.

45. Ryan Bort, "Elon Musk Can't Stop Peddling Putin Propaganda," *Rolling Stone*, October 17, 2022, www.rollingstone.com/politics/politics-news/elon-musk-peddling-putin-propaganda-ukraine-crimea-1234612343/.

46. Michael E. Mann, *The Hockey Stick and the Climate Wars: Dispatches from the Front Lines* (Columbia University Press, 2014).

47. W. Bragman and A. Kotch, "How the Koch Network Hijacked the War on Covid," Exposed by CMD (Center for Media and Democracy),

December 22, 2021, www.exposedbycmd.org/2021/12/22/how-the-koch-network -hijacked-the-war-on-covid/.

48. A. B. Keene et al., "Association of Surge Conditions with Mortality Among Critically Ill Patients with COVID-19," *Journal of Intensive Care Medicine* 37, no. 4 (2022): 500–509, https://doi.org/10.1177/08850666211067509, PMID: 34939474, PMCID: PMC8926920.

49. Bragman and Kotch, "How the Koch Network Hijacked the War on Covid."

50. Bragman and Kotch, "How the Koch Network Hijacked the War on Covid." See also A. Rowell, "Koch-Funded Denial Group 'Shaping' White House Strategy on COVID-19," Oil Change International, October 16, 2020, https: //priceofoil.org/2020/10/16/koch-funded-climate-denial-group-shaping-white -house-strategy-on-covid-19/; and G. Yamey and D. H. Gorski, "Covid-19 and the New Merchants of Doubt," *BMJ Opinion*, September 13, 2021, https://blogs.bmj .com/bmj/2021/09/13/covid-19-and-the-new-merchants-of-doubt/.

51. M. Zenone et al., "Analyzing Natural Herd Immunity Media Discourse in the United Kingdom and the United States," *PLOS Global Public Health* 2, no. 1 (2022): e0000078, https://doi.org/10.1371/journal.pgph.0000078, PMID: 36962077, PMCID: PMC10021579; "WHO Chief Says Herd Immunity Approach to Pandemic 'Unethical,'" *Guardian*, October 12, 2020, www.theguardian.com /world/2020/oct/12/who-chief-says-herd-immunity-approach-to-pandemic -unethical; T. S. Brett and P. Rohani, "Transmission Dynamics Reveal the Imprac- ticality of COVID-19 Herd Immunity Strategies," *Proceedings of the National Academy of Sciences* 117, no. 41 (2020): 25897–25903, https://doi.org/10.1073 /pnas.2008087117.

52. "WHO Chief Says Herd Immunity Approach to Pandemic 'Unethical.'"

53. Bragman and Kotch, "How the Koch Network Hijacked the War on Covid."

54. Merrill Matthews, "Trump's Revenge: Naming Dr. Bhattacharya to Head the NIH," *Hill*, December 10, 2024, https://thehill.com/opinion/5031472 -trump-nominates-nih-revenge/.

55. Bragman and Kotch, "How the Koch Network Hijacked the War on Covid."

56. F. Ege, G. Mellace, and S. Menon, "The Unseen Toll: Excess Mortality During Covid-19 Lockdowns," *Scientific Reports* 13, no. 1 (2023): 18745, https: //doi.org/10.1038/s41598-023-45934-2, PMID: 37907531, PMCID: PMC10618514.

57. E. Green, "The Christian Liberal-Arts School at the Heart of the Culture Wars," *New Yorker*, April 3, 2023, www.newyorker.com/magazine/2023/04/10 /the-christian-liberal-arts-school-at-the-heart-of-the-culture-wars.

58. Bragman and Kotch, "How the Koch Network Hijacked the War on Covid."

59. Michael E. Mann (@MichaelEMann), "Grok knows," X, November 10, 2024, 4:36 p.m., https://x.com/MichaelEMann/status/1855972194926358981.

60. B. Bambrough, "Elon Musk Confirms 'Progress' on Wild X Crypto Rumors—Triggering a Dogecoin Price Surge to Rival Bitcoin, Ethereum and XRP," *Forbes*, December 6, 2023, www.forbes.com/sites/digital-assets/2023/12/06 /elon-musk-confirms-progress-on-wild-x-crypto-rumors-triggering-a-dogecoin -price-surge-to-rival-bitcoin-ethereum-xrp/?sh=6566aa845bb3; A. Shephard, "Elon

Musk Is the *New Republic*'s 2023 Scoundrel of the Year," *New Republic*, December 27, 2024, https://newrepublic.com/article/177695/elon-musk-scoundrel-year-2023 -new-republic; P. Markolin, "Elon Musk's War Against Science, Evidence, and Objective Truth," *Byline Times*, January 16, 2024, https://bylinetimes.com/2024 /01/16/elon-musks-war-against-science-evidence-and-objective-truth/.

61. Musk Foundation, www.muskfoundation.org/, accessed January 21, 2024.

62. Richard Lawler, "Every Ridiculous Thing We Learned Today About Elon Musk's Plan to Take over Twitter," *Verge*, April 29, 2022, www.theverge .com/2022/4/29/23049172/elon-musk-buys-twitter-thiel-farts-jack-dorsey.

63. David Corn, "How Dangerous Is Peter Thiel?," *Mother Jones*, November 26, 2021, www.motherjones.com/politics/2021/11/how-dangerous-is-peter-thiel/.

64. Shephard, "Elon Musk Is the *New Republic*'s 2023 Scoundrel of the Year."

65. Damian Carrington, "Alarm at Exodus of Climate Voices on Twitter After Musk Takeover," *Guardian*, August 15, 2023, www.theguardian.com /environment/2023/aug/15/twitter-exodus-climate-green-voices-musk-takeover.

66. Markolin, "Elon Musk's War Against Science, Evidence, and Objective Truth."

67. Carrington, "Alarm at Exodus of Climate Voices on Twitter After Musk Takeover."

68. Dana Nuccitelli, "Rupert Murdoch Doesn't Understand Climate Change Basics, and That's a Problem," *Guardian*, July 14, 2021, www.theguardian.com /environment/climate-consensus-97-per-cent/2014/jul/14/rupert-murdoch-doesnt -understand-climate-basics; Rupert Murdoch (@rupertmurdoch), "Just flying over N Atlantic 300 miles of ice. Global warming!," Twitter (now X), February 27, 2015, 2:18 p.m., https://twitter.com/rupertmurdoch/status/571389009651486720.

69. Nuccitelli, "Rupert Murdoch Doesn't Understand Climate Change Basics."

70. Kenneth Li, "Murdoch Receives COVID-19 Vaccine as Fox News Host Casts Suspicion on Campaign," Reuters, December 18, 2020, www.reuters.com /article/us-health-coronavirus-murdoch-idUSKBN28S2J7/.

71. Matthew Mosley, "This James Bond Villain Is 007's Most Underrated Enemy," *Collider*, June 5, 2023, https://collider.com/most-underrated-james-bond -villain-tomorrow-never-dies-jonathan-pryce/.

72. David Bauder, "Study: Fox Viewers More Likely to Believe COVID Falsehoods," Associated Press, November 10, 2021, https://apnews.com/article /coronavirus-pandemic-media-misinformation-health-b59e98ca50f37ddeea 217903915a53fc.

73. Hotez, "Journey in Science"; P. J. Hotez, "Vaccine Preventable Disease and Vaccine Hesitancy," *Medical Clinics of North America* 107, no. 6 (2023): 979–987, https://doi.org/10.1016/j.mcna.2023.05.012, PMID: 37806729.

74. F. C. Elliott, *The Birth of the Texas Medical Center: A Personal Account* (Texas A&M University Press, 2004); W. H. Kellar, *Enduring Legacy: The M.D. Anderson Foundation and the Texas Medical Center* (Texas A&M University Press, 2014); W. T. Butler and D. L. Ware, *Arming for Battle Against Disease Through Education, Research and Patient Care at Baylor College of Medicine*, 5 vols. (Baylor College of Medicine, 2011).

75. B. Burrough, *The Big Rich: The Rise and Fall of the Greatest Texas Oil Fortunes* (Penguin Books, 2009).

76. Butler and Ware, "Arming for Battle against Disease."

77. Burrough, *Big Rich*; B. Porterfield, "H. L. Hunt's Long Goodbye," *Texas Monthly*, March 1975, www.texasmonthly.com/news-politics/h-l-hunts-long-goodbye/.

78. Burrough, *Big Rich*.

79. L. Badash, "Science and McCarthyism," *Minerva* 38, no. 1 (2000): 53–80, www.jstor.org/stable/41821155.

80. C. Tolan et al., "How Two Texas Megadonors Have Turbocharged the State's Far-Right Shift," CNN, July 24, 2022, www.cnn.com/2022/07/24/politics/texas-far-right-politics-invs/index.html.

81. R. Gold, "The Billionaire Bully Who Wants to Turn Texas into a Christian Theocracy," *Texas Monthly*, March 2024, www.texasmonthly.com/news-politics/billionaire-tim-dunn-runs-texas/.

82. D. Seidman, "The Money Behind Empower Texans," Public Accountability Initiative, September 25, 2019, https://public-accountability.org/report/the-money-behind-empower-texans/; M. Smith, "Committee Mobilizes to Defend Vaccine Exemptions in Texas," *Texas Tribune*, April 2, 2016, www.texastribune.org/2016/04/02/pac-mobilizes-defend-vaccine-exemptions-texas/.

83. L. Reigstad, "The Anti-vaccination Movement Is Gaining Ground in Texas," *Texas Monthly*, September 1, 2016, www.texasmonthly.com/the-daily-post/anti-vaccination-movement-gaining-ground-texas/; S. Novack, "The Antivaccine Lobby's Revenge Candidate Is Running in North Dallas," *Texas Observer*, October 19, 2018, www.texasobserver.org/the-anti-vaccine-lobbys-revenge-candidate-is-running-in-north-dallas/; S. Novack, "How Anti-vaxxers Are Injecting Themselves into the Texas Republican Primaries," *Texas Observer*, February 28, 2018, www.texasobserver.org/anti-vaxxers-injecting-texas-republican-primaries/.

84. P. J. Hotez, "America's Deadly Flirtation with Antiscience and the Medical Freedom Movement," *Journal of Clinical Investigation* 131, no. 7 (2021): e149072, https://doi.org/10.1172/JCI149072, PMID: 33630759, PMCID: PMC8011881; P. J. Hotez, *The Deadly Rise of Anti-science: A Scientist's Warning* (Johns Hopkins University Press, 2023).

85. Reigstad, "Anti-vaccination Movement Is Gaining Ground in Texas"; Novack, "Antivaccine Lobby's Revenge Candidate Is Running in North Dallas"; Novack, "How Anti-vaxxers Are Injecting Themselves into the Texas Republican Primaries."

86. Seidman, "Money Behind Empower Texans."

87. L. H. Sun, "Trump Energizes the Anti-vaccine Movement in Texas," *Texas Tribune*, February 20, 2017, www.texastribune.org/2017/02/20/trump-energizes-anti-vaccine-movement-texas/; C. Tomlinson, "Texas Oilmen Pushing Right-Wing Extremism on GOP Employ Antisemitic Allies, Investigations Show," *Houston Chronicle*, November 3, 2023, www.houstonchronicle.com/business/columnists/tomlinson/article/texas-dunn-wilks-politics-oilmen-18461978.php.

88. Seidman, "Money Behind Empower Texans."

89. P. J. Hotez, *Vaccines Did Not Cause Rachel's Autism: My Journey as a Vaccine Scientist, Pediatrician, and Autism Dad* (Johns Hopkins University Press, 2018); Jonathan Stickland (@RepStickland), "You are bought and paid for by the biggest special interest in politics. Do our state a favor and mind your own business. Parental rights mean more to us than your self enriching 'science.' #txlege," Twitter (now X), May 7, 2019, 11:52 a.m., https://twitter.com/RepStickland

/status/1125790483895279617; Jonathan Stickland (@RepStickland), "Make the case for your sorcery to consumers on your own dime. Like every other business. Quit using the heavy hand of government to make your business profitable through mandates and immunity. It's disgusting," Twitter (now X), May 7, 2019, 2:32 p.m., https://twitter.com/RepStickland/status/1125830765961457664; R. Pearson, "Yes, I Am a Sorceress," *Texas Observer*, May 16, 2019, www.texasobserver.org/yes-i-am-a-sorceress/; D. Paul, "GOP State Legislator Attacks Vaccine Scientist on Twitter: Accusing Him of Self-Enrichment, 'Sorcery,'" *Washington Post*, May 8, 2019, www.washingtonpost.com/health/2019/05/08/gop-legislator-attacks-top-vaccine-scientist-twitter-accusing-him-self-enrichment-sorcery/.

90. Hotez, "Journey in Science"; K. P. Puchner et al., "Vaccine Value Profile for Hookworm," *Vaccine* 42, no. 19 (2024): S25–S41, https://doi.org/10.1016/j.vaccine.2023.05.013, PMID: 37863671; P. J. Hotez et al., "From Concept to Delivery: A Yeast-Expressed Recombinant Protein-Based COVID-19 Vaccine Technology Suitable for Global Access," *Expert Review of Vaccines* 22, no. 1 (2023): 495–500, https://doi.org/10.1080/14760584.2023.2217917, PMID: 37252854; P. J. Hotez, "Global Vaccine Access Demands Combating Both Inequity and Hesitancy," *Health Affairs* 42, no. 12 (2023): 1681–1688, https://doi.org/10.1377/hlthaff.2023.00775, PMID: 38048497; M. Goozner, "How to Stop the Drug Companies from Hiking Prices for Covid Vaccines," *Washington Monthly*, March 6, 2023, https://washingtonmonthly.com/2023/03/06/how-to-stop-the-drug-companies-from-hiking-prices-for-covid-vaccines/; J. Palca, "Whatever Happened to the New No-Patent COVID Vaccine Touted as a Global Game Changer?," *NPR Goats and Soda*, August 31, 2023, www.npr.org/transcripts/1119947342.

91. Goozner, "How to Stop the Drug Companies from Hiking Prices for Covid Vaccines"; Palca, "Whatever Happened to the New No-Patent COVID Vaccine?"

92. Texas Department of State Health Services (DSHS), "Conscientious Exemption Data, 2022–23," n.d., www.dshs.texas.gov/immunization-unit/immunization-coverage-levels/conscientious-exemptions-data-vaccination.

93. Texas Homeschool Coalition, "History of Homeschooling in Texas— Getting Started," https://thsc.org/texas-homeschooling/, accessed December 29, 2023.

94. P. J. Hotez, "The Great Texas COVID Tragedy," *PLOS Global Public Health* 2, no. 10 (2022): e0001173, https://doi.org/10.1371/journal.pgph.0001173, PMID: 36962661, PMCID: PMC10021311.

95. Hotez, *Deadly Rise of Anti-science*; Hotez, "Great Texas COVID Tragedy."

96. D. Holmes, "They Clapped for Death at CPAC," *Esquire*, July 12, 2021, www.esquire.com/news-politics/a37001629/cpac-vaccination-goal-biden-miss-clap/.

97. Sheldon Whitehouse, "Time to Wake Up 277: Donors Trust," speech, December 9, 2020, www.whitehouse.senate.gov/news/speeches/time-to-wake-up-277-donors-trust/.

98. Hailey Fuchs, "Two Anonymous $425 Million Donations Give Dark Money Conservative Group a Massive Haul," *Politico*, November 16, 2022, www.politico.com/news/2022/11/16/two-anonymous-425-million-donations-gives-dark-money-conservative-group-a-massive-haul-00067493.

99. Julie Kelly, "Hockey Sticks, Changing Goal Posts, and Hysteria," AmericanGreatness.com, March 31, 2020; https://amgreatness.com/2020/03/31/hockey-sticks-changing-goal-posts-and-hysteria/.

100. Alex Kotch, "Right-Wing Megadonors Are Financing Media Operations to Promote Their Ideologies," PR Watch, January 27, 2020, www.prwatch.org /news/2020/01/13531/right-wing-megadonors-are-financing-media-operations -promote-their-ideologies.

101. "Heritage Foundation," *DeSmog*, n.d., www.desmog.com/heritage -foundation/.

102. Heritage Foundation (@Heritage), "When you hear a pharmacist or physician (or academic dean) parrot the malignant trope of 'Ivermectin doesn't work for Covid' or that there is 'no evidence' to support ivermectin's use for Covid-19, send them Dr. Gortler's review of a statistical analysis of both positive and negative studies, collectively showing a statistically significant improvement associated with ivermectin for 1) mortality, 2) ventilation, 3) ICU, 4) hospitalization, 5) recovery, 6) cases, and 7) viral clearance. The only people who will refuse to read and give these findings due consideration, are those whose personal or political ideologies won't allow them to objectively consider clinical/scientific data. Dr. Gortler writes: 'Science doesn't care about consensus. In fact, many of the world's biggest scientific advancements were the result of questioning an established consensus,'" X, June 11, 2024, 1:52 p.m., https://x.com/heritage/status/1800586976262758882?s =43&t=KnVZfaEFPXgfy-Hsr5cuhg.

103. Joe Bastardi, "The Arrogance of Authority in Covid and Climate," *CFACT*, February 8, 2022, www.cfact.org/2022/02/08/the-arrogance-of -authority-in-covid-and-climate/.

104. "Whitehouse Illuminates Dark-Money Network's Project to Undermine Democracy," August 4, 2022, www.whitehouse.senate.gov/news/release /whitehouse-illuminates-dark-money-networks-project-to-undermine-democracy/.

105. "Whitehouse, Chu Introduce Bill Ending Billionaires' Tax-Free Giveaways to Dark Money Groups," February 6, 2024, www.whitehouse.senate.gov /news/release/whitehouse-chu-introduce-bill-ending-billionaires-tax-free -giveaways-to-dark-money-groups/.

106. "Whitehouse, Cicilline Reintroduce DISCLOSE Act to End Corrupting Influence of Dark Money in American Democracy," February 17, 2023, www.whitehouse.senate.gov/news/release/whitehouse-cicilline-reintroduce -disclose-act-to-end-corrupting-influence-of-dark-money-in-american-democracy/.

107. Keerti Gopal, "Despite Climate Concerns, Young Voter Turnout Slumped and Its Support Split Between the Parties," *Inside Climate News*, November 8, 2024, https://insideclimatenews.org/news/08112024/young-voter -turnout-slumped-support-split-between-parties/.

CHAPTER 3: THE PETROSTATES

1. G. Ibadoghlu and R. Sadigov, "The Economics of Petro-authoritarianism: Post-Soviet Transitions and Democratization," *Resources Policy* 85 (2023): 103752.

2. C. Daggett, "Petro-masculinity and the Politics of Climate Refusal," *Autonomy*, May 1, 2022, https://autonomy.work/portfolio/petro-masculinity -climate-refusal/.

3. E. Ashford, "The Problem with Being a Petrostate," *Foreign Policy*, June 19, 2022, https://foreignpolicy.com/2022/06/19/petrostates-oil-production-weapon -foreign-policy-war-economy/.

4. N. Ferris, "What Four Years of 'Non-existent' Climate Action Has Done

to Brazil," *Energy Monitor*, September 29, 2022, www.energymonitor.ai/policy /bolsonaro-what-four-years-of-non-existent-climate-action-has-done-to-brazil /?cf-view.

5. P. J. Hotez, "Anti-science Kills: From Soviet Embrace of Pseudoscience to Accelerated Attacks on US Biomedicine," *PLOS Biology* 19, no. 1 (2021): e3001068, https://doi.org/10.1371/journal.pbio.3001068, PMID: 33507935, PMCID: PMC7842901.

6. Hotez, "Anti-science Kills"; P. J. Hotez, *The Deadly Rise of Anti-science: A Scientist's Warning* (Johns Hopkins University Press, 2023).

7. R. J. Rummel, *Lethal Politics: Soviet Genocide and Mass Murder Since 1917* (Routledge, 1990).

8. Olga Ivshina, "Invisible Losses: Tens of Thousands Fighting for Russia Are Dying Unnoticed on the Frontline in Ukraine," BBC, February 22, 2025, https: //www.bbc.com/news/articles/cgkm7lly61do.

9. Hotez, "Anti-science Kills"; Hotez, *Deadly Rise of Anti-science*; S. Ings, *Stalin and the Scientists: A History of Triumph and Tragedy, 1905–1953* (Grove Press, 2016).

10. Hotez, *Deadly Rise of Anti-science*.

11. Ings, *Stalin and the Scientists*; P. Pringle, *The Murder of Nikolai Vavilov* (Simon and Schuster, 2008).

12. R. Lourie, *Sakharov, a Biography* (Brandeis University Press, 2002).

13. B. Grozovski, "Putin's War Costs: Changing Russia's Economy," *The Russia File* (blog), Wilson Center, January 17, 2023, www.wilsoncenter.org/blog-post /putins-war-costs-changing-russias-economy.

14. "Putin Says Climate Change Is Not Man-Made and We Should Adapt to It, Not Try to Stop It," Agence France-Presse, March 31, 2017, www.scmp.com /news/world/russia-central-asia/article/2083650/trump-vladimir-putin-says -climate-change-not-man-made.

15. Andrew Roth, "Western Leaders Point Finger at Putin After Alexei Navalny's Death in Jail," *Guardian*, February 16, 2024, www.theguardian.com /world/2024/feb/16/russian-activist-and-putin-critic-alexei-navalny-dies-in-prison.

16. www.thedailybeast.com/trump-campaign-changed-ukraine-platform -lied-about-it/.

17. Aja Romano, "Twitter Released 9 Million Tweets from One Russian Troll Farm. Here's What We Learned," *Vox*, October 19, 2018, www.vox .com/2018/10/19/17990946/twitter-russian-trolls-bots-election-tampering.

18. Robert Windrem, "Russians Launched Pro–Jill Stein Social Media Blitz to Help Trump Win Election, Reports Say," NBC News, December 22, 2018, www .nbcnews.com/politics/national-security/russians-launched-pro-jill-stein-social -media-blitz-help-trump-n951166.

19. Robert Windrem, "Guess Who Came to Dinner with Flynn and Putin," NBC News, April 18, 2017, www.nbcnews.com/news/world/guess-who-came -dinner-flynn-putin-n742696.

20. Brad Plumer, "On Climate Change, the Difference Between Trump and Clinton Is Really Quite Simple," *Vox*, November 4, 2016, www.vox.com/science -and-health/2016/10/10/13227682/trump-clinton-climate-energy-difference.

21. Rebecca Leber, "Many Young Voters Don't See a Difference Between Clinton and Trump on Climate," *Grist*, July 31, 2016, https://grist.org/election

-2016/many-young-voters-dont-see-a-difference-between-clinton-and-trump-on -climate/.

22. Michael E. Mann, *The Hockey Stick and the Climate Wars: Dispatches from the Front Lines* (Columbia University Press, 2014).

23. Kathryn Watson, "How Did WikiLeaks Become Associated with Russia?," CBS News, November 15, 2017, www.cbsnews.com/news/how-did-wikileaks -become-associated-with-russia/.

24. See Iggy Ostanin, "Exclusive: 'Climategate' Email Hacking Was Carried Out from Russia, in Effort to Undermine Action on Global Warming," *Medium*, July 1, 2019, https://medium.com/@iggyostanin/exclusive-climategate -email-hacking-was-carried-out-from-russia-in-effort-to-undermine-action -78b19bc3ca5a. Ostanin is an award-winning investigative reporter.

25. Mann, *Hockey Stick and the Climate Wars*; David Folkenflik, "The Saudi Prince, the Mosque and Fox News," NPR, September 1, 2010, www.npr .org/2010/09/01/129584557/the-saudi-prince-the-mosque-and-fox-news.

26. The evidence is laid out in Michael Mann and Tom Toles, *The Madhouse Effect* (Columbia University Press, 2016), 164–166.

27. Craig Timberg and Tony Romm, "Russian Trolls Sought to Inflame Debate over Climate Change, Fracking, Dakota Pipeline," *Chicago Tribune*, March 1, 2018, www.chicagotribune.com/nation-world/ct-russian-trolls-climate -change-20180301-story.html.

28. See Simon Rosenberg (@SimonWDC), "MAGA/Trump/RFK/JDVance/ Tucker are Russian-backed wrecking balls trying to destroy America from within. Undermining our public health systems has long been a focus of Russian info ops in the US," X, November 2, 2024, 8:00 p.m., https://x.com/SimonWDC /status/1852863273625944348; and Simon Rosenberg (@SimonWDC), "Undermining public health in America—sowing discord and confusion and causing more Americans to die—has been a central goal of Russian info ops in the US for years. These three men are all the most pro-Putin people in America, w/Tucker on Putin's payroll. Eyes open all," X, November 4, 2024, 9:34 a.m., https://x.com /SimonWDC/status/1853445613108461801.

29. Brian Barrett, "The Manosphere Won," *Wired*, November 6, 2024, www .wired.com/story/donald-trump-manosphere-won/.

30. Keerti Gopal, "Despite Climate Concerns, Young Voter Turnout Slumped and Its Support Split Between the Parties," *Inside Climate News*, November 8, 2024, https://insideclimatenews.org/news/08112024/young-voter -turnout-slumped-support-split-between-parties/.

31. Mary Whitfill Roeloffs, "Who Are Tim Pool and Benny Johnson? What to Know About the Six Right-Wing Commentators DOJ Alleges Were Funded by Russia," *Forbes*, September 5, 2024, www.forbes.com/sites/maryroeloffs /2024/09/05/who-are-tim-pool-and-benny-johnson-what-to-know-about-the-six -right-wing-commentators-doj-alleges-were-funded-by-russia/.

32. Geoff Dembicki, "Inside the Anti-climate Culture War Led by Jordon Peterson and Project 2025, *DeSmog*, September 9, 2024, www.desmog.com/2024/09/09 /inside-the-anti-climate-culture-war-led-by-jordan-peterson-and-project-2025/.

33. Manuel Roig-Franzia et al., "How the 'Bad Boys of Brexit' Forged Ties with Russia and the Trump Campaign—and Came Under Investigators' Scrutiny," *Washington Post*, June 28, 2018, www.washingtonpost.com/politics/how-the

-bad-boys-of-brexit-forged-ties-with-russia-and-the-trump-campaign—and
-came-under-investigators-scrutiny/2018/06/28/6e3a5e9c-7656-11e8-b4b7-
308400242c2e_story.html.

34. Richard Collett-White, Chloe Farand, and Mat Hope, "Meet the Brexit Party's Climate Science Deniers," *DeSmogBlog* (blog), May 1, 2019, www.desmog .co.uk/2019/05/01/brexit-party-climate-science-deniers.

35. Roman Goncharenko, "France's 'Yellow Vests' and the Russian Trolls That Encourage Them," Deutsche Welle, December 15, 2018, www.dw.com/en /frances-yellow-vests-and-the-russian-trolls-that-encourage-them/a-46753388.

36. Emily Atkin, "France's Yellow Vest Protesters Want to Fight Climate Change: Trump Says the Violence Is Proof That People Oppose Environmental Protection. He Couldn't Be More Wrong," *New Republic*, December 10, 2018, https:// newrepublic.com/article/152585/frances-yellow-vest-protesters-want-to-fight -climate-change.

37. Alexander Panetta, "Notorious Russian Troll Farm Targeted Trudeau, Canadian Oil in Online Campaigns," *Toronto Star*, March 18, 2018, www.thestar .com/news/canada/2018/03/18/notorious-russian-troll-farm-targeted-trudeau -canadian-oil-in-online-campaigns.html.

38. Ahmed Al-Rawi and Yasmin Jiwani, "Russian Twitter Trolls Stoke Anti-immigrant Lies Ahead of Canadian Election," *Vancouver Sun*, August 7, 2019, https://vancouversun.com/opinion/op-ed/ahmed-al-rawi-and-yasmin-jiwani -russian-twitter-trolls-stoke-anti-immigrant-lies-ahead-of-canadian-election.

39. W. J. Broad, "Putin's Long War Against American Science," *New York Times*, April 13, 2020, updated June 16, 2021, www.nytimes.com/2020/04/13 /science/putin-russia-disinformation-health-coronavirus.html.

40. M. R. Gordon and D. Volz, "Russian Disinformation Campaign Aims to Undermine Confidence in Pfizer, Other Covid-19 Vaccines, U.S. Officials Say," *Wall Street Journal*, March 7, 2021, www.wsj.com/articles/russian-disinformation -campaign-aims-to-undermine-confidence-in-pfizer-other-covid-19-vaccines -u-s-officials-say-11615129200.

41. Y. Roshchina, S. Roshchin, and K. Rozhkova, "Determinants of COVID-19 Vaccine Hesitancy and Resistance in Russia," *Vaccine* 40, no. 39 (2022): 5739–5747, https://doi.org/10.1016/j.vaccine.2022.08.042, PMID: 36050249, PMCID: PMC9411140.

42. S. Scherbov et al., "COVID-19 and Excess Mortality in Russia: Regional Estimates of Life Expectancy Losses in 2020 and Excess Deaths in 2021," *PLOS One* 17, no. 11 (2022): e0275967, https://doi.org/10.1371/journal.pone.0275967, PMID: 36322565, PMCID: PMC9629588.

43. J. Frevele, "*Morning Joe* Crew 'Connects the Dots' and Claims the Republican Party Is 'Close to Being a Cult to Vladimir Putin,'" *Mediaite*, February 13, 2024, www.mediaite.com/politics/morning-joe-crew-connects-the-dots-and-claims-the -republican-party-is-close-to-being-a-cult-to-vladimir-putin/.

44. Roth, "Western Leaders Point Finger at Putin"; Stephanie Kirchgaessner, "US Finds Saudi Crown Prince Approved Khashoggi Murder but Does Not Sanction Him," *Guardian*, February 26, 2021, www.theguardian.com/world/2021/feb /26/jamal-khashoggi-mohammed-bin-salman-us-report.

45. According to Kenneth Li, "Alwaleed Backs James Murdoch," *Financial Times*, January 21, 2010. Prince Alwaleed bin Talal al-Saud of the Saudi royal

family of Saudi Arabia, through his Kingdom Holding Company, owns 7 percent of News Corp.'s shares, making Kingdom Holdings the second-largest shareholder.

46. Lisa Lerer, "Hacking into the Mind of the CRU Climate Change Hacker," *Guardian*, February 5, 2010.

47. Lisa Lerer, "Saudi Arabia Calls for 'Climategate' Investigation," *Politico*, December 7, 2009.

48. Richard Black, "Climate E-mail Hack 'Will Impact on Copenhagen Summit,'" BBC News, December 3, 2009.

49. Joe Lo, "How Russia Won a 'Dangerous Loophole' for Fossil Gas at Cop28," *Climate Change News*, December 15, 2023, www.climatechangenews.com/2023/12/15/how-russia-won-a-dangerous-loophole-for-fossil-gas-at-cop28/.

50. Lisa Friedman, Brad Plumer, and Vivian Nereim, "Saudia Arabia Is Trying to Block a Global Deal to End Fossil Fuels, Negotiators Say," *New York Times*, December 10, 2023, www.nytimes.com/2023/12/10/climate/saudi-arabia-cop28-fossil-fuels.html.

51. Andrew Stanton, "Elon Musk's Twitter Takeover Faces Backlash over Saudi Financing," *Newsweek*, October 29, 2022, www.newsweek.com/elon-musks-twitter-takeover-faces-backlash-over-saudi-financing-1755606.

52. Ivan Levingston and Ben Bartenstein, "Musk Twitter Bid Counts Secretive $5 Billion Fund Among Backers," *Bloomberg*, June 9, 2022, https://www.bloomberg.com/news/articles/2022-06-09/musk-twitter-bid-counts-secretive—5-billion-fund-among-backers.

53. See Quinn Slobodian, "Does Trump Want America to Look More Like Saudi Arabia?," *New York Times*, March 15, 2025, www.nytimes.com/2025/03/15/opinion/trump-saudi-arabia-america.html; and Michael Hirsh, "How Russian Money Helped Save Trump's Business," *Foreign Policy*, December 21, 2018, https://foreignpolicy.com/2018/12/21/how-russian-money-helped-save-trumps-business/.

54. Jonathan Watts and Ben Doherty, "US and Russia Ally with Saudi Arabia to Water Down Climate Pledge," *Guardian*, December 10, 2018, www.theguardian.com/environment/2018/dec/09/us-russia-ally-saudi-arabia-water-down-climate-pledges-un.

55. Michael E. Mann and Susan Joy Hassol, "COP28 Has Become a Shameless Exercise in the Fight Against Climate Change. But Can We Afford to Walk Out?," *Los Angeles Times*, December 11, 2023, www.latimes.com/opinion/story/2023-12-11/climate-summit-dubai-cop28.

56. William James, Kate Abnett, and Simon Jessop, "COP29 Host Azerbaijan Hits Out at West in Defence of Oil and Gas," Reuters, November 12, 2024, www.reuters.com/business/environment/cop29-pay-up-or-face-climate-led-disaster-humanity-warns-un-chief-2024-11-12/.

57. Gloria Dickie, "COP29 Host Azerbaijan Promoted Fossil Fuel Deals Ahead of Climate Summit, NGO Says," Reuters, November 8, 2024, www.reuters.com/business/environment/cop29-host-azerbaijan-promoted-fossil-fuel-deals-ahead-climate-summit-ngo-says-2024-11-08/.

58. "COP29: Baku Call Launched on Climate Action for Peace, Relief, and Recovery," News on Air, November 15, 2024, www.newsonair.gov.in/cop29-baku-call-launched-on-climate-action-for-peace-relief-and-recovery/.

59. "Coronavirus: Bolsonaro Downplays Threat of Pandemic to Brazil," BBC,

March 25, 2020, www.bbc.com/news/world-latin-america-52040205; B. Dupeyron and C. Segatto, "Just Like Trump, Brazil's Bolsonaro Puts the Economy Ahead of His People During Coronavirus," *Conversation*, April 20, 2020, https://theconversation.com/just-like-trump-brazils-bolsonaro-puts-the-economy-ahead-of-his-people-during-coronavirus-136351.

60. M. K. Sott, M. S. Bender, and K. da Silva Baum, "Covid-19 Outbreak in Brazil: Health, Social, Political, and Economic Implications," *International Journal of Health Services* 52, no. 4 (2022): 442–454, https://doi.org/10.1177/00207314221122658, PMID: 36062608, PMCID: PMC9445630; P. Reeves, "Brazil COVID-19: 'Humanitarian Crisis' with More than 3,000 Deaths a Day," NPR [serial on the internet], April 15, 2021, www.npr.org/2021/04/15/987741403/brazil-covid-19-humanitarian-crisis-with-more-than-3-000-deaths-a-day.

61. R. Pedroso, "Brazil's Bolsonaro Says He Will Not Be Vaccinated Against COVID-19," CNN [serial on the internet], October 13, 2021, www.cnn.com/2021/10/13/americas/bolsonaro-no-vaccine-intl/index.html.

62. P. J. Hotez, "Global Vaccine Access Demands Combating Both Inequity and Hesitancy," *Health Affairs* 42, no. 12 (2023): 1681–1688, https://doi.org/10.1377/hlthaff.2023.00775, PMID: 38048497.

63. G. J. Seara-Morais et al., "The Pervasive Association Between Political Ideology and COVID-19 Vaccine Uptake in Brazil: An Ecologic Study," *BMC Public Health* 23, no. 1 (2023): 1606, https://doi.org/10.1186/s12889-023-16409-w, PMID: 37612648, PMCID: PMC10464231.

64. Médecins sans Frontières, "Failed COVID-19 Response Drives Brazil to Humanitarian Catastrophe," press release, April 15, 2021, www.msf.org/failed-coronavirus-response-drives-brazil-humanitarian-catastrophe.

65. J. Mendoza, "COVID-19 Deaths in Latin America, 2023, by Country and Number of Deaths Due to the Novel Coronavirus (COVID-19) in Latin America and the Caribbean as of December 6, 2023," Statista, December 13, 2023, www.statista.com/statistics/1103965/latin-america-caribbean-coronavirus-deaths/.

66. A. Cheatham and D. Roy, "Venezuela: The Rise and Fall of a Petrostate," Council on Foreign Relations, December 22, 2023, www.cfr.org/backgrounder/venezuela-crisis.

67. P. J. Hotez, *Preventing the Next Pandemic: Vaccine Diplomacy in a Time of Anti-science* (Johns Hopkins University Press, 2021).

68. P. J. Hotez et al., "Venezuela and Its Rising Vector-Borne Neglected Diseases," *PLOS Neglected Tropical Diseases* 11, no. 6 (2017): e0005423, https://doi.org/10.1371/journal.pntd.0005423, PMID: 28662038, PMCID: PMC5490936; A. E. Paniz-Mondolfi et al., "Resurgence of Vaccine-Preventable Diseases in Venezuela as a Regional Public Health Threat in the Americas," *Emerging Infectious Diseases* 25, no. 4 (2019): 625–632, https://doi.org/10.3201/eid2504.181305, PMID: 30698523, PMCID: PMC6433037; M. E. Grillet et al., "Venezuela's Humanitarian Crisis, Resurgence of Vector-Borne Diseases, and Implications for Spillover in the Region," *Lancet Infectious Diseases* 19, no. 5 (2019): e149–e161, https://doi.org/10.1016/S1473-3099(18)30757-6, errata in *Lancet Infectious Diseases* (February 27, 2019), PMID: 30799251.

69. Hotez, *Preventing the Next Pandemic*; A. E. Paniz-Mondolfi et al., "Venezuela's Upheaval Threatens Yanomami," *Science* 365, no. 6455 (2019): 766–767, https://doi.org/10.1126/science.aay6003, PMID: 31439788.

70. Deyanira Garcia Zea, "Brain Drain in Venezuela: The Scope of the Human Capital Crisis," *Human Resource Development International* 23, no. 2 (2020): 188–195, https://doi.org/10.1080/13678868.2019.1708156; J. Requena, "Economy Crisis: Venezuela's Brain Drain Is Accelerating," *Nature* 536, no. 7617 (2016): 396, https://doi.org/10.1038/536396d, PMID: 27558053.

71. R. Perez Ortega, "Scientists Rush to Defend Venezuelan Colleagues Threatened over Coronavirus Study," *Science* (June 2, 2020), www.science.org /content/article/scientists-rush-defend-venezuelan-colleagues-threatened-over -coronavirus-study.

72. R. Stone, "Healing Venezuela," *Science* 375, no. 6585 (2022): 1082–1084, https://doi.org/10.1126/science.adb1875, PMID: 35271340.

73. Simon Bolivar Foundation, "About," www.simonbolivarfoundation.org /grants/index.html, accessed February 4, 2024.

74. Texas Economic Development Corporation, "Why Texas Is the Best State for Business," https://businessintexas.com/why-texas/, accessed December 30, 2023.

75. B. Mulder, "Fact-Check: Is the Texas Oil and Gas Industry 35% of the State Economy?," *Austin American-Statesman*, December 20, 2020, www .statesman.com/story/news/politics/politifact/2020/12/22/fact-check-texas-oil-and -gas-industry-35-state-economy/4009134001/.

76. P. Stone, "Texas Fracking Billionaire Brothers Fuel Rightwing Media with Millions of Dollars," *Guardian*, September 5, 2023, www.theguardian.com /us-news/2023/sep/05/texas-fracking-billionaire-brothers-prageru-daily-wire.

77. Mann, *Hockey Stick and the Climate Wars*.

78. "CorbettAdministrationIgnoringPennsylvania'sClimateChangeLaw,"State Impact Pennsylvania/NPR, https://stateimpact.npr.org/pennsylvania/2013/07/18 /corbett-administration-ignoring-pennsylvanias-climate-change-law/.

79. Mark Levy, "Pennsylvania Court Permanently Blocks Effort to Make Power Plants Pay for Greenhouse Gas Emissions," Associated Press, November 1, 2023, https://apnews.com/article/climate-change-pennsylvania-power-plants -energy-shapiro-e6315cb32a7a7be57ddcad9ad0567e4b.

80. "ALEC: 50 Years of Attacking Environmental Protection and Democracy," Greenpeace, October 3, 2023, www.greenpeace.org/usa/alec-50-years-of -attacking-environmental-protection-and-democracy/.

81. Tori Otten, "Ex-Senator Jim Inhofe Retired Due to Long Covid, Says at Least Five Other Congress Members Also Have It," *New Republic*, February 24, 2023, https://newrepublic.com/post/170783/jim-inhofe -retired-due-long-covid-says-least-5-congressmembers-have-it.

82. Suzanne Goldenberg, "US Congressman Cites Biblical Flood to Dispute Human Link to Climate Change," *Guardian*, April 10, 2013, www.theguardian. com/environment/blog/2013/apr/11/republican-biblical-flood-climate-change.

83. Mann, *Hockey Stick and the Climate Wars*; Mann and Toles, *The Madhouse Effect*.

84. C. Mooney, *The Republican War on Science* (Basic Books, 2005).

85. Mann, *Hockey Stick and the Climate Wars*.

86. S. Gross, "What Is the Trump Administration's Track Record on the Environment?," Brookings Institution, August 4, 2020, www.brookings.edu/articles /what-is-the-trump-administrations-track-record-on-the-environment/.

87. J. Daley, "U.S. Exits Paris Climate Accord After Trump Stalls Global Warming Action for Four Years," *Scientific American*, November 4, 2020, www.scientificamerican.com/article/u-s-exits-paris-climate-accord-after-trump-stalls-global-warming-action-for-four-years/.

88. A. J. Blinken, "The United States Officially Rejoins the Paris Agreement," US State Department press statement, February 19, 2021, www.state.gov/the-united-states-officially-rejoins-the-paris-agreement/.

89. John Larsen et al., "A Turning Point for US Climate Progress: Assessing the Climate and Clean Energy Provisions in the Inflation Reduction Act," Rhodium Group, August 12, 2022, https://rhg.com/research/climate-clean-energy-inflation-reduction-act/.

90. "Analysis: Trump Election Win Could Add 4bn Tonnes to US Emissions by 2030," Carbon Brief, March 6, 2024, www.carbonbrief.org/analysis-trump-election-win-could-add-4bn-tonnes-to-us-emissions-by-2030/.

91. Collin Eaton and Benoit Morenne, "Exxon Says Trump Should Keep U.S. in Paris Climate Pact," *Wall Street Journal*, November 12, 2024, www.wsj.com/business/energy-oil/exxon-says-trump-should-keep-u-s-in-paris-climate-pact-3d8de471.

92. Matthew Dalton, "Trump Victory Leaves China Calling the Shots at COP29 Climate Negotiations," *Wall Street Journal*, November 11, 2024, www.wsj.com/world/trump-victory-leaves-china-calling-the-shots-at-cop29-climate-negotiations-f4161f7f.

93. Lisa Friedman, "A Republican 2024 Climate Strategy: More Drilling, Less Clean Energy," *New York Times*, August 4, 2023, www.nytimes.com/2023/08/04/climate/republicans-climate-project2025.html.

94. Arianna Skibell, "Trump's 30-Day Climate Assault," *Politico*, February 20, 2025, www.politico.com/newsletters/power-switch/2025/02/20/trumps-30-day-climate-assault-00205201.

95. A. Sheikh, C. Robertson, and B. Taylor, "BNT162b2 and ChAdOx1 nCoV-19 Vaccine Effectiveness Against Death from the Delta Variant," *New England Journal of Medicine* 385, no. 23 (2021): 2195–2197, https://doi.org/10.1056/NEJMc2113864, PMID: 34670038, PMCID: PMC8552534; C. R. Tamandjou Tchuem et al., "Vaccine Effectiveness and Duration of Protection of COVID-19 mRNA Vaccines Against Delta and Omicron BA.1 Symptomatic and Severe COVID-19 Outcomes in Adults Aged 50 Years and Over in France," *Vaccine* 41, no. 13 (2023): 2280–2288, https://doi.org/10.1016/j.vaccine.2023.02.062, PMID: 36870880, PMCID: PMC9968619; P. Tang et al., "BNT162b2 and mRNA-1273 COVID-19 Vaccine Effectiveness Against the SARS-CoV-2 Delta Variant in Qatar," *Nature Medicine* 27, no. 12 (2021): 2136–2143, https://doi.org/10.1038/s41591-021-01583-4, PMID: 34728831; J. Lopez Bernal et al., "Effectiveness of Covid-19 Vaccines Against the B.1.617.2 (Delta) Variant," *New England Journal of Medicine* 385, no. 7 (2021): 585–594, https://doi.org/10.1056/NEJMoa2108891, errata in *New England Journal of Medicine* 388, no. 7 (2023): 672, PMID: 34289274, PMCID: PMC8314739.

96. Hotez, *Deadly Rise of Anti-science*.

97. Hotez, *Deadly Rise of Anti-science*.

98. D. Leonhard, "Red Covid," *New York Times*, September 27, 2021, updated October 1, 2021, www.nytimes.com/2021/09/27/briefing/covid-red-states-vaccinations.html; D. Leonhardt, "U.S. Covid Deaths Get Even Redder (the

Morning Newsletter)," *New York Times*, November 8, 2021, updated November 24, 2021, www.nytimes.com/2021/11/08/briefing/covid-death-toll-red-america.html.

99. D. Holmes, "They Clapped for Death at CPAC," *Esquire*, July 12, 2021, www.esquire.com/news-politics/a37001629/cpac-vaccination-goal-biden-miss-clap/.

100. Hotez, *Deadly Rise of Anti-science*.

101. D. Castronuovo, "Cawthorn: Biden Door-to-Door Vaccine Strategy Could Be Used to 'Take' Guns, Bibles," *Hill*, July 9, 2021, https://thehill.com/homenews/house/562372-cawthorn-biden-door-to-door-vaccine-strategy-could-be-used-to-take-guns-bibles/.

102. A. Smith, "Conservative Hostility to Biden Vaccine Push Surges with Covid Cases on the Rise," NBC News, July 19, 2021, www.nbcnews.com/politics/politics-news/conservative-hostility-biden-vaccine-push-surges-covid-cases-rise-n1273692.

103. P. LeBlanc, "Marjorie Taylor Greene Compares Biden Vaccine Push to Nazi-Era 'Brown Shirts' Weeks After Apologizing for Holocaust Comments," CNN, July 7, 2021, www.cnn.com/2021/07/07/politics/marjorie-taylor-greene-brown-shirts-vaccine/index.html.

104. P. J. Hotez, "Global Vaccinations: New Urgency to Surmount a Triple Threat of Illness, Antiscience, and Anti-Semitism," *Rambam Maimonides Medical Journal* 14, no. 1 (2023): e0004, https://doi.org/10.5041/RMMJ.10491, PMID: 36719666, PMCID: PMC9888484.

105. G. C. Altschuler, "The Clear and Present Danger of Jim Jordan & Co.," *Hill*, December 12, 2021, https://thehill.com/opinion/campaign/585413-the-clear-and-present-danger-of-jim-jordan-co; A. Henderson, "'A Race to the Bottom': House GOP Slammed for 'Really Disgraceful' Anti-vaccine Tweet," *Salon*, January 1, 2022, www.salon.com/2022/01/01/a-race-to-the-bottom-slammed-for-really-disgraceful-anti-vaccine-tweet/; A. Solender, "GOP Rep. Brooks Pushes Anti-vaccine Talking Points in Letter to Biden," *Forbes*, July 19, 2021, www.forbes.com/sites/andrewsolender/2021/07/19/gop-rep-brooks-pushes-anti-vaccine-talking-points-in-letter-to-biden/?sh=7e171e2354be; A. Grayer, L. Fox, and S. Fortinsky, "Not All Republicans Are Embracing McConnell's Vaccine Push. Read What Some Had to Say When Asked This Week," CNN, July 22, 2021, www.cnn.com/2021/07/22/politics/house-republicans-vaccination-rates/index.html.

106. H. Lybrand and T. Subramani, "Fact-Checking Sen. Ron Johnson's Anti-vaccine Misinformation," CNN, May 10, 2021, www.cnn.com/2021/05/07/politics/ron-johnson-vaccine-misinformation-fact-check/index.html; H. Redman, "In Interview, Sen. Johnson Says It 'May Be True' That COVID Vaccines Cause AIDS," *Wisconsin Examiner*, May 3, 2022, https://wisconsinexaminer.com/brief/in-interview-sen-johnson-says-it-may-be-true-that-covid-vaccines-cause-aids/; K. Dickeson, "Sen. Ron Johnson Hosts COVID-19 Panel Focusing on Vaccine Skepticism; Local Doctors Respond," NBC 26 Green Bay (WI), January 24, 2022, www.nbc26.com/news/local-news/sen-ron-johnson-hosts-covid-19-panel-focusing-on-vaccine-skepticism-local-doctors-respond; S. G. Stolberg, "Anti-vaccine Doctor Has Been Invited to Testify Before Senate Committee," *New York Times*, December 6, 2020, updated April 26, 2021, www.nytimes.com/2020/12/06/us/politics/anti-vax-scientist-senate-hearing.html; C. Cillizza, "What Rand Paul Gets Wrong on Vaccines," CNN Politics, May 24, 2021, www.cnn.com/2021/05/24

/politics/rand-paul-vaccines-covid-19/index.html; Rand Paul, "Rand Paul: The Science Proves People with Natural Immunity Should Skip COVID Vaccines," *Louisville Courier-Journal*, May 27, 2021, www.courier-journal.com/story/opinion/2021/05/27 /rand-paul-says-people-natural-covid-immunity-should-skip-vaccine/7468051002/.

107. Kat So, "Climate Deniers of the 118th Congress," American Progress, July 18, 2024, www.americanprogress.org/article/climate-deniers-of-the-118th -congress/.

108. Rand Paul (@RandPaul), "Perhaps Peter Hotez won't debate on Joe Rogan show because he fears his connection to Chinese military scientists will be exposed? Dr. Peter Hotez's Funding Linked to Controversial Chinese Military Scientists at Wuhan Lab," Twitter (now X), June 27, 2023, 10:32 a.m., https://twitter .com/RandPaul/status/1673700802316181506; Ron Johnson (@SenRonJohnson), "More fear-mongering coming from a card-carrying member of the Covid Cartel. Will he promote more lockdowns and another mRNA vaccine to end the 'Disease X' pandemic?," X, November 29, 2023, 7:00 p.m., https://x.com/SenRonJohnson /status/1730013980511371423; Rand Paul, *Deception: The Great Covid Cover-up* (Regnery, 2023).

109. Ron Johnson (@SenRonJohnson), "More fear-mongering coming from a card-carrying member of the Covid Cartel."

110. Paul (@RandPaul), "Perhaps Peter Hotez won't debate on Joe Rogan show."

111. Paul, *Deception*.

112. N. Confessore, "Hillary Was Right," *American Prospect*, November 14, 2001, https://prospect.org/features/hillary-right/.

113. A. J. Pennyfarthing, "Ron DeSantis Fed COVID Crow by Doctor He'd Ridiculed on Fox News," *Daily Kos*, July 15, 2021, www.dailykos.com/stories /2021/7/15/2040161/-Ron-DeSantis-fed-COVID-crow-by-doctor-he-d-ridiculed -on-Fox-News.

114. "'Florida Has Defeated Fauci-ism,' DeSantis Tells CPA Convention," *Orlando Sentinel*, February 24, 2022, www.orlandosentinel.com/2022/02/24 /florida-has-defeated-fauci-ism-desantis-tells-cpac-convention/.

115. D. Wood and G. Brumfiel, "Pro-Trump Counties Now Have Far Higher COVID Death Rates. Misinformation Is to Blame," NPR, December 5, 2021, www.npr.org/sections/health-shots/2021/12/05/1059828993/data-vaccine -misinformation-trump-counties-covid-death-rate; C. Gaba, "Challenge Accepted: The Elephant in the Room, Now Age-Adjusted," ACA Signups, February 2, 2022, https://acasignups.net/22/02/03/challenge-accepted-elephant-room-now-age -adjusted; C. Gaba, "Weekly Update: County-Level #COVID19 Vaccination Levels by 2020 Partisan Lean," ACA Signups, February 14, 2022, https://acasig-nups.net/22/02/14/weekly-update-county-level-covid19-vaccination-levels-2020- partisan-lean; C. Gaba and A. Stokes, "Exclusive: 'Non-COVID' Excess Death Rates Ran 21x Higher in Reddest Counties than Bluest in 2021," ACA Signups, May 5, 2022, https://acasignups.net/22/05/09/exclusive-non-covid-excess-death-rates -ran-21x-higher-reddest-counties-bluest-2021; J. Wallace, P. Goldsmith-Pinkham, and J. L. Schwartz, "Excess Death Rates for Republican and Democratic Registered Voters in Florida and Ohio During the COVID-19 Pandemic," *JAMA Internal Medicine* 183, no. 9 (2023): 916–923, https://doi.org/10.1001/jamaint-ernmed.2023.1154, PMID: 37486680, PMCID: PMC10366951.

116. Hotez, *Deadly Rise of Anti-science*; C. Gaba, "My Own Crude Estimate: Vaccine Refusal Has Likely Killed 180K–235K Americans to Date," ACA Signups, March 10, 2022, https://acasignups.net/22/03/10/my-own-crude-estimate -vaccine-refusal-has-likely-killed-180k-235k-americans-date; K. Amin et al., "COVID-19 Mortality Preventable by Vaccines," Peterson-KFF Health System Tracker, April 21, 2022, www.kff.org/coronavirus-covid-19/issue-brief/covid-19 -continues-to-be-a-leading-cause-of-death-in-the-u-s/.

117. K. M. Jia et al., "Estimated Preventable COVID-19-Associated Deaths Due to Non-vaccination in the United States," *European Journal of Epidemiology* 38, no. 11 (2023): 1125–1128, https://doi.org/10.1007/s10654-023-01006-3, PMID: 37093505, PMCID: PMC10123459.

118. Wood and Brumfiel, "Pro-Trump Counties Now Have Far Higher COVID Death Rates."

119. David Corn, "New Revelations Emerge on How Donald Trump Killed 400,000 (or More) Americans," *Mother Jones*, November 17, 2021, www .motherjones.com/politics/2021/11/new-revelations-emerge-on-how-donald -trump-killed-400000-coronavirus-pandemic/.

120. P. J. Hotez, "Anti-science Conspiracies Pose New Threats to US Biomedicine in 2023," *FASEB BioAdvances* 5, no. 6 (2023): 228–232, https://doi .org/10.1096/fba.2023-00032, PMID: 37287866, PMCID: PMC10242190.

121. K. Fung, "Marjorie Taylor Greene Wants to Make Certain Vaccines Illegal to Fund," *Newsweek*, January 17, 2024, www.newsweek.com/marjorie -taylor-greene-wants-certain-vaccines-illegal-fund-1861475.

122. P. Sah et al., "Estimating the Impact of Vaccination on Reducing COVID-19 Burden in the United States: December 2020 to March 2022," *Journal of Global Health* (September 3, 2022): 03062, https://doi.org/10.7189 /jogh.12.03062, PMID: 36056814, PMCID: PMC9441009.

123. Hotez, "Anti-science Conspiracies Pose New Threats."

124. G. Jacobs (@BioinfoTools), "Occasionally I remind ppl much of the COVID-19 'lab leak' story from the USA is political theatre. This from the (Republican-led) Select Subcommittee on the Coronavirus Pandemic [down arrow] 'Explosive', 'Get your popcorn ready folks . . .' is not the language of committees or investigation," Twitter (now X), June 30, 2023, 7:35 a.m., https://twitter.com /BioinfoTools/status/1674551435235840000/photo/1, referring to House Select Subcommittee on the Coronavirus (@COVIDSelect), "Explosive emails from Dr. David Morens—Senior Scientific Advisor @NIAIDNews & Dr. Fauci ally— reveal a potential violation of federal record keeping laws. Get your popcorn ready folks . . . ," Twitter (now X), June 30, 2023, 7:35 a.m.

125. Katherine Fung, "Marjorie Taylor Greene Wants to Make Certain Vaccines Illegal to Fund," *Newsweek*, January 17, 2024, www.newsweek.com/marjorie -taylor-greene-wants-certain-vaccines-illegal-fund-1861475.

126. C. Campanile, "Dr. Anthony Fauci Deserves to Go to Prison over 'Dishonest' on COVID-19 Origins: Sen. Rand Paul," *New York Post*, January 14, 2024, https://nypost.com/2024/01/14/news/dr-anthony-fauci-deserves-to-go-to-prison -over-dishonesty-on-covid-19-origins-sen-rand-paul/.

127. Dan Diamond and Alex Horton, "Navy Demoted Ronny Jackson After Probe into White House Behavior," *Texas Tribune*, March 7, 2024, www .texastribune.org/2024/03/07/ronny-jackson-white-house-navy-demoted/.

128. S. MacFarlane, "Lawmakers Questioned Fauci About 'Lab Leak' COVID Theory in Marathon Closed-Door Congressional Interview," CBS News, January 16, 2024.

129. M. Worobey et al., "The Huanan Seafood Wholesale Market in Wuhan Was the Early Epicenter of the COVID-19 Pandemic," *Science* 377, no. 6609 (2022): 951–959, https://doi.org/10.1126/science.abp8715, PMID: 35881010, PMCID: PMC9348750; J. E. Pekar et al., "The Molecular Epidemiology of Multiple Zoonotic Origins of SARS-CoV-2," *Science* 377, no. 6609 (2022): 960–966, https://doi.org/10.1126/science.abp8337, errata in *Science* 382, no. 6667 (2023): eadl0585, PMID: 35881005, PMCID: PMC9348752; J. E. Pekar et al., "The Recency and Geographical Origins of the Bat Viruses Ancestral to SARS-CoV and SARS-CoV-2," *bioRxiv* [preprint] (July 12, 2023): 2023.07.12.548617, https://doi.org/10.1101/2023.07.12.548617, PMID: 37502985, PMCID: PMC10369958; A. Crits-Christoph et al., "Genetic Tracing of Market Wildlife and Viruses at the Epicenter of the COVID-19 Pandemic," *bioRxiv* [preprint] (September 14, 2023): 2023.09.13.557637, https://doi.org/10.1101/2023.09.13.557637, PMID: 37745602, PMCID: PMC10515900; A. Crits-Christoph et al., "Genetic Tracing of Market Wildlife and Viruses at the Epicenter of the COVID-19 Pandemic," *Cell* 187, no. 19 (2024): 5468–5482.e11, https://doi.org/10.1016/j.cell.2024.08.010, PMID: 39303692, PMCID: PMC11427129.

130. Director of National Intelligence, "Key Takeaways," www.dni.gov/files/ODNI/documents/assessments/Unclassified-Summary-of-Assessment-on-COVID-19-Origins.pdf, accessed January 1, 2024.

131. J. Tollefson, "Trump's Presidential Push Renews Fears for US Science," *Nature* (January 29, 2024), https://doi.org/10.1038/d41586-024-00258-7, e-publication ahead of print, PMID: 38287160.

132. Jason Breslow, "Trump Secretly Sent Putin COVID-19 Tests During Pandemic Shortage, a New Book Reports," NPR, October 8, 2024, www.npr.org/2024/10/08/nx-s1-5146501/trump-putin-covid-tests.

133. Laura Barrón-López and Jackson Hudgins, "Trump Vows to Defund Schools Requiring Vaccines for Students If He's Reelected," PBS News, June 24, 2024, www.pbs.org/newshour/show/trump-vows-to-defund-schools-requiring-vaccines-for-students-if-hes-reelected.

134. Meridith McGraw and Chelsea Cirruzzo, "Trump to Select Robert F. Kennedy Jr. to Lead HHS," *Politico*, November 14, 2024, www.politico.com/news/2024/11/14/robert-f-kennedy-jr-trump-hhs-secretary-pick-00188617.

135. Interview with RFK Jr. archived here: Camus (@newstart_2024), "RFK Jr. Threatens to Prosecute Medical Journals for Fraud Under Racketeering Laws: 'They are presenting themselves to medical professionals as an arbiter of truth and as a neutral referee and a reliable referee of the truth.' 'And they know that those medical professionals are relying on journal articles to treat patients and that if they tell a lie, if they're committing fraud, that they can injure and kill people.' 'So I believe they can be prosecuted and not only can they be prosecuted for those injuries, but they can be prosecuted on the racketeering statutes for promoting fraud. So I will, I'm going to do that as soon as I get in there," X, December 16, 2024, 4:16 p.m., https://x.com/newstart_2024/status/1868767103467323867.

136. Ron Filipkowski (@RonFilipkowski), "Junior Is Now Selling

Ivermectin," X, February 12, 2024, 7:59 p.m., https://x.com/RonFilipkowski/status/1757207953461977333?s=20.

137. K. Bales, "Prominent Anti-vaxxer's Snake Oil Store Has a Propaganda Outlet," *Who What Why*, February 12, 2024, https://whowhatwhy.org/science/health-medicine/prominent-anti-vaxxers-snake-oil-store-has-a-propaganda-outlet/.

138. Christina Pagel, "Censor, Purge, Defund: How Trump Is Following the Autocratic Playbook on Science and Universities," *Diving into Data and Decision Making*, March 10, 2025, https://christinapagel.substack.com/p/censor-purge-defund-how-trump-following.

139. Michael Sainato and Dharna Noor, "Musk's 'Efficiency' Agency Site Adds Data from Controversial Rightwing Thinktank," *Guardian*, February 12, 2025, www.theguardian.com/us-news/2025/feb/12/trump-musk-doge-website.

140. Igor Bobic, "Trump Administration Orders 'Catastrophic' Funding Cuts for Science Research," *Huffpost*, February 8, 20252, www.huffpost.com/entry/donald-trump-elon-musk-nih-cuts_n_67a780f3e4b0c0ca26442926.

141. Kai Kupferschmidt, "Pentagon Abruptly Ends All Funding for Social Science Research," *Science* (March 10, 2025), www.science.org/content/article/pentagon-abruptly-ends-all-funding-social-science-research.

142. Paul Voosen, "NASA to Eliminate Chief Scientist Position," *Science* (March 10, 2025), www.science.org/content/article/nasa-eliminate-chief-scientist-position.

143. Oliver Milman with visuals by Andrew Witherspoon, "Scientists Brace 'for the Worst' as Trump Purges Climate Mentions from Websites," *Guardian*, February 4, 2025, www.theguardian.com/us-news/2025/feb/04/trump-climate-change-federal-websites.

144. William J. Broad, "Why Trump Picked a Science Adviser Who Isn't a Scientist," *New York Times*, January 29, 2025, www.nytimes.com/2025/01/29/science/trump-science-advisor-michael-kratsios.html.

145. Kimberly M. S. Cartier, "Crowds Stand Up for Science Across the United States," *Eos*, March 7, 2025, https://eos.org/articles/crowds-stand-up-for-science-across-the-united-states.

146. Sam Stein, "The DOGE Brain Drain Has Begun," *Bulwark*, February 19, 2025, www.thebulwark.com/p/the-doge-brain-drain-has-begun-science-government-cuts-budget-researchers.

147. N. Jankowicz, "The Coming Flood of Disinformation: How Washington Gave Up on the Fight Against Falsehood," *Foreign Affairs*, February 7, 2024, www.foreignaffairs.com/united-states/coming-flood-disinformation.

148. P. J. Hotez, "Opinion: Scientists Have Become Sitting Ducks. We Need Leaders to Step Up and Defend Us," *Los Angeles Times*, September 22, 2023, www.latimes.com/opinion/story/2023-09-22/antiscience-climate-change-covid-vaccines-antivax.

149. Editorial Board, "Ken Cuccinelli's Climate-Change Witch Hunt," *Washington Post*, March 11, 2012, www.washingtonpost.com/opinions/ken-cuccinellis-climate-change-witch-hunt/2012/03/08/gIQApmdu5R_story.html.

150. G. Steward, "It's Clear the Anti-science Movement Has a Tight Grip on the Alberta Government," *Toronto Star*, January 2, 2024, www.thestar.com/opinion/contributors/it-s-clear-the-anti-science-movement-has-a-tight-grip-on-the-alberta-government/article_d350d58a-a0e5-11ee-8d77-87baa2dc1f86.html.

151. P. J. Hotez, "Will Anti-vaccine Activism in the USA Reverse Global Goals?," *Nature Reviews Immunology* 22, no. 9 (2022): 525–526, https://doi.org /10.1038/s41577-022-00770-9, PMID: 35915141, PMCID: PMC9340755.

152. Elena Giordano and Karl Mathiesen, "Beset by Fire and Heat, Meloni's Government Flirts with Climate Denial," *Politico*, August 2, 2023, www.politico.eu/arti cle/bgiorgia-meloni-italy-heat-summer-wildfires-climate-change-climate-denial/.

153. Shaun Walker, "Migration v Climate: Europe's New Political Divide," *Guardian*, December 2, 2019, www.theguardian.com/environment/2019/dec/02 /migration-v-climate-europes-new-political-divide.

154. Ben Hubbard, "Rich Gulf States Have Huge Ambitions. Will Extreme Heat Hold Them Back?," June 27, 2024, www.nytimes.com/2024/06/27/world /middleeast/gulf-states-heat-climate-change.html?smid=nytcore-ios-share &referringSource=articleShare&sgrp=c-cb.

155. "Noam Chomsky on Midterms: Republican Party Is the 'Most Dangerous Organization in Human History,'" *Democracy Now!*, November 5, 2018, www .youtube.com/watch?v=jeHOCId5T-w.

156. Lucas Ropek, "Trump's Defense Secretary Hegseth Order Cyber Command to 'Stand Down' on All Russia Operations," *Gizmodo*, March 2, 2025, https: //gizmodo.com/trumps-defense-secretary-hegseth-orders-cyber-command-to -stand-down-on-all-russia-operations-2000570343.

157. Justin Ling, "Pentagon Cuts Threaten Programs That Secure Loose Nukes and Weapons of Mass Destruction," *Wired*, March 6, 2025, www.wired .com/story/pentagon-cuts-nukes-chemical-weapons-wmd/.

CHAPTER 4: THE PROS

1. The origins of the term are this *Washington Post* op-ed by Michael E. Mann and Tom Toles. We here have written about a restricted set of the individuals named and expanded on the earlier discussion: www.washingtonpost .com/posteverything/wp/2016/09/16/deniers-club-meet-the-people-clouding -the-climate-change-debate/.

2. Michael E. Mann, *Our Fragile Moment: How Lessons from Earth's Past Can Help Us Survive the Climate Crisis* (PublicAffairs, 2023).

3. Michael E. Mann, *The Hockey Stick and the Climate Wars: Dispatches from the Front Lines* (Columbia University Press, 2014).

4. Dan Harris et al., "Global Warming Denier: Fraud or 'Realist'?," ABC News, March 23, 2008.

5. Fred Singer, "The Sea Is Rising, but Not Because of Climate Change," *Wall Street Journal*, May 15, 2018, www.wsj.com/articles/the-sea-is -rising-but-not-because-of-climate-change-1526423254.

6. Andrea L. Dutton and Michael E. Mann, "Water's Rising Because It's Getting Warmer," *Wall Street Journal*, May 22, 2018, www.wsj.com/articles/waters -rising-because-its-getting-warmer-1527007433.

7. "Steve Milloy," *DeSmog*, www.desmog.com/steve-milloy/, accessed March 7, 2024.

8. The ads were sponsored by Milloy's "Green Hell Blog."

9. The episode was covered in a UCLA Environmental Law online article. See Cara Horowitz, "What's It Like to Be Climate Scientist Michael Mann? Think Bounty (Not the Good Kind)," *Legal Planet*, February 13, 2012, http://legalplanet

.wordpress.com/2012/02/13/whats-it-like-to-be-climate-scientist-michael-mann-think-bounty-not-the-good-kind/.

10. Steven J. Milloy, "Freaky-Frog Fraud," Fox News, November 8, 2002, www.foxnews.com/story/2002/11/08/freaky-frog-fraud.html, quoted in Sara Jerving, "Syngenta's Paid Third Party Pundits Spin the 'News' on Atrazine," *PRWatch*, February 7, 2012, http://dev.prwatch.org/news/2012/02/11276/syngentas-paid-third-party-pundits-spin-news-atrazine.

11. Steve Milloy (@JunkScience), "Red Pope calls for 'CULTURAL REVOLUTION.' Been there, done that," Twitter (now X), June 18, 2015, 8:55 a.m., https://twitter.com/JunkScience/status/611517554964721664.

12. Steve Milloy (@JunkScience), "Democrat dirty tricks were able to drive the heroic Joe McCarthy to an early grave. Democrat dirty tricks forced Nixon to resign. But with @realDonaldTrump, Democrat dirty tricks will backfire and destroy the Democrat party (or what's left of it after Obama)," Twitter (now X), February 3, 2018, 9:00 p.m., https://twitter.com/JunkScience/status/959969983383121925.

13. Clive Hamilton, "Silencing the Scientists: The Rise of Right-Wing Populism," *Our World*, March 2, 2011, http://ourworld.unu.edu/en/silencing-the-scientists-the-rise-of-right-wing-populism/#authordata.

14. "Committee for a Constructive Tomorrow," SourceWatch, www.sourcewatch.org/index.php/Committee_for_a_Constructive_Tomorrow, accessed March 7, 2024.

15. "Marc Morano: Biden's Green Energy Agenda Is in Complete Collapse," *Jesse Waters Primetime*, November 1, 2023, www.foxnews.com/video/6340327348112.

16. "Biden Forcing an 'Energy Transition' on a Nation That's Not Ready: Mark Morano," *Varney & Co.*, January 10, 2024, www.foxbusiness.com/video/6344622191112.

17. "Biden's Halting of Liquified Natural Gas Projects Is 'Utter Nonsense': Marc Morano," *Fox and Friends*, January 27, 2024, www.foxnews.com/video/6345709173112.

18. Michael E. Mann, *The New Climate War: The Fight to Take Back Our Planet* (PublicAffairs, 2021).

19. Bjorn Lomborg, "Who's Afraid of Climate Change?," *Project Syndicate*, August 11, 2010, www.project-syndicate.org/commentary/who-s-afraid-of-climate-change.

20. "Michael Mann Responds on Wildfires and Climate Change: Bjorn Lomborg Should Know That We Can Only Adapt So Much," *Wall Street Journal*, August 6, 2023, www.wsj.com/articles/michael-mann-bjorn-lomborg-wildfires-climate-change-f1e15053; Alyssa Lukpat, "The World Bakes Under Extreme Heat: Warming Oceans and Heat Domes Are Contributing to One of the Hottest Summers on Record," *Wall Street Journal*, July 19, 2023, www.wsj.com/articles/extreme-heat-waves-across-the-world-photos-7cc1544d?msockid=0f0cb8ba251364260a3baa7324f3656e.

21. L. A. Grossman, *Choose Your Medicine: Freedom of Therapeutic Choice in America* (Oxford University Press, 2021); M. A. Flannery, "The Early Botanical Medical Movement as a Reflection of Life, Liberty, and Literacy in Jacksonian America," *Journal of the Medical Library Association* 90, no. 4 (2002): 442–454, PMID: 12398251, PMCID: PMC128961.

22. Grossman, *Choose Your Medicine*; Flannery, "Early Botanical Medical Movement."

23. Alexis de Tocqueville, *Democracy in America*, specially edited and abridged for the modern reader by Richard D. Heffner (Mentor Book from New American Library, 1956).

24. S. Thomson, "Dr. Samuel Thomson's Method of Treating Small Pox," *Contagion: Historical Views of Diseases and Epidemics*, Harvard Library, Curiosity Collections, https://curiosity.lib.harvard.edu/contagion/catalog/36-99005759229 0203941, accessed January 2, 2024; P. Cash and Y. Higomoto, "Further Information Concerning Dr. Benjamin Waterhouse's Appointment as Harvard's First Professor of Medicine," *Journal of the History of Medicine and Allied Sciences* 49, no. 3 (1994): 419–428, https://doi.org/10.1093/jhmas49.3.419, PMID: 8083511; "Benjamin Waterhouse, 1754–1846," Harvard Library, Curiosity Collections, https://curiosity.lib.harvard.edu/contagion/feature/benjamin-waterhouse-1754-1846, accessed January 2, 2024.

25. Grossman, *Choose Your Medicine*; A. Berman, "Wooster Beach and the Early Eclectics," *Medical Bulletin* (Ann Arbor) 24, no. 7 (1958): 277–286, PMID: 13581363.

26. A. Flexner, "Medical Education in the United States and Canada: A Report to the Carnegie Foundation for the Advancement of Teaching, Bulletin Number Four," 1910, D. E. Updike, Merrymount Press, Boston, privately printed for members of the Classics of Medicine Library, 1990.

27. Grossman, *Choose Your Medicine*; L. A. Grossman, "The Origins of American Health Libertarianism," *Yale Journal of Health Policy, Law, and Ethics* 13, no. 1 (2013): 76–134, PMID: 23815041; P. J. Hotez, "America's Deadly Flirtation with Antiscience and the Medical Freedom Movement," *Journal of Clinical Investigation* 131, no. 7 (2021): e149072, https://doi.org/10.1172/JCI149072, PMID: 33630759, PMCID: PMC8011881.

28. Grossman, *Choose Your Medicine*; Grossman, "Origins of American Health Libertarianism"; Hotez, "America's Deadly Flirtation with Antiscience."

29. Grossman, *Choose Your Medicine*.

30. Hotez, "America's Deadly Flirtation with Antiscience."

31. A. J. Wakefield et al., "Ileal-Lymphoid-Nodular Hyperplasia, Non-specific Colitis, and Pervasive Developmental Disorder in Children," *Lancet* 351, no. 9103 (1998): 637–664, https://doi.org/10.1016/s0140-6736(97)11096-0, retraction in *Lancet* 375, no. 9713 (2010): 445, errata in *Lancet* 363, no. 9411 (2004): 750, PMID: 9500320.

32. F. Godlee, J. Smith, and H. Marcovitch, "Wakefield's Article Linking MMR Vaccine and Autism Was Fraudulent," *BMJ* 342 (2011): c7452, https://doi.org/10.1136/bmj.c7452, PMID: 21209060.

33. L. Eggertson, "*Lancet* Retracts 12-Year-Old Article Linking Autism to MMR Vaccines," *CMAJ* 182, no. 4 (2010): E199–E200, https://doi.org/10.1503/cmaj.109-3179, PMID: 20142376, PMCID: PMC2831678.

34. Godlee, Smith, and Marcovitch, "Wakefield's Article Linking MMR Vaccine and Autism Was Fraudulent."

35. Z. Kmietowicz, "Wakefield Is Struck Off for the 'Serious and Wide-Ranging Findings Against Him,'" *BMJ* 340 (2010): c2803, https://doi.org/10.1136/bmj.c2803, PMID: 20498165.

36. P. J. Hotez, *Vaccines Did Not Cause Rachel's Autism: My Journey as a Vaccine Scientist, Pediatrician, and Autism Dad* (Johns Hopkins University Press, 2018).

37. K. Lauerman, "Correcting Our Record: We've Removed an Explosive 2005 Report by Robert F. Kennedy Jr. About Autism and Vaccines. Here's Why," *Salon*, January 16, 2011, www.salon.com/2011/01/16/dangerous_immunity/, accessed January 3, 2024; S. Bernard et al., "Autism: A Novel Form of Mercury Poisoning," *Medical Hypotheses* 56, no. 4 (2001): 462–471, https://doi.org/10.1054/mehy.2000.1281, PMID: 11339848.

38. Lauerman, "Correcting Our Record"; S. Mnookin, *The Panic Virus: The True Story Behind the Vaccine-Autism Controversy* (Simon & Schuster, 2012).

39. Hotez, *Vaccines Did Not Cause Rachel's Autism*.

40. B. S. Gadad et al., "Administration of Thimerosal-Containing Vaccines to Infant Rhesus Macaques Does Not Result in Autism-Like Behavior or Neuropathology," *Proceedings of the National Academy of Sciences of the United States of America* 112, no. 40 (2015): 12498–12503, https://doi.org/10.1073/pnas.1500968112, errata in *Proceedings of the National Academy of Sciences of the United States of America* 112, no. 49 (2015): E6827, PMID: 26417083, PMCID: PMC4603476.

41. Hotez, *Vaccines Did Not Cause Rachel's Autism*.

42. Hotez, *Vaccines Did Not Cause Rachel's Autism*; Mnookin, *Panic Virus*.

43. X. Meng et al., "Assembloid CRISPR Screens Reveal Impact of Disease Genes in Human Neurodevelopment," *Nature* 622, no. 7982 (2023): 359–366, https://doi.org/10.1038/s41586-023-06564-w, PMID: 37758944, PMCID: PMC10567561.

44. Hotez, *Vaccines Did Not Cause Rachel's Autism*.

45. Hotez, *Vaccines Did Not Cause Rachel's Autism*.

46. L. Beil, "Peter Hotez vs. Measles and the Anti-vaccination Movement," *Texas Monthly*, November 22, 2017, www.texasmonthly.com/news-politics/scientist-stop-measles-texas/.

47. Center for Countering Digital Hate, "About," https://counterhate.com/about/, accessed January 4, 2024; Center for Countering Digital Hate, "The Disinformation Dozen," https://counterhate.com/research/the-disinformation-dozen/, accessed February 3, 2024; I. Ahmed, "Dismantling the Anti-vaxx Industry," *Nature Medicine* 27, no. 3 (2021): 366, https://doi.org/10.1038/s41591-021-01260-6, PMID: 33723446; Center for Countering Digital Hate, "Anti-vaxx Misinformation," https://counterhate.com/topic/anti-vaxx-misinformation/, accessed January 4, 2024.

48. Center for Countering Digital Hate, "About."

49. Center for Countering Digital Hate, "The Disinformation Dozen"; Ahmed, "Dismantling the Anti-vaxx Industry."

50. Center for Countering Digital Hate, "Anti-vaxx Misinformation."

51. Center for Countering Digital Hate, "Substack & Anti-vaxx Newsletters," https://counterhate.com/research/substack-anti-vaxx-newsletters/, accessed February 3, 2024.

52. S. Frenkel, "The Most Influential Spreader of Coronavirus Misinformation Online," *New York Times*, July 24, 2021, updated November 25, 2021, www.nytimes.com/2021/07/24/technology/joseph-mercola-coronavirus-misinformation-online.html.

53. Frenkel, "Most Influential Spreader of Coronavirus Misinformation Online."

54. R. Abrams and J. Hoffman (reporters), L. Moftah (producer/director), *Dr. Joseph Mercola: The Misinformation "Superspreader,"* New York Times, August 16, 2022, www.nytimes.com/2022/08/16/NYT-Presents/joseph-mercola-coronavirus-misinformation.html.

55. Center for Countering Digital Hate, "*Substack* & Anti-vaxx Newsletters."

56. "Anti-vaccine Activist RFK Jr. Launches Presidential Campaign," Associated Press, April 19, 2023, https://apnews.com/article/kennedy-biden-president-2024-democrat-dd9d6ecf17b54f4b32e7b17778540431; J. McDonald, "Fact Checking Robert F. Kennedy Jr.," FactCheck.org, August 9, 2023, www.factcheck.org/2023/08/scicheck-factchecking-robert-f-kennedy-jr/.

57. Anna Merlan, "Robert F. Kennedy Jr. Reveals Plan to Fire 600 Federal Health Care Workers," *Mother Jones*, November 12, 2024, www.motherjones.com/politics/2024/11/robert-f-kennedy-jr-national-institutes-of-health/; Will Stone, Pien Huang, and Rob Stein, "Staff at CDC and NIH Are Reeling as Trump Administration Cuts Workforce," NPR, February 14, 2025, www.npr.org/sections/shots-health-news/2025/02/14/nx-s1-5297913/cdc-layoffs-hhs-trump-doge.

58. "Pope Francis Lays Smackdown on Climate Deniers; RFK, Jr. Interview—*The Ring of Fire*," July 14, 2015, www.youtube.com/watch?v=e0rE0ERgYPA.

59. Dan Farber, "RFK Jr. Joins the War on Climate Scientists," *Legal Planet*, February 27, 2024, https://legal-planet.org/2024/02/27/rfk-jr-joins-the-war-on-climate-scientists/.

60. A. Allen, "Vaccine Skeptic RFK Jr. Says He'll Chair Vaccine Commission for Trump," *Politico*, January 10, 2017, www.politico.com/blogs/donald-trump-administration/2017/01/trump-meets-robert-f-kennedy-jr-233417.

61. McDonald, "FactChecking Robert F. Kennedy Jr."

62. D. Gorski, "Robert F. Kennedy, Jr. Hosts a Trainwreck of an Anti-vaccine Forum in Harlem," *Science-Based Medicine*, October 21, 2019, https://sciencebasedmedicine.org/tag/harlem-vaccine-forum/; M. R. Smith and A. Swenson, "RFK Jr. Spent Years Stoking Fear and Mistrust of Vaccines. These People Were Hurt by His Work," Associated Press, October 18, 2023, https://apnews.com/article/rfk-kennedy-election-vaccines-2ccde2df146f57b5e8c26e8494f0a16a.

63. F. Manjoo, "It's Not Possible to 'Win' an Argument with Robert F. Kennedy Jr.," *New York Times*, June 23, 2023, www.nytimes.com/2023/06/23/opinion/rfk-jr-joe-rogan.html; S. Jones, "The Intelligencer: Anti-vaxxers Don't Want a Debate; They Want a Spectacle," *New York Magazine*, June 24, 2023, https://nymag.com/intelligencer/2023/06/robert-f-kennedy-jr-vs-peter-hotez-would-not-be-a-debate.html.

64. K. Weir, "How Robert F. Kennedy Jr. Became the Anti-vaxxer Icon of America's Nightmares," *Vanity Fair*, May 13, 2021, www.vanityfair.com/news/2021/05/how-robert-f-kennedy-jr-became-anti-vaxxer-icon-nightmare.

65. McDonald, "FactChecking Robert F. Kennedy Jr."

66. McDonald, "FactChecking Robert F. Kennedy Jr."

67. Robert F. Kennedy Jr., *The Real Anthony Fauci: Bill Gates, Big Pharma, and the Global War on Democracy and Public Health* (Children's Health Defense and Skyhorse, 2021).

68. P. J. Hotez, "Mounting Antiscience Aggression in the United States," *PLOS Biology* 19, no. 7 (2021): e3001369, https://doi.org/10.1371/journal.pbio.3001369, PMID: 34319972, PMCID: PMC8351985.

69. K. P. Puchner et al., "Vaccine Value Profile for Hookworm," *Vaccine* (October 18, 2023): S0264-410X(23)00540-6, https://doi.org/10.1016/j.vaccine.2023.05.013, e-publication ahead of print, PMID: 37863671.

70. J. Palca, "Whatever Happened to the New No-Patent COVID Vaccine Touted as a Global Game Changer?," *NPR Goats and Soda*, August 31, 2022, www.npr.org/sections/goatsandsoda/2022/08/31/1119947342/whatever-happened-to-the-new-no-patent-covid-vaccine-touted-as-a-global-game-cha; M. Goozner, "How to Stop the Drug Companies from Hiking Prices for Covid Vaccines," *Washington Monthly*, March 6, 2023, https://washingtonmonthly.com/2023/03/06/how-to-stop-the-drug-companies-from-hiking-prices-for-covid-vaccines/.

71. W. Bragman and A. Kotch, "America's Biggest Charities Bankrolled RFK Jr.'s Anti-vax Outfit," *Rolling Stone*, October 19, 2023, www.rollingstone.com/politics/politics-features/rfk-jr-anti-vax-orgs-funding-major-charities-1234857214/.

72. M. Alfred, "RFK Jr. Hires Leading Anti-vaxxer as New Comms Director," *Daily Beast*, January 2, 2024, www.thedailybeast.com/rfk-jr-hires-leading-anti-vaxxer-del-bigtree-as-new-comms-director.

73. Bragman and Kotch, "America's Biggest Charities Bankrolled RFK Jr.'s Anti-vax Outfit."

74. C. T. Bramante et al., "Randomized Trial of Metformin, Ivermectin, and Fluvoxamine for Covid-19," *New England Journal of Medicine* 387, no. 7 (2022): 599–610, https://doi.org/10.1056/NEJMoa2201662, PMID: 36070710, PMCID: PMC9945922; G. Reis et al., "Effect of Early Treatment with Ivermectin Among Patients with Covid-19," *New England Journal of Medicine* 386, no. 18 (2022): 1721–1731, https://doi.org/10.1056/NEJMoa2115869, PMID: 35353979, PMCID: PMC9006771; C. Temple, R. Hoang, and R. G. Hendrickson, "Toxic Effects from Ivermectin Use Associated with Prevention and Treatment of Covid-19," *New England Journal of Medicine* 385, no. 23 (2021): 2197–2198, https://doi.org/10.1056/NEJMc2114907, PMID: 34670041, PMCID: PMC8552535; S. Naggie et al., "Accelerating Covid-19 Therapeutic Interventions and Vaccines (ACTIV)-6 Study Group and Investigators: Effect of Higher-Dose Ivermectin for 6 Days vs. Placebo on Time to Sustained Recovery in Outpatients with COVID-19, a Randomized Clinical Trial," *JAMA* 329, no. 11 (2023): 888–897, https://doi.org/10.1001/jama.2023.1650, PMID: 36807465, PMCID: PMC9941969; E. López-Medina et al., "Effect of Ivermectin on Time to Resolution of Symptoms Among Adults with Mild COVID-19: A Randomized Clinical Trial," *JAMA* 325, no. 14 (2021): 1426–1435, https://doi.org/10.1001/jama.2021.3071, PMID: 33662102, PMCID: PMC7934083.

75. A. Pradelle et al., "Deaths Induced by Compassionate Use of Hydroxychloroquine During the First COVID-19 Wave: An Estimate," *Biomedicine and Pharmacotherapy* 171 (2024): 116055, https://doi.org/10.1016/j.biopha.2023.116055, PMID: 38171239.

76. V. Bergengruen, "'What Price Was My Father's Life Worth?': Right-Wing Doctors Are Still Peddling Dubious COVID Drug," *Time*, May 15, 2023, https://time.com/6278831/ivermectin-americas-frontline-doctors-lawsuits-covid-19/.

77. M. Lee, "Network of Right-Wing Health Care Providers Is Making

Millions Off Hydroxychloroquine and Ivermectin, Hacked Data Reveals," *Intercept*, September 28, 2021, https://theintercept.com/2021/09/28/covid-telehealth -hydroxychloroquine-ivermectin-hacked/.

78. Bergengruen, "'What Price Was My Father's Life Worth?'"

79. B. Bachman and T. Rooney, "Florida's New Surgeon General, Dr. Joseph Ladapo, Has Ties to Fringe Group Pushing Bogus COVID Cures," *Salon*, September 22, 2021, www.salon.com/2021/09/22/floridas-new-surgeon-general-dr -joseph-ladapo-has-ties-to-fringe-group-pushing-bogus-cures/.

80. Florida Health, "Florida State Surgeon General Calls for Halt in the Use of COVID-19 mRNA Vaccines," January 3, 2024, www.floridahealth.gov /newsroom/2024/01/20240103-halt-use-covid19-mrna-vaccines.pr.html; R. Luscombe, "'Scientific Nonsense': Experts Dismiss Florida Official's Covid Vaccine Remarks," *Guardian*, January 4, 2024, www.theguardian.com/us-news/2024 /jan/04/florida-surgeon-general-covid-vaccine-misinformation-joseph-ladapo.

81. Jeff Storobinsky (@jeffstorobinsky), "1.5.24 1250 AM ET Ali Velshi @AliVelshi pinch-hitting for @WagnerTonight hosting Professor Peter Hotez MD, PHD @PeterHotez and author of The Deadly Rise of Anti-Science, A Scientist's warning Brilliant Segment Full Segment (5 minutes)," Twitter (now X), January 5, 2024, 12:54 a.m., https://twitter.com/jeffstorobinsky/status/1743148864620601777.

82. D. Diamond, L. Weber, and J. Dawsey, "Florida Surgeon General Calls for Halt on mRNA Covid Vaccines, Citing Debunked Claim," *Washington Post*, January 3, 2024, www.washingtonpost.com/health/2024/01/03/florida -surgeon-general-ladapo-covid-vaccine/.

83. Storobinsky (@jeffstorobinsky), "1.5.24 1250 AM ET Ali Velshi @AliVelshi pinch-hitting for @WagnerTonight."

84. Simone Gold (@drsimonegold), "Deadly Disinformation Dozen #5: Dr. Peter Hotez Dr. Hotez spread disinformation about the efficacy and safety of the COVID vaccine. Dr. Hotez, like many others, must be held accountable. Who else agrees?," X, November 13, 2023, 4:09 p.m., https://x.com/drsimonegold /status/1724172676057518425.

85. L. Romero, "Dr. Simone Gold, Leading Anti-vax Figure, Sentenced for Storming Capitol on Jan. 6," ABC News, June 16, 2022, https://abcnews.go.com /US/dr-simone-gold-leading-anti-vax-figure-sentenced/story?id=85445732; Cheryl Clark, "Simone Gold, MD, Pardoned by Trump: America's Frontline Doctors Founder Already Served Time in Prison," *MedPage Today*, January 23, 2025, www .medpagetoday.com/special-reports/features/113902.

86. O. Goldhill, "Encouraged by Right-Wing Doctor Groups, Desperate Patients Turn to Ivermectin for Long Covid," *STAT News*, July 26, 2022, www .statnews.com/2022/07/26/ivermectin-has-become-a-popular-treatment-for-long -covid-with-a-push-from-doctors-with-ties-to-right-wing-political-groups/.

87. Bragman and Kotch, "America's Biggest Charities Bankrolled RFK Jr.'s Anti-vax Outfit."

88. Goldhill, "Encouraged by Right-Wing Doctor Groups"; Kristina Fiore, "ABIM Revokes Certification of Another Doctor Who Made Controversial COVID Claims," *MedPage Today*, January 2, 2025; M. DePeau-Wilson, "Two Doctors Will Lose ABIM Certification for Allegedly Spreading False Information," *MedPage Today*, August 10, 2023, www.medpagetoday.com/special-reports /exclusives/105835, accessed January 23, 2024.

89. J. Jarry, "Dr. Peter McCullough's Libertarian Medical Train Makes a Pit Stop in East Palestine: McCullough's Cash-Only Wellness Company Doesn't Want the Government or the Pharmaceutical Industry to Tread on Its Profits," McGill University Office of Science and Society, March 10, 2023, www .mcgill.ca/oss/article/covid-19-medical-critical-thinking/dr-peter-mcculloughs -libertarian-medical-train-makes-pit-stop-east-palestine.

90. W. Bragman, "Fueled by Mystery Donors, COVID Conspiracy Group Grew Budget Last Year," *Important Context News*, December 14, 2023, www .importantcontext.news/p/fueled-by-mystery-donors-covid-conspiracy; N. Nattrass, "Promoting Conspiracy Theory: From AIDS to COVID-19," *Global Public Health* 18, no. 1 (2023): 2172199, https://doi.org/10.1080/17441692.2023.217 2199, PMID: 36749932; W. Bragman, "Who Is Funding the Brownstone Institute?," *Important Context News*, July 24, 2023, www.importantcontext.news/p /who-is-funding-the-brownstone-institute; D. Gorski, "The 'Spiritual Child of the Great Barrington Declaration' Promotes Antivaccine Misinformation," *Science Based Medicine*, January 24, 2022, https://sciencebasedmedicine.org/the-spiritual -child-of-the-great-barrington-declaration-promotes-antivaccine-misinformation/; Brownstone Institute, https://brownstone.org/articles/, accessed January 23, 2024.

91. Bragman, "Fueled by Mystery Donors, COVID Conspiracy Group Grew Budget Last Year"; Bragman, "Who Is Funding the Brownstone Institute?"

92. Nattrass, "Promoting Conspiracy Theory"; Bragman, "Who Is Funding the Brownstone Institute?" D. Gorski, "The Ultimate COVID-19 Antivax Conspiracy Theory, Courtesy of the Brownstone Institute and Jeffrey Tucker," Science-Based Medicine, May 27, 2024, https://sciencebasedmedicine.org/the -ultimate-covid-19-antivax-conspiracy-theory-courtesy-of-the-brownstone -institute-and-jeffery-tucker/.

93. Nattrass, "Promoting Conspiracy Theory."

94. Brownstone Institute, https://brownstone.org/articles/.

95. M. J. Harris et al., "Perceived Experts Are Prevalent and Influential Within an Antivaccine Community on Twitter," *PNAS Nexus* 3, no. 2 (2024): pgae007, https://doi.org/10.1093/pnasnexus/pgae007, PMID: 38328781, PMCID: PMC10847722.

96. Bragman and Kotch, "America's Biggest Charities Bankrolled RFK Jr.'s Anti-vax Outfit."

CHAPTER 5: THE PROPAGANDISTS

1. A. Shephard, "Elon Musk Is the *New Republic*'s 2023 Scoundrel of the Year," *New Republic*, December 27, 2023, https://newrepublic.com/article/177695 /elon-musk-scoundrel-year-2023-new-republic.

2. Geoff Dembicki, "Inside the Anti-Climate Culture War Led by Jordan Peterson and Project 2025," *DeSmog*, September 9, 2024, www.desmog .com/2024/09/09/inside-the-anti-climate-culture-war-led-by-jordan-peterson -and-project-2025/; Alex Boyd, "Jordan Peterson Snaps Back over Russia Claims Made by Justin Trudeau," *Toronto Star*, October 18, 2024, www.thestar.com /politics/federal/jordan-peterson-snaps-back-over-russia-claims-made-by-justin- trudeau/article_7b8d8d24-8d94-11ef-afb4-5b2236e17082.html; Batey, "Tucker Carlson Funded by Russian Propaganda Machine"; Jordon B. Peterson, "Meaning

Through Responsibility: The Heritage Foundation & Dr. Kevin Roberts," You-Tube, www.youtube.com/watch?v=pR6DodpTrXI.

3. P. Markolin, "The Manipulation Playbook of Anti-science Actors, Part 3: Disparage, Disorient, Dispute," *Medium*, November 17, 2023, https://protagonist -science.medium.com/disparage-disorient-dispute-ff494e7f3ec4.

4. Naomi Oreskes and Erik M. Conway, *Merchants of Doubt: How a Handful of Scientists Obscured the Truth on Issues from Tobacco Smoke to Climate Change* (Bloomsbury, 2010); A. Lantian et al., "Maybe a Free Thinker but Not a Critical One: High Conspiracy Belief Is Associated with Low Critical Thinking Ability," *Applied Cognitive Psychology* 35, no. 3 (2021): 674–684, https://onlinelibrary.wiley .com/doi/10.1002/acp.3790.

5. Steven Lee Myers and Sheera Frenkel, "G.O.P. Targets Researchers Who Study Disinformation Ahead of 2024 Election," *New York Times*, June 19, 2023, www.nytimes.com/2023/06/19/technology/gop-disinformation-researchers-2024 -election.html.

6. Lantian et al., "Maybe a Free Thinker but Not a Critical One."

7. G. Sparks, A. Kirzinger, and M. Brodie, "KFF COVID-19 Vaccine Monitor: Profile of the Unvaccinated," Kaiser Family Foundation, June 11, 2021, www .kff.org/coronavirus-covid-19/poll-finding/kff-covid-19-vaccine-monitor-profile -of-the-unvaccinated/; P. J. Hotez, *The Deadly Rise of Anti-science: A Scientist's Warning* (Johns Hopkins University Press, 2023).

8. Moira Fagan and Christine Huang, "A Look at How People Around the World View Climate Change," Pew Research Center, April 18, 2019, www .pewresearch.org/short-reads/2019/04/18/a-look-at-how-people-around-the-world -view-climate-change/.

9. Hunter Bassler, "Water Conspiracy Theories Are Coming to Arizona. Here's Why Experts Say They Will Flood the State with Lies," KPNX (Phoenix), November 17, 2022, updated January 31, 2023, www.12news.com/article/news /local/water-wars/arizona-drought-water-conspiracy-theories-how-to-stop-them /75-50988a0d-bce0-421c-b85a-d46b44fa7380.

10. Ashley Cleek, "Russian Scholar Warns of 'Secret' U.S. Climate Change Weapon," July 30, 2010, www.rferl.org/a/Russian_Scholar_Warns_Of_Secret _US_Climate_Change_Weapon/2114381.html.

11. PragerU (@prageru), "The Good News about Climate Change @curryja," X, April 19, 2024, 2:00 p.m., https://x.com/prageru/status/1781382259628519711.

12. Judith Curry (@curryja), "Torrential rains in Dubai—unintended consequence of cloud seeding?," X, April 16, 2024, 2:30 p.m., https://x.com/curryja /status/1780302797973659936.

13. Seth Borenstein and Brittany Peterson, "Here's Why Experts Don't Think Cloud Seeding Played a Role in Dubai's Downpour," Associated Press, April 17, 2024, https://apnews.com/article/dubai-united-arab-emirates -oman-flooding-cloud-seeding-2f8c12854017e11ac7438579646b3758.

14. Michael E. Mann, "It's Not Rocket Science: Climate Change Was Behind This Summer's Extreme Weather," November 2, 2018, www.washingtonpost .com/opinions/its-not-rocket-science-climate-change-was-behind-this-summers -extreme-weather/2018/11/02/b8852584-dea9-11e8-b3f0-62607289efee_story .html.

15. David Klepper, "Russia Amplified Hurricane Disinformation to Drive

Americans Apart, Researchers Find," Associated Press, October 24, 2024, https://apnews.com/article/russia-hurricane-disinformation-fema-9e37c73ab8ffa2a2d338797a1a827e57.

16. Kinsey Crowley and Chris Mueller, "No, the Government Didn't Create the Hurricanes. President Biden Addresses MTG Conspiracy," *USA Today*, October 10, 2024, updated January 17, 2025, www.usatoday.com/story/news/politics/elections/2024/10/10/government-controls-weather-marjorie-taylor-greene-conspiracy/75607365007/.

17. Alejandra Borunda and Rachel Waldholz, "Climate Change Made Helene More Dangerous. It Also Makes Similar Storms More Likely," NPR, October 9, 2024, www.npr.org/2024/10/09/nx-s1-5144216/climate-change-hurricane-helene.

18. Sommer Brugal, "Climate Change References Removed from Florida Textbooks, Authors Say," *Axios*, July 11, 2024, www.axios.com/local/miami/2024/07/11/climate-change-florida-textbooks.

19. Tori Otten, "DeSantis Rejected $350 Million in Climate Funding Before Hurricane Idalia," *New Republic*, August 30, 2023, https://newrepublic.com/post/175301/desantis-rejected-350-million-climate-funding-florida-hurricane-idalia.

20. Scott Waldman, "Climate Misinformation Spreads on Musk's Twitter," *E&E News*, December 23, 2022, www.eenews.net/articles/climate-misinformation-spreads-on-musks-twitter/.

21. Michael J. I. Brown, "Another Attack on the Bureau, but Top Politicians Have Stopped Listening to Climate Change Denial," *Conversation*, August 4, 2017, https://theconversation.com/another-attack-on-the-bureau-but-top-politicians-have-stopped-listening-to-climate-change-denial-81993.

22. Nitasha Kaul and Tom Buchanan, "Misogyny, Authoritarianism, and Climate Change," *Analyses of Social Issues and Public Policy* 23, no. 2 (2023): 308–333, https://spssi.onlinelibrary.wiley.com/doi/full/10.1111/asap.12347.

23. "UN Agenda 2030," Directorate of Programme Co-ordination, n.d., www.coe.int/en/web/programmes/un-2030-agenda.

24. See, for example, Brett D. Schaefer, "The United Nations' Agenda 2030 and the Sustainable Development Goals Fall Flat," Heritage Foundation, January 31, 2023, www.heritage.org/global-politics/commentary/the-united-nations-agenda-2030-and-the-sustainable-development-goals; and Brett D. Schaefer, "The U.N. Sustainable Development Goals Are Beyond Saving," Heritage Foundation, September 26, 2023, www.heritage.org/global-politics/commentary/the-un-sustainable-development-goals-are-beyond-saving.

25. "Heritage Foundation," SourceWatch, n.d., www.sourcewatch.org/index.php/Heritage_Foundation.

26. This passage first appeared in M. Mann, "Project 2025: The Right-Wing Conspiracy to Torpedo Global Climate Action," *Bulletin of the Atomic Scientists*, August 16, 2024, https://thebulletin.org/2024/08/project-2025-the-right-wing-conspiracy-to-torpedo-global-climate-action/.

27. "Brownstone Institute," *DeSmog*, n.d., www.desmog.com/brownstone-institute/.

28. Toby Rogers (@uTobian), "Wait, hold the f*ck up. Michael Mann (creator of the global warming 'hockey stick graph') spoke at an EcoHealth Alliance conference this week!? The same EcoHealth Alliance that laundered the money to the Wuhan Institute of Virology to create SARS-CoV-2. And Hotez

reposted this!??," X, April 20, 2024, 9:29 p.m., https://x.com/uTobian/status
/1781857637631742162.

29. Toby Rogers, "The Teams Are Set for World War III," *Todayville*, April
22, 2024, www.todayville.com/the-teams-are-set-for-world-war-iii/; Toby Rogers,
"This Really Is World War III," *uTobian* (blog), April 21, 2024, https://tobyrogers
.substack.com/p/this-really-is-world-war-iii.

30. Marc Morano, "Watch: Morano on Fox Talks Biden's Access to
'COVID-Like' Powers I He Declares a National 'Climate Emergency'—
'Dictatorial Powers,'" Climate Depot, April 21, 2024, www.climatedepot.com/2024
/04/21/watch-morano-on-fox-talks-bidens-access-to-covid-like-powers-if-he
-declares-a-national-climate-emergency/?mc_cid=dbcc19ee65.

31. Daniel Zuidijk and Olivia Rudgard, "What Are 15-Minute Cities and
Why Is Britain's Conservative Party Suddenly Talking About Them?," Bloomberg,
October 3, 2023, www.bloomberg.com/news/articles/2023-10-03/15-minute
-cities-what-are-they-and-why-are-they-controversial.

32. Zuidijk and Rudgard, "What Are 15-Minute Cities?"

33. "Climate Change Lockdowns Disguised as '15-Minute Cities'
Under Agenda 2030," *Red Pill Conspiracy* (blog), December 22, 2022, https:
//redpillconspiracy.substack.com/p/climate-change-lockdowns-disguised; Isabella
Fertel, "Fact Check: False Claim '15-Minute Cities' Are Actually 'Climate Lockdowns,'" *USA Today*, February 6, 2023, www.usatoday.com/story/news/factcheck
/2023/02/06/fact-check-false-claim-15-minute-cities-plan-confine-residents
/11179132002/.

34. Joey Garrison, "White House Blames Trump Administration and
Republicans over East Palestine, Ohio Spill," *USA Today*, February 22, 2023,
www.usatoday.com/story/news/politics/2023/02/22/white-house-blame-trump
-gop-east-palestine-spill/11322623002/.

35. Jacob Vaughn, "At Dallas QAnon Convention, Conspiracy Theories Run Wild and Reporters Kicked Out," *Dallas Observer*, June 1, 2021, www
.dallasobserver.com/news/dallas-qanon-convention-conspiracy-theories-coups
-pricey-auctions-and-reporters-ejected-12028633.

36. Patriot Voice, https://thepatriotvoice.us/.

37. "Ohio Train Derailment Fuels 15-Minute City Conspiracy Theories,"
RTL Today, March 16, 2023, https://today.rtl.lu/news/fact-check/a/2041548
.html.

38. Gregor Semieniuk et al., "Stranded Fossil-Fuel Assets Translate to Major
Losses for Investors in Advanced Economies," *Nature Climate Change* 12 (2022):
532–538, www.nature.com/articles/s41558-022-01356-y.

39. Ceri Parker, "End Fossil Fuel Subsidies and Reset the Economy for
a Better World—IMF Head," World Economic Forum, June 3, 2020, www
.weforum.org/agenda/2020/06/end-fossil-fuel-subsidies-economy-imf-georgieva
-great-reset-climate/.

40. Claire Giangravé, "Archbishop Viganò Pushes Conspiracy Theories About
Ukraine and Russia in 10,000-Word Letter," *America: The Jesuit Review*, March 7,
2022, www.americamagazine.org/faith/2022/03/07/vigano-ukraine-242526.

41. "'The Great Reset' Conspiracy Flourishes amid Continued Pandemic,"
ADL, December 29, 2020, www.adl.org/resources/blog/great-reset-conspiracy
-flourishes-amid-continued-pandemic.

42. "Trudeau UN Speech Sparks 'Great Reset' Conspiracy," AFP Canada, November 19, 2020, https://factcheck.afp.com/trudeau-un-speech-sparks-great-reset-conspiracy.

43. Tucker Carlson, "Tucker Carlson: The Elites Want COVID-19 Lockdowns to Usher in 'Great Reset' and That Should Terrify You," Fox News, November 16, 2020, www.foxnews.com/opinion/tucker-carlson-coronavirus-pandemic-lockdowns-great-reset.

44. "Alex Jones Accuses the Left of 'Having a Fetish' for 'Hating Putin,'" *Independent*, October 11, 2022, https://news.yahoo.com/alex-jones-accuses-left-having-122021933.html; "Alex Jones Promotes Conspiracy Theory Alexei Navanly [*sic*] Was Murdered by 'Western Intelligence Agencies,'" *Media Matters*, February 16, 2024, www.mediamatters.org/alex-jones/alex-jones-promotes-conspiracy-theory-alexei-navanly-was-murdered-western-intelligence.

45. "'The Great Reset' Conspiracy Flourishes amid Continued Pandemic."

46. Gabrielle Chan, "Pauline Hanson Says Australians Want Vladimir Putin's Style of Leadership," *Guardian*, March 4, 2017, www.theguardian.com/australia-news/2017/mar/05/pauline-hanson-says-australians-want-vladimir-putins-style-of-leadership.

47. https://web.archive.org/web/20201117174547/https://spectator.us/great-reset-davos-klaus-schwab/.

48. https://web.archive.org/web/20201117174547/https://spectator.us/great-reset-davos-klaus-schwab/.

49. Craig Timberg and Tony Romm, "Russian Trolls Sought to Inflame Debate over Climate Change, Fracking, Dakota Pipeline," *Chicago Tribune*, March 1, 2018, www.chicagotribune.com/nation-world/ct-russian-trolls-climate-change-20180301-story.html.

50. Sandra Laville and David Pegg, "Fossil Fuel Firms' Social Media Fightback Against Climate Action," *Guardian*, October 11, 2019, www.theguardian.com/environment/2019/oct/10/fossil-fuel-firms-social-media-fightback-against-climate-action.

51. Oliver Milman, "Revealed: Quarter of All Tweets About Climate Crisis Produced by Bots," *Guardian*, February 21, 2020, www.theguardian.com/technology/2020/feb/21/climate-tweets-twitter-bots-analysis.

52. Douglas Fisher and the Daily Climate, "Cyber Bullying Intensifies as Climate Data Questioned," *Scientific American*, March 1, 2010.

53. "Who Is Behind the Climate Change and Carbon Trading Scams," Stormfront.org, March 30, 2010, www.stormfront.org/forum/showthread.php?p=7961482.

54. Richard Black, "Scientist Leaves Behind a Climate of Abuse," BBC, July 20, 2020, www.bbc.co.uk/blogs/thereporters/richardblack/2010/07/i_didnt_know_stephen_schneider.html.

55. "Stormfront Website Posters Have Murdered Almost 100 People, Watchdog Group Says," ABC News, April 17, 2014, https://abcnews.go.com/US/stormfront-website-posters-murdered-100-people-watchdog-group/story?id=23365815.

56. See "Paul Driessen," SourceWatch, www.sourcewatch.org/index.php?title=Paul_Driessen, accessed May 3, 2024.

57. Each of the groups named has been funded by some combination of

ExxonMobil, Scaife Foundation, or Koch Industries funding. For ExxonMobil, see SourceWatch.org page on "Committee for a Constructive Tomorrow" (archived May 3, 2011), www.sourcewatch.org/index.php?title=Committee_for_a _Constructive_Tomorrow; and "Koch Industries: Still Fueling Climate Denial" (April 2011, published by Greenpeace USA). For Scaife Foundation, see Source-Watch.org page on "Center for the Defense of Free Enterprise" (archived May 3, 2011), www.sourcewatch.org/index.php?title=Center_for_the_Defense_of_Free _Enterprise; For Frontiers of Freedom and the Atlas Economic Research Foundation, see the Greenpeace reports on Koch Industries cited for ExxonMobil.

58. "Paul Driessen," SourceWatch.

59. The piece first appeared on townhall.com, October 15, 2009, Paul Driessen, "None Dare Call It Fraud," http://townhall.com/columnists/pauldriessen /2009/10/15/none_dare_call_it_fraud, accessed May 3, 2011.

60. Arshad R. Zargar, "Summer Heat Hits Asia Early, Killing Dozens as One Expert Calls It the 'Most Extreme Event' in Climate History," CBS News, May 2, 2024, www.cbsnews.com/news/heat-wave-asia-2024-deaths-india -severe-weather-climate-change/.

61. See SourceWatch pages for Anthony Watts and Heartland Institute, www .sourcewatch.org/index.php/Anthony_Watts and www.sourcewatch.org/index .php/Heartland_Institute.

62. Eric Worrall, "Katharine Hayhoe Attacks Greta Thunberg's Climate 'Shaming' Crusade," *WattsUpWithThat* (blog), August 19, 2019, https: //wattsupwiththat.com/2019/08/19/katharine-hayhoe-attacks-greta-thunbergs -climate-shaming-crusade/.

63. Katharine Hayhoe (@Khayhoe), "I don't normally bother to call out liars on twitter but I will here, as they're trying to invent a disagreement to drive a wedge between us. @GretaThunberg is not personally shaming anyone: she is acting according to her principles. What I really said: www.cbc.ca/radio /thecurrent/the-current-for-august-19-2019-1.5251826/shaming-people-into -fighting-climate-change-won-t-work-says-scientist-1.5251832," Twitter (now X), August 20, 2019, 12:34 p.m., https://twitter.com/KHayhoe/status/116385187492 1005057.

64. Maxine Joselow, "Quitting Burgers and Plants Won't Stop Warming, Experts Say," *Climatewire*, December 6, 2019, www.eenews.net/stories/1061734031.

65. Seth Borenstein, "Climate Scientists Try to Cut Their Own Carbon Footprints," Associated Press, December 8, 2019, https://apnews.com/dde2bf108411ecd 973de60bfda5250aa.

66. Katharine Hayhoe (@katharinehayhoe.com), "I've long suspected it, but it was officially confirmed today. Doomerism/personal guilters have wrapped so far around, they are now co-posting with climate dismissives. The below are responses to a post by a doomer ridiculing my newsletter. @michaelemann.bsky.social I bet you have similar proof," Bluesky, March 27, 2024, 2:44 p.m., https://bsky.app /profile/katharinehayhoe.com/post/3kop57uklk522.

67. Eliot Jacobson, "On Being a 'Doomer,'" *Watching the World Go Bye* (blog), June 26, 2023, https://climatecasino.net/2023/06/on-being-a-doomer/.

68. https://twitter.com/sandlwise01/status/1709686068935553376 (now deleted account).

69. The exchange is here: Kevin Anderson (@KevinClimate), "But doesn't the

Act require, if the minister asks (eg. Claire Perry's letter), that the CCC revisit its recommendations if there are material changes to the science (eg. AR5 & SR1.5 budgets) &/or the international political environment (eg. Paris 1.5-2°C)?," Twitter (now X), January 28, 2020, 4:18 a.m., https://twitter.com/KevinClimate /status/1222080140383080448, responding to Chris Stark (@StarkClimate), "It's not a politically dogmatic stone. It's the UK's Climate Change Act, which we're obliged follow," Twitter (now X), January 28, 2020, 3:58 a.m.

70. Seth Borenstein, "Pioneering Scientist Says Global Warming Is Accelerating. Some Experts Call His Claims Overheated," Phys.org, November 3, 2023, https://phys.org/news/2023-11-scientist-global-experts-overheated.html.

71. James Edward Hansen (@DrJamesEHansen), "The United Nations and COP28 are lying. They know the 1.5C and 2C global warming targets are dead. Young people can and should take charge of their future. More later. See: 'A Miracle Will Occur' Is Not Sensible Climate Policy," X, December 7, 2023, 10:13 a.m., https://x.com/DrJamesEHansen/status/1732780395211657481.

72. Michael E. Mann (@MichaelEMann), "It's OK to express pessimism about our will to act fast enough and urgently enough to avert 1.5C or 2C warming. It's NOT OK to wrongly conflate that pessimism with the false claim that it is physically or technologically impossible for us to do so at this point," X, May 8, 2024, 4:20 p.m., https://x.com/MichaelEMann/status/1788302998424207593.

73. See, for example, Katharine Hayhoe (@Khayhoe), "Also because those of us who relentlessly share positive news are often labelled as liars and hopium peddlers by our own allies," X, April 15, 2024, 1:31 p.m., https://x.com/KHayhoe /status/1779925427017965894.

74. Michael E. Mann (@MichaelEMann), "And I appreciate your efforts to help us leave a livable planet for my grandkids," X, February 11, 2020, 6:31 p.m., https://x.com/MichaelEMann/status/1227374498124451842.

75. Naomi Wolf (@naomirwolf), "@MichaelEMann has a new book, Jan 2021. Building perhaps on case he made in classic The Hockey Stick and the Climate Wars. PRE-ORDER pls. It helps authors to do so. If we want independent thought, in a climate of powerful wars on independent thinkers, vote w resources," Twitter (now X), September 24, 2020, 10:13 p.m., https://twitter.com/naomirwolf /status/1309315026726129666.

76. Eve Andrews, "The Real Fear Behind Climate Conspiracy Theories," *Grist*, April 6, 2018, https://grist.org/culture/the-real-fear-behind-climate-conspiracy -theories/.

77. Liza Featherstone, "The Madness of Naomi Wolf," *New Republic*, June 10, 2021, https://newrepublic.com/article/162702/naomi-wolf-madness-feminist -icon-antivaxxer.

78. Rebecca Onion, "A Modern Feminist Classic Changed My Life. Was It Actually Garbage?," *Slate*, March 30, 2021, https://slate.com/human-interest/2021/03 /naomi-wolf-beauty-myth-feminism-conspiracy-theories.html.

79. Jem Bendell, "It's Not Too Late to Stop Being a Tool of Oppression," *Thoughts on Collapse Readiness and Recovery*, November 21, 2022, https: //jembendell.com/2022/11/21/its-not-too-late-to-stop-being-a-tool-of-oppression/.

80. Michael E. Mann (@MichaelEMann), "That's a good point!," X, April 11, 2020, 9:21 p.m., https://x.com/MichaelEMann/status/1249145581374668800.

81. See, for example, this thread: Michael E. Mann (@MichaelEMann),

"That's not how the 1.5C (or 2C) warming threshold has been defined in scientific assessments. It doesn't refer to interannual fluctuations that might cross the threshold temporarily. It refers to the TREND line crossing that threshold," X, February 7, 2021, 1:35 p.m., https://x.com/MichaelEMann/status/1358484507557580802.

82. Roda Group, www.rodagroup.com/companies.html.

83. Hotez, *Deadly Rise of Anti-science*; Office of the Surgeon General, US Department of Health and Human Services, "Confronting Health Misinformation: The Surgeon General's Advisory on Building a Healthy Information Environment," July 15, 2021, www.hhs.gov/sites/default/files/surgeon-general-misinformation-advisory.pdf.

84. Hotez, *Deadly Rise of Anti-science*; Office of the Surgeon General, US Department of Health and Human Services, "Confronting Health Misinformation."

85. Virality Project, "Memes, Magnets and Microchips: Narrative Dynamics Around COVID-19 Vaccines," Stanford Digital Repository, 2022, https://purl.stanford.edu/mx395xj8490 and https://cyber.fsi.stanford.edu/io/news/virality-project-final-report, accessed January 8, 2024.

86. S. Owermohle, "Supreme Court to Weight Whether Covid Misinformation Is Protected Speech," *STAT News*, February 6, 2024, www.statnews.com/2024/02/06/supreme-court-covid-misinformation-public-health-free-speech/.

87. Ian Millhiser, "The Supreme Courts Hands an Embarrassing Defeat to America's Trumpiest Court," *Vox*, June 26, 2024, www.vox.com/scotus/357111/supreme-court-murthy-missouri-fifth-circuit-jawboning-first-amendment.

88. C. T. Bramante et al., "Randomized Trial of Metformin, Ivermectin, and Fluvoxamine for Covid-19," *New England Journal of Medicine* 387, no. 7 (2022): 599–610, https://doi.org/10.1056/NEJMoa2201662, PMID: 36070710, PMCID: PMC9945922; G. Reis et al., "Effect of Early Treatment with Ivermectin Among Patients with Covid-19," *New England Journal of Medicine* 386, no. 18 (2022): 1721–1731, https://doi.org/10.1056/NEJMoa2115869, PMID: 35353979, PMCID: PMC9006771; C. Temple, R. Hoang, and R. G. Hendrickson, "Toxic Effects from Ivermectin Use Associated with Prevention and Treatment of Covid-19." *New England Journal of Medicine* 385, no. 23 (2021): 2197–2198, https://doi.org/10.1056/NEJMc2114907, PMID: 34670041, PMCID: PMC8552535; S. Naggie et al., "Accelerating Covid-19 Therapeutic Interventions and Vaccines (ACTIV)-6 Study Group and Investigators: Effect of Higher-Dose Ivermectin for 6 Days vs Placebo on Time to Sustained Recovery in Outpatients with COVID-19, a Randomized Clinical Trial," *JAMA* 329, no. 11 (2023): 888–897, https://doi.org/10.1001/jama.2023.1650, PMID: 36807465, PMCID: PMC9941969; E. López-Medina et al., "Effect of Ivermectin on Time to Resolution of Symptoms Among Adults with Mild COVID-19: A Randomized Clinical Trial," *JAMA* 325, no. 14 (2021): 1426–1435, https://doi.org/10.1001/jama.2021.3071, PMID: 33662102, PMCID: PMC7934083; RECOVERY Collaborative Group et al., "Effect of Hydroxychloroquine in Hospitalized Patients with Covid-19," *New England Journal of Medicine* 383, no. 21 (2020): 2030–2040, https://doi.org/10.1056/NEJMoa2022926, PMID: 33031652, PMCID: PMC7556338; D. R. Boulware et al., "A Randomized Trial of Hydroxychloroquine as Postexposure Prophylaxis for Covid-19," *New England Journal of Medicine* 383, no. 6 (2020): 517–525, https://doi.org/10.1056/NEJMoa2016638, PMID: 32492293, PMCID: PMC7289276; C. P. Skipper et al.,

"Hydroxychloroquine in Nonhospitalized Adults with Early COVID-19: A Randomized Trial," *Annals of Internal Medicine* 173, no. 8 (2020): 623–631, https://doi .org/10.7326/M20-4207, errata in *Annals of Internal Medicine* 174, no. 3 (2021): 435, PMID: 32673060, PMCID: PMC7384270; O. Mitjà et al., "A Cluster-Randomized Trial of Hydroxychloroquine for Prevention of Covid-19," *New England Journal of Medicine* 384, no. 5 (2021): 417–427, https://doi.org/10.1056/NEJMoa2021801, PMID: 33289973, PMCID: PMC7722693; A. B. Cavalcanti et al., "Hydroxychloroquine With or Without Azithromycin in Mild-to-Moderate Covid-19," *New England Journal of Medicine* 383, no. 21 (2020): 2041–2052, https://doi .org/10.1056/NEJMoa2019014, errata in *New England Journal of Medicine* 383, no. 21 (2020): e119, PMID: 32706953, PMCID: PMC7397242.

89. A. Pradelle et al., "Deaths Induced by Compassionate Use of Hydroxychloroquine During the First COVID-19 Wave: An Estimate," *Biomedicine & Pharmacotherapy* 171 (2024): 116055, https://doi.org/10.1016/j.biopha.2023.116055, PMID: 38171239.

90. K. Niburski and O. Niburski, "Impact of Trump's Promotion of Unproven COVID-19 Treatments and Subsequent Internet Trends: Observational Study," *Journal of Medical Internet Research* 22, no. 11 (2020): e20044, https://doi. org/10.2196/20044, PMID: 33151895, PMCID: PMC7685699; K. Bales, "How Musk Sold MAGA on HCQ—and Opened the COVID-19 Disinformation Floodgates," *Who What Why*, January 8, 2024, https://whowhatwhy.org/culture /journalism-media/how-musk-sold-maga-on-hcq-and-opened-the-covid-19 -disinformation-floodgates/.

91. Peter Navarro (@RealPNavarro), "At the White House, I had a million tablets of hydroxy that could have saved thousands of lives but @cnn crusaded against it to beat @realDonaldTrump. Negligent homicide at a minimum. @fda was also implicated in hydroxy suppression," X, September 24, 2023, 3:14 p.m., https://x .com/RealPNavarro/status/1706024344797716608.

92. J. McDonald, "Trump Hypes Potential COVID-19 Drugs, but Evidence So Far Is Slim," FactCheck.org, March 25, 2020, updated April 7, 2020, www.factcheck .org/2020/03/trump-hypes-potential-covid-19-drugs-but-evidence-so-far-is-slim/.

93. H. H. Thorp, "Underpromise, Overdeliver," *Science* 367, no. 6485 (2020): 1405, https://doi.org/10.1126/science.abb8492, PMID: 32205459.

94. Elon Musk (@elonmusk), "Maybe worth considering chloroquine for C19 docs.google.com/document/d/e2 . . . ," Twitter (now X), March 16, 2020, 4:31 p.m., https://twitter.com/elonmusk/status/1239650597906898947.

95. Bales, "How Musk Sold MAGA on HCQ"; Musk (@elonmusk), "Maybe worth considering chloroquine for C19."

96. Bales, "How Musk Sold MAGA on HCQ"; C. Ferguson, "This Tech Millionaire Went from Covid Trial Funder to Misinformation Superspreader," *MIT Technology Review*, October 5, 2021, www.technologyreview.com/2021/10/05/1036408 /silicon-valley-millionaire-steve-kirsch-covid-vaccine-misinformation/.

97. Skipper et al., Hydroxychloroquine in Nonhospitalized Adults with Early COVID-19"; Bales, "How Musk Sold MAGA on HCQ."

98. RECOVERY Collaborative Group et al., "Effect of Hydroxychloroquine in Hospitalized Patients with Covid-19"; Boulware et al., "Randomized Trial of Hydroxychloroquine"; Skipper et al., "Hydroxychloroquine in Nonhospitalized Adults with Early COVID-19"; Mitjà et al., "Cluster-Randomized Trial of

Hydroxychloroquine"; Cavalcanti et al., "Hydroxychloroquine With or Without Azithromycin."

99. Bales, "How Musk Sold MAGA on HCQ."

100. K. Butler, "He Was Just Trying to Study COVID Treatments. Ivermectin Zealots Sent Hate Mail Calling Him a Nazi," *Mother Jones*, August 24, 2021, www.motherjones.com/politics/2021/08/ivermectin-hcq-fluvoxamine-covid-boulware/.

101. US Food and Drug Administration, "Coronavirus (COVID-19) Update: FDA Revokes Emergency Use Authorization for Chloroquine and Hydroxychloroquine," June 15, 2020, www.fda.gov/news-events/press-announcements/coronavirus-covid-19-update-fda-revokes-emergency-use-authorization-chloroquine-and.

102. US Food and Drug Administration, "Why You Should Not Use Ivermectin to Treat or Prevent COVID-19," www.fda.gov/consumers/consumer-updates/why-you-should-not-use-ivermectin-treat-or-prevent-covid-19, accessed January 25, 2024.

103. Bales, "How Musk Sold MAGA on HCQ."

104. D. Milbank, "Pro-lifers, RIP. The Pro-death Movement Is Born," *Washington Post*, January 24, 2022, www.washingtonpost.com/opinions/2022/01/24/march-life-anti-vaccine-protest-hypocrisy/.

105. Bales, "How Musk Sold MAGA on HCQ"; J. Visser, "Elon Musk Makes Very Normal Justin Trudeau Is Worse than Hitler Tweet," *Vice*, February 17, 2022, www.vice.com/en/article/n7nq9g/elon-musk-justin-trudeau-hitler-tweet.

106. Bales, "How Musk Sold MAGA on HCQ"; C. Hoard et al., "Elon Musk Continues to Cater to Far-Right Twitter Accounts Promoting Bigotry, Extremism, and Misinformation," *Media Matters for America*, March 27, 2023, www.congress.gov/118/meeting/house/115561/documents/HHRG-118-IF16-20230328-SD035.pdf; C. Jewers, "'I'm Back!': Accounts for Vaccine Creator Dr. Robert Malone and Cardiologist Dr. Peter McCullough Are the Latest to Be Reinstated to Twitter," *Daily Mail*, December 13, 2022, www.dailymail.co.uk/news/article-11532743/Twitter-reinstates-accounts-Dr-Robert-Malone-Dr-Peter-McCullough.html; N. Bose and K. Singh, "Musk Hosts Twitter Event for Anti-vaxx Democratic Candidate RFK Jr.," Reuters, June 5, 2023, www.reuters.com/world/us/musk-hosts-twitter-event-anti-vaxx-democratic-candidate-rfk-jr-2023-06-05/.

107. Elon Musk (@elonmusk), "My pronouns are Prosecute/Fauci," Twitter (now X), December 11, 2022, 5:58 a.m., https://twitter.com/elonmusk/status/1601894132573605888; B. Y. Lee, "Elon Musk Tweets 'My Pronouns Are Prosecute/Fauci,' Here's the Response," *Forbes*, December 11, 2022, www.forbes.com/sites/brucelee/2022/12/11/elon-musk-tweets-my-pronouns-are-prosecutefauci-heres-the-response/?sh=30d2cfa6531d.

108. Hotez, *Deadly Rise of Anti-science*; W. Carless, "With Trump in the Rearview Mirror, Proud Boys Offer Muscle at Rallies Against Vaccine Mandates, Masks," *USA Today*, September 8, 2021, www.usatoday.com/story/news/nation/2021/09/08/proud-boys-join-protests-against-covid-vaccine-mandates-masks/5703785001/?gnt-cfr=1.

109. Bales, "How Musk Sold MAGA on HCQ."

110. Bales, "How Musk Sold MAGA on HCQ"; Grace Chong (@gc22gc), "Hotez is a criminal . . . ," Twitter (now X), June 12, 2023, 10:09 p.m., https://twitter.com/gc22gc/status/1668440433410842624.

111. A. Merlan, "Spotify Has Stopped Even Sort of Trying to Stem Joe Rogan's Vaccine Misinformation," *Motherboard Tech by Vice*, June 16, 2024, www.vice .com/en/article/k7zz9z/spotify-rogan-rfk-vaccine-misinformation-policy.

112. J. Rogan, *The Joe Rogan Experience*, https://open.spotify.com/show /4rOoJ6Egrf8K2IrywzwOMk, accessed February 5, 2024.

113. S. Bond, "What the Joe Rogan Podcast Controversy Says About the Online Misinformation System," NPR, January 21, 2022, www.npr .org/2022/01/21/1074442185/joe-rogan-doctor-covid-podcast-spotify -misinformation.

114. C. Eller, "Neil Young and Joni Mitchell Put Protest over Profit," *Variety*, February 2, 2022, https://variety.com/2022/digital/columns/neil-young-joni -mitchell-spotify-joe-rogan-1235169514/.

115. Anastasia Tsioulcas, "Joe Rogan Has Responded to the Protests Against Spotify over His Podcast," NPR, January 31, 2022, www.npr.org/2022/01/31/1076891070 /joe-rogan-responds-spotify-podcast-covid-misinformation.

116. A. Merlan, "Joe Rogan, Elon Musk Instigate Harassment Campaign Against Vaccine Scientist." *Motherboard Tech by Vice*, June 20, 2024, www.vice .com/en/article/93kkp7/rogan-musk-rfk-hotez-harassment-vaccines.

117. Wikipedia, s.v. "Gish gallop," last modified March 5, 2025, https: //en.wikipedia.org/wiki/Gish_gallop.

118. Michael E. Mann (@MichaelEMann), "Sorry @JoeRogan: I don't debate deniers and disinformers. Here's the low-down on Steven Koonin via @SciAM: scientificamerican.com/article/that-obama-scientist-climate-skeptic-youve -been-hearing-about/," Twitter (now X), February 12, 2022, 11:17 a.m., https: //twitter.com/MichaelEMann/status/1492533309796196363.

119. Joe Rogan (@joerogan), "Peter, if you claim what RFKjr is saying is 'misinformation' I am offering you $100,000.00 to the charity of your choice if you're willing to debate him on my show with no time limit," X, June 17, 2023, 6:27 p.m., https://x.com/joerogan/status/1670196590928068609?lang=en.

120. Elon Musk (@elonmusk), "He's afraid of a public debate, because he knows he's wrong," Twitter (now X), June 17, 2023, 7:58 p.m., https://twitter.com /elonmusk/status/1670219488485154816.

121. Merlan, "Joe Rogan, Elon Musk Instigate Harassment Campaign."

122. Elon Musk (@elonmusk), "Maybe @PeterHotez just hates charity [shrug emoji]," Twitter (now X), June 17, 2023, 10:15 p.m., https://twitter.com/elonmusk /status/1670253846902259715.

123. Merlan, "Joe Rogan, Elon Musk Instigate Harassment Campaign."

124. K. Powell, "Does It Take Too Long to Publish Research?," *Nature* 530, no. 7589: 148–151, https://doi.org/10.1038/530148a, PMID: 26863966.

125. A. Tinniswood, *The Royal Society & the Invention of Modern Science* (Basic Books, 2019); A. Lightman, *The Discoveries: Great Breakthroughs in 20th Century Science, Including the Original Papers* (Vintage Books, 2005); J. Gribbin, *The Scientists: A History of Science Told Through the Lives of Its Greatest Inventors* (Random House, 2002); W. Bynum, *A Little History of Science* (Yale University Press, 2012).

126. R. Skibba, "Einstein, Bohr and the War over Quantum Theory," *Nature* 555, no. 7698 (2018): 582–584, https://doi.org/10.1038/d41586-018-03793-2, PMID: 32099168.

127. F. Manjoo, "It's Not Possible to 'Win' an Argument with Robert F. Kennedy

Jr.," *New York Times*, June 23, 2023, www.nytimes.com/2023/06/23/opinion
/rfk-jr-joe-rogan.html.

128. U. Samarasekera, "Peter Hotez: Physician-Scientist-Warrior Combating
Anti-science," *Lancet* 403, no. 10422 (2024): 134.

129. R. Brand, "'Debate RFK for $2.6M!': Rogan vs. Hotez 'Misinformation'
Clash Goes VIRAL," www.youtube.com/watch?v=Igjp-6rIluw, accessed January
12, 2024.

130. J. Dore, "Joe Rogan Challenges Pro-vaxx Doctor to . . . ," *Deezer*, June
23, 2023, www.deezer.com/mx/episode/522639315, accessed January 12, 2024.

131. Peter Hotez (@PeterHotez), "Just matter of time before InfoWars/Alex
Jones joined the party. 'Big Pharma Globalist Gremlin' quite a mouthful. Never
mind we don't take BigPharma money but make low-cost patent-free vaccines
for global health bypassing BigPharma. And 1 billion people awakened? Impressive," Twitter (now X), June 22, 2023, 4:18 p.m., https://twitter.com/PeterHotez
/status/1671976009346826241.

132. M. Florio, "Aaron Rodgers Embraces Effort by Robert F. Kennedy Jr.
to Debate Dr. Peter Hotez," NBC Sports, June 22, 2023, www.nbcsports.com
/nfl/profootballtalk/rumor-mill/news/aaron-rodgers-embraces-effort-by-robert-f
-kennedy-jr-to-debate-dr-peter-hotez.

133. R. Douthat, "Go Ahead. Debate Robert F. Kennedy Jr.," *New York Times*,
June 24, 2023, www.nytimes.com/2023/06/24/opinion/rfk-jr-joe-rogan-debate
.html.

134. Y. Rosenberg, "The Most Shocking Aspect of RFK Jr.'s Antisemitism," *Atlantic*, July 16, 2023, www.theatlantic.com/ideas/archive/2023/07/rfk-
kennedy-covid-anti-semitism/674727/.

135. "Twitter Ends Enforcement of COVID Misinformation Policy," Associated Press, November 29, 2022, https://apnews.com/article/twitter-ends-covid
-misinformation-policy-cc232c9ce0f193c505bbc63bf57ecad6; S. Sule et al.,
"Communication of COVID-19 Misinformation on Social Media by Physicians in
the US," *JAMA Network Open* 6, no. 8 (2023): e2328928, https:///doi.org/10.1001
/jamanetworkopen.2023.28928.

136. Peter Hotez (@PeterHotez), "Interesting how the same bot message
'Hotez is just another satanic agent of the Luciferan globalists' appears under 3
different accounts at the same time. You would think they want to mix it up a bit
so it looks more credible? This must be what X means by 'free speech,'" X, January 2, 2024, 7:30 a.m., https://x.com/PeterHotez/status/1742161372849287451
/photo/1.

137. R. DiResta, "The Supply of Disinformation Will Soon Be Infinite,"
Atlantic, September 20, 2020, www.theatlantic.com/ideas/archive/2020/09/future
-propaganda-will-be-computer-generated/616400/; S. Feuerriegel et al., "Research
Can Help to Tackle AI-Generated Disinformation," *Nature Human Behavior* 7,
no. 11 (2023): 1818–1821, https://doi.org/10.1038/s41562-023-01726-2, PMID:
37985906.

138. B. D. Menz et al., "Health Disinformation Use Case Highlighting
the Urgent Need for Artificial Intelligence Vigilance: Weapons of Mass Disinformation," *JAMA Internal Medicine* 184, no. 1 (2024): 92-96, https://doi
.org/10.1001/jamainternmed.2023.5947, PMID: 37955873; P. J. Hotez, "Health
Disinformation—Gaining Strength, Becoming Infinite," *JAMA Internal Medicine*

184, no. 1 (2024): 96–97, https://doi.org/10.1001/jamainternmed.2023.5946, PMID: 37955920.

139. T. Hsu, "Fake and Explicit Images of Taylor Swift Started on 4chan, Study Says," *New York Times*, February 5, 2024, www.nytimes.com/2024/02/05/business/media/taylor-swift-ai-fake-images.html.

140. S. Cubbon, "Banned Sites and Pro-Russian Networks Are Driving Anti-Pfizer Vaccine Disinformation," *First Draft News*, March 31, 2021, https://firstdraftnews.org/articles/anti-pfizer-vaccine-narratives/.

141. Office of the Surgeon General, US Department of Health and Human Services, "Confronting Health Misinformation."

142. "EU Warns Musk That Twitter Faces Ban over Content Moderation—FT," Reuters, November 30, 2022, www.reuters.com/technology/eu-warns-musk-that-twitter-faces-ban-over-content-moderation-ft-2022-11-30/.

143. N. Jankowicz, "The Coming Flood of Disinformation," *Foreign Affairs*, February 7, 2024, www.foreignaffairs.com/united-states/coming-flood-disinformation.

144. Menz et al., "Health Disinformation Use Case"; Hotez, "Health Disinformation—Gaining Strength, Becoming Infinite."

145. "AI Watch: Global Regulatory Tracker—European Union," White & Case, February 27, 2025, www.whitecase.com/insight-our-thinking/ai-watch-global-regulatory-tracker-european-union.

146. John Cook, "Inoculation Theory: Using Misinformation to Fight Misinformation," *Conversation*, May 14, 2017, https://theconversation.com/inoculation-theory-using-misinformation-to-fight-misinformation-77545.

147. Bobi Ivanov and Kimberly A. Parker, "Mitigating the Effects of the Coronavirus Outbreak," in *Communicating Science in Times of Crisis: The COVID-19 Pandemic*, edited by H. Dan O'Hair and Mary John O'Hair (Wiley, 2021), https://onlinelibrary.wiley.com/doi/abs/10.1002/9781119751809.ch13.

148. P. J. Hotez, "Global Vaccinations: New Urgency to Surmount a Triple Threat of Illness, Antiscience, and Anti-Semitism," *Rambam Maimonides Medical Journal* 14, no. 1 (2023): e0004, https://doi.org/10.5041/RMMJ.10491, PMID: 36719666, PMCID: PMC9888484; P. J. Hotez, "On Antiscience and Antisemitism," *Perspectives in Biology and Medicine* 66, no. 3 (2023): 420–436 .

149. Hotez, "Global Vaccinations"; Jennifer Bardi, "A Virulent Antisemitism: An Interview with Dr. Peter Hotez," *Moment*, December 6, 2022, https://momentmag.com/a-virulent-antisemitism-interview-with-dr-peter-hotez/?srsltid=AfmBOoqqlxw3qFdVow7RtAFjc33GvhMbDZdJiukr7lh5JK662HsPhEsu.

150. Hotez, "On Antiscience and Antisemitism"; P. Ball, *Serving the Reich: The Struggle for the Soul of Physics Under Hitler* (University of Chicago Press, 2014).

151. Hotez, "On Antiscience and Antisemitism."

152. Q. Jurecic, "How Elon Musk Broke Twitter as He Turned It into X," *Washington Post*, February 13, 2024, www.washingtonpost.com/books/2024/02/13/zoe-schiffer-extremely-hardcore-twitter-elon-musk-reveiw/.

CHAPTER 6: THE PRESS

1. Caleb Pershan, "The Bot That Saw the *Times*," *Columbia Journalism Review*, May 23, 2022, www.cjr.org/the_profile/nyt-pitchbot.php.

2. Doug J. Balloon (@DougJBallon), "There are many topics on which people

can reasonably disagree: climate change, vaccination, and January 6, for example. But every decent person should admit that the college protests, like the budget deficit and Bill Clinton's extramarital affairs, are objectively bad," X, May 6, 2024, 8:57 a.m., https://x.com/DougJBalloon/status/1787466651027304794.

3. M. Wolff, *The Fall: The End of Fox News and the Murdoch Dynasty* (Henry Holt, 2023).

4. D. Folkenflik, "Rupert Murdoch Says Fox Stars 'Endorsed' Lies About 2020. He Chose Not to Stop Them," *NPR Morning Edition*, February 28, 2023, www.npr.org/2023/02/28/1159819849/fox-news-dominion-voting-rupert-murdoch-2020-election-fraud.

5. C. Malone, "The Fallout of Fox News' Public Shaming," *New Yorker*, March 15, 2023, www.newyorker.com/news/annals-of-communications/the-fallout-of-fox-news-public-shaming.

6. S. A. Thompson, "How Russian Media Uses Fox News to Make Its Case," *New York Times*, April 15, 2022, www.nytimes.com/2022/04/15/technology/russia-media-fox-news.html.

7. Michelle Castillo, "Roger Ailes' Knack for Turning Politics into Entertainment Changed Cable News," CNBC, May 18, 2017, www.cnbc.com/2017/05/18/how-foxs-ailes-changed-cable-news.html.

8. "Rupert Murdoch Mocks Global Warming with Icy Photo, Enrages Twitter—Again," *Hollywood Reporter*, February 27, 2015, www.hollywoodreporter.com/news/rupert-murdoch-mocks-global-warming-778302.

9. Dana Nuccitelli, "Rupert Murdoch Doesn't Understand Climate Change Basics, and That's a Problem," *Guardian*, July 14, 2014, www.theguardian.com/environment/climate-consensus-97-per-cent/2014/jul/14/rupert-murdoch-doesnt-understand-climate-basics.

10. P. J. Hotez, *The Deadly Rise of Anti-science: A Scientist's Warning* (Johns Hopkins University Press, 2023); M. Dowd, "James Murdoch, Rebellious Scion," *New York Times*, October 10, 2020, www.nytimes.com/2020/10/10/style/james-murdoch-maureen-dowd.html.

11. Jim Waterson, "James Murdoch Criticises Father's News Outlets for Climate Crisis Denial," *Guardian*, January 14, 2020, www.theguardian.com/media/2020/jan/14/james-murdoch-criticises-fathers-news-outlets-for-climate-crisis-denial.

12. Ross Garnaut, *The Garnaut Review, 2011: Australia in the Global Response to Climate Change* (Cambridge University Press, 2012), www.cambridge.org/core/books/garnaut-review-2011/394F7386E5CD788EDAF06FBD039A7969.

13. Thomas Catenacci, "Al Gore Has History of Climate Predictions, Statements Proven False," Fox News, January 22, 2023, www.foxnews.com/politics/al-gore-history-climate-predictions-statements-proven-false.

14. "Jesse Waters: Al Gore Is Psychologically Obsessed with Control," *The Five*, December 6, 2023, www.foxnews.com/video/6342487228112.

15. Chuck DeVore, "Texas Town's Environmental Narcissism Makes Al Gore Happy While Sticking Its Citizens with the Bill," Fox News, January 29, 2019, www.foxnews.com/opinion/texas-towns-environmental-narcissism-makes-al-gore-happy-while-sticking-its-citizens-with-the-bill; "Al Gore Goes Vegan," Fox News, November 26, 2013, updated November 21, 2016, www.foxnews.com/food-drink/al-gore-goes-vegan.

16. Rita Panahi, "Hollywood Hypocrite's Global Warming Sermon," *Herald Sun*

(Melbourne), October 7, 2016, www.heraldsun.com.au/blogs/rita-panahi/hollywood
-hypocrites-global-warming-sermon/news-story/b4cc2e4b6034c032998fb3c
13e6df4a6.

17. Andrea Peyser, "Leo DiCaprio Isn't the Only Climate Change Hypocrite," *New York Post*, May 26, 2016, https://nypost.com/2016/05/26/leo-di-caprio
-isnt-the-only-climate-change-hypocrite/.

18. "Greta Thunberg," Fox News, www.foxnews.com/category/person/greta
-thunberg (archived May 20, 2024).

19. Christopher Carbone, "Teen Climate Crusader Greta Thunberg Com-
pletes Carbon-Free Voyage by Yacht from Europe to New York City," Fox News,
August 28, 2019, www.foxnews.com/science/teen-climate-crusader-greta-thunberg
-completes-carbon-free-voyage-by-yacht-from-europe-to-new-york-city.

20. Andrew Bolt, "Look, in the Sky! A Hypocrite Called McKibben," *Herald Sun* (Melbourne), April 8, 2013, www.heraldsun.com.au/blogs/andrew-bolt
/look-in-the-sky-a—hypocrite-called-mckibben/news-story/164435dd3d4447
ba60bcea92c349edda.

21. Isabel Vincent and Melissa Klein, "Gas-Guzzling Car Rides Expose AOC's
Hypocrisy amid Green New Deal Pledge," *New York Post*, March 2, 2019, https:
//nypost.com/2019/03/02/gas-guzzling-car-rides-expose-aocs-hypocrisy-amid
-green-new-deal-pledge/.

22. Clover Moore (@CloverMoore), "These tactics seek to distract us from the
reality that our government is presiding over a period of shameful inaction. Vested
interests are failing us while saying: 'We're not the problem; look at the hypocrites
over there!' Shame," Twitter (now X), December 5, 2019, 1:47 a.m., https://twitter
.com/CloverMoore/status/1202479544172630016.

23. Robert Bryce, "Backlash Against Wind and Solar Projects Is Real, It's
Global and It's Growing," Fox News, April 5, 2024, www.foxnews.com/opinion
/backlash-against-wind-solar-projects-real-global-growing.

24. Fiona Harvey, "'Massive Disinformation Campaign' in Slowing Global
Transition to Green Energy," *Guardian*, August 8, 2024, www.theguardian.com
/environment/article/2024/aug/08/fossil-fuel-industry-using-disinformation
-campaign-to-slow-green-transition-says-un.

25. "The Daily Caller," SourceWatch, n.d., www.sourcewatch.org/index.php
/The_Daily_Caller.

26. Denise Robbins, "Study: How Mainstream Media Misled on the Success of the Clean Energy Loan Program," *Media Matters*, April 10, 2014, www
.mediamatters.org/new-york-times/study-how-mainstream-media-misled-success
-clean-energy-loan-program.

27. Sam Dorman, "AOC Accused of Soviet-Style Propaganda with Green New
Deal 'Art Series,'" Fox News, August 30, 2019, www.foxnews.com/media/aoc-green
-new-deal-art-series-propaganda; Daniel Turner, "Daniel Turner: Stealth Version
of AOC 'Green New Deal' Now the Law in New Mexico, Voters Be Damned,"
Fox News, May 27, 2019, www.foxnews.com/opinion/daniel-turner-stealth
-version-of-aoc-green-new-deal-now-the-law-in-new-mexico-voters-be-
damned.

28. Dominique Jackson, "*The Daily Show* Brutally Ridicules Fox's Sean
Hannity for Whining AOC's Green New Deal Will Deprive Him of Hamburgers," *Daily Beast*, February 14, 2019, www.rawstory.com/2019/02/watch

-the-daily-shows-trevor-noah-brutally-mocks-sean-hannity-over-thinking-aoc
-wants-outlaw-hamburgers-with-green-new-deal/.

29. Antonia Noori Farzan, "The Latest Right-Wing Attack on Democrats: 'They Want to Take Away Your Hamburgers,'" *Washington Post*, March 1, 2019, www.washingtonpost.com/nation/2019/03/01/latest-right-wing-attack-democrats-they-want-take-away-your-hamburgers/.

30. Tom Jacobs, "Did Fox News Quash Republican Support for the Green New Deal?," *Pacific Standard*, May 13, 2019, https://psmag.com/economics/did-fox-news-quash-republican-support-for-the-green-new-deal.

31. "Solar Energy Plants in Tortoises' Desert Habitat Pit Green Against Green," Fox News, February 20, 2014, www.foxnews.com/us/solar-energy-plants-in-tortoises-desert-habitat-pit-green-against-green; Associated Press, "Environmental Concerns Threaten Solar Power Expansion in California Desert," Fox News, April 18, 2009, updated January 14, 2015, www.foxnews.com/story/environmental-concerns-threaten-solar-power-expansion-in-california-desert; Alex Pappas, "Massive East Coast Solar Project Generates Fury from Neighbors," Fox News, February 15, 2019, www.foxnews.com/politics/massive-east-coast-solar-project-generates-fury-from-neighbors-in-virginia; "World's Largest Solar Plant Scorching Birds in Nevada Desert," Fox News, February 15, 2014, updated November 29, 2015, www.foxnews.com/us/worlds-largest-solar-plant-scorching-birds-in-nevada-desert.

32. Elliott Negin, "The Wind Energy Threat to Birds Is Overblown," *Live Science*, December 3, 2013, www.livescience.com/41644-wind-energy-threat-to-birds-overblown.html.

33. Adam Morton, Jordyn Beazley, and Ariel Bogle, "How a False Claim About Wind Turbines Killing Whales Is Spinning Out of Control in Coastal Australia," *Guardian*, November 11, 2023, www.theguardian.com/environment/2023/nov/12/how-a-false-claim-about-wind-turbines-killing-whales-is-spinning-out-of-control-in-coastal-australia.

34. "Whales and Climate Change: Big Risks to the Ocean's Biggest Species," NOAA Fisheries, last updated June 23, 2022, www.fisheries.noaa.gov/national/climate/whales-and-climate-change-big-risks-oceans-biggest-species.

35. Max Greenberg, "Fox News' Wind Power Hypochondria," *Media Matters*, February 26, 2013, www.mediamatters.org/fox-news/fox-news-wind-power-hypochondria.

36. Lee Moran, "Fox News' Jesse Watters Gets Schooled over Nonsensical Winter Solar Panels Claim," *Huffington Post*, February 1, 2019, www.huffpost.com/entry/fox-news-jesse-watters-solar-panels_n_5c540aa7e4b043e25b1b2168.

37. Troy Matthews (@Troy_in_Tahoe), "Fox: Plane skidded off runway because Democrats want you afraid to fly to reduce carbon pollution," X, March 8, 2024, 12:36 p.m., https://x.com/Troy_in_Tahoe/status/1766155978637217801.

38. Will Oremus, "Fox News Claims Solar Won't Work in America Because It's Not Sunny Like Germany," *Slate*, February 7, 2013, https://slate.com/technology/2013/02/fox-news-expert-on-solar-energy-germany-gets-a-lot-more-sun-than-we-do-video.html.

39. Max Greenberg, "Fox Cedes Solar Industry to Germany Special Programs Climate & Energy," *Media Matters*, February 7, 2013, www.mediamatters.org/fox-friends/fox-cedes-solar-industry-germany.

40. Catalina Jaramillo, "What to Know About Trump's Executive Order on Wind Energy," FactCheck.org, February 5, 2025, www.factcheck.org/2025/02/what-to-know-about-trumps-executive-order-on-wind-energy/.

41. Hotez, *Deadly Rise of Anti-science*; M. Knott, "Former Murdoch Exec Slams Fox News over Vaccine Misinformation," *Sydney Morning Herald*, July 18, 2021, www.smh.com.au/world/north-america/former-fox-news-exec-slams-network-over-vaccine-misinformation-20210717-p58ain.html.

42. Hotez, *Deadly Rise of Anti-science*.

43. M. M. Grynbaum, "Fox News Stars Trumpeted a Malaria Drug, Until They Didn't," *New York Times*, April 22, 2020, updated October 2, 2020, www.nytimes.com/2020/04/22/business/media/virus-fox-news-hydroxychloroquine.html.

44. A. J. Katz, "Top Cable News Shows of 2021: *Tucker Carlson Tonight* Is No. 1 in All Measurements for First Time Ever," *AdWeek TV Newser*, January 3, 2022, www.adweek.com/tvnewser/top-cable-news-shows-of-2021-tucker-carlson-tonight-is-no-1-in-all-categories-for-first-time-ever/496940/.

45. A. Syal, "Hydroxychloroquine for COVID-19: Scientists Say It's Time to Stop Promoting the Drug," NBC News, July 30, 2020, www.nbcnews.com/health/health-news/hydroxychloroquine-covid-19-scientists-say-it-s-time-stop-promoting-n1235347.

46. *The Daily Show* (@TheDailyShow), "Hannity. Rush. Dobbs. Ingraham. Pirro. Nunes. Tammy. Geraldo. Doocy. Hegseth. Schlapp. Siegel. Watters. Dr. Drew. Henry. Ainsley. Gaetz. Inhofe. Pence. Kudlow. Conway. Trump. Today, we salute the Heroes of the Pandumbic," Twitter (now X), April 3, 2020, 2:45 p.m., https://twitter.com/TheDailyShow/status/1246146713523453957.

47. Judi Ketteler, "Rush Limbaugh Died from Lung Cancer After Denying Smoking's Risks. Why'd He Believe His Lie?," NBC News, February 20, 2021, www.nbcnews.com/think/opinion/rush-limbaugh-died-lung-cancer-after-denying-smoking-s-risk-ncna1258395.

48. Benny Peiser and Andrew Montford, "Coronavirus Lessons from the Asteroid That Didn't Hit Earth," *Wall Street Journal*, April 1, 2020, www.wsj.com/articles/coronavirus-lessons-from-the-asteroid-that-didnt-hit-earth-11585780465?mod=e2two.

49. Bobby Lewis and Kayla Gogarty, "Pro-Trump Media Have Ramped Up Attacks Against Dr. Anthony Fauci," *Media Matters*, March 24, 2020, www.mediamatters.org/coronavirus-covid-19/pro-trump-media-have-ramped-attacks-against-dr-anthony-fauci.

50. See video clip and my comment here: Michael E. Mann (@MichaelEMann), "OMG. This is so inhuman and horrific," Twitter (now X), March 23, 2020, 8:49 p.m., https://twitter.com/MichaelEMann/status/1242252283557093377.

51. Matthew Chapman, "Internet Explodes as Fox's Brit Hume Says Its 'Entirely Reasonable' to Let Grandparents Die for the Stock Market," *Raw Story*, March 24, 2020, www.rawstory.com/2020/03/internet-explodes-as-foxs-brit-hume-says-its-entirely-reasonable-to-let-grandparents-die-for-the-stock-market/.

52. Bill Mitchell (@mitchellvii), "While death is sad for the living left behind, for the dying, it is merely a passage out of this physical body to a spiritual existence, free of this mortal coil. If one turns off the radio, the music is still there. For all we know, the dead weep for us," Twitter (now X), April 4, 2020, 10:21 p.m., https://twitter.com/mitchellvii/status/1246623932767141890.

53. The remarkable parallels between the various stages of denial and inactivism with both climate change and inactivism were explored in an exchange between Mike and *Sydney Morning Herald* environmental journalists Peter Hannam. See Michael E. Mann (@MichaelEMann), "Hmm. Where have I seen this before? [puzzled emoji] #MadhouseEffect @TomTolesToons," Twitter (now X), April 5, 2020, 5:08 p.m., https://twitter.com/MichaelEMann/status/1246907494221451270; and Peter Hannam (@p_hannam), "'Merely a passage out of this physical body . . . ,'" Twitter (now X), April 5, 2020, 5:02 p.m., https://twitter.com/p_hannam/status /1246906143932215297.

54. https://twitter.com/IceSheetMike/status/1242808580350177287 (now deleted account).

55. Dan Rather (@DanRather), "After years when we should have learned of the dangers of 'false equivalence' it baffles me that we are seeing a framing that pits the health of our citizens against some vague notion of getting back to work. We must do better," Twitter (now X), March 25, 2020, 12:10 p.m., https://twitter.com /DanRather/status/1242846264963678209.

56. Hotez, *Deadly Rise of Anti-science*; R. Savillo and T. Monroe, "Fox's Efforts to Undermine Vaccines Has Only Worsened," *Media Matters for America*, August 19, 2021, www.mediamatters.org/fox-news/foxs-effort-undermine-vaccines -has-only-worsened.

57. M. Roy, "Delta Variant Already Dominant in U.S., CDC Estimates Show," Reuters, July 7, 2021, www.reuters.com/world/us/delta-variant-already -dominant-us-cdc-estimates-show-2021-07-07/.

58. Hotez, *Deadly Rise of Anti-science*.

59. Savillo and Monroe, "Fox's Efforts to Undermine Vaccines Has Only Worsened"; E. Kleefeld, "Fox News Boss Lachlan Murdoch Supports Tucker Carlson's Misinformation Against the COVID-19 Vaccines," *Media Matters*, May 19, 2021, www.mediamatters.org/lachlan-murdoch/fox-news-boss-lachlan -murdoch-supports-tucker-carlsons-misinformation-against-covid.

60. M. Makary, "We'll Have Herd Immunity by April," *Wall Street Journal*, February 18, 2021, www.wsj.com/articles/well-have-herd-immunity-by-april -11613669731; J. A. Ladapo and H. A. Risch, "Are Covid Vaccines Riskier than Advertised?," *Wall Street Journal*, June 22, 2021, www.wsj.com/articles/are-covid -vaccines-riskier-than-advertised-11624381749; E. Bendavid and J. Bhattacharya, "Is the Coronavirus as Deadly as They Say?," *Wall Street Journal*, May 24, 2020, www.wsj.com/articles/is-the-coronavirus-as-deadly-as-they-say-11585088464.

61. J. Howard, *We Want Them Infected: How the Failed Quest for Herd Immunity Led Doctors to Embrace the Anti-vaccine Movement and Blinded Americans to the Threat of COVID* (Redhawk, 2023).

62. Makary, "We'll Have Herd Immunity by April."

63. Savillo and Monroe, "Fox's Efforts to Undermine Vaccines Has Only Worsened."

64. R. Martin and A. Sadowski, "The Dishonest Doctors Who Were Fox News' Most Frequent Medical Guests in 2021," *Media Matters for America*, December 30, 2021, www.mediamatters.org/fox-news/dishonest-doctors-who-were-fox-news -most-frequent-medical-guests-2021-0.

65. Savillo and Monroe, "Fox's Efforts to Undermine Vaccines Has Only Worsened"; Martin and Sadowski, "Dishonest Doctors."

66. M. Kulldorf, S. Gupta, and J. Battacharya, "The Great Barrington Declaration," https://gbdeclaration.org/, accessed January 15, 2024.

67. M. Zenone et al., "Analyzing Natural Herd Immunity Media Discourse in the United Kingdom and the United States," *PLOS Global Public Health* 2, no. 1 (2022): e0000078, https://doi.org/10.1371/journal.pgph.0000078, PMID: 36962077, PMCID: PMC10021579; "WHO Chief Says Herd Immunity Approach to Pandemic 'Unethical,'" *Guardian*, October 12, 2020, www.theguardian.com /world/2020/oct/12/who-chief-says-herd-immunity-approach-to-pandemic-unethical; T. S. Brett and P. Rohani, "Transmission Dynamics Reveal the Impracticality of COVID-19 Herd Immunity Strategies," *Proceedings of the National Academy of Sciences* 117, no. 41 (2020): 25897–25903, https://doi.org/10.1073 /pnas.2008087117.

68. L. Bruggeman and L. Cathey, "Former Stanford Colleagues Warn Dr. Scott Atlas Fosters 'Falsehoods and Misrepresentations of Science,'" ABC News, September 10, 2020, https://abcnews.go.com/Politics/stanford-colleagues-warn -dr-scott-atlas-fosters-falsehoods/story?id=72926212; Editorial Board, "How One Doctor Wrecked the Pandemic Response," *Washington Post*, June 29, 2022, www .washingtonpost.com/opinions/2022/06/29/scott-atlas-covid-herd-immunity -strategy/.

69. S. Atlas, "When Will Academia Account for Its Covid Failures?," *Wall Street Journal*, December 29, 2022, www.wsj.com/articles/when-will -academia-account-for-its-covid-failures-pandemic-lockdowns-stanford-ivy-league -elite-narrative-ideology-11672346923.

70. T. Caulfield, "Current Affairs: Lies, Damn Lies, and Tucker Carlson," *Walrus*, January 25, 2024, https://thewalrus.ca/tucker-carlson-alberta/; K. Yandell, "Tucker Carlson Video Spreads Falsehoods on COVID-19 Vaccines, WHO Accord," FactCheck.org, January 12, 2024, www.factcheck.org/2024/01/scicheck -tucker-carlson-video-spreads-falsehoods-on-covid-19-vaccines-who-accord/.

71. Caulfield, "Current Affairs."

72. Yandell, "Tucker Carlson Video Spreads Falsehoods."

73. P. Sah et al., "Estimating the Impact of Vaccination on Reducing COVID-19 Burden in the United States: December 2020 to March 2022," *Journal of Global Health* (September 3, 2022), https://doi.org/10.7189/jogh.12.03062, PMID: 36056814, PMCID: PMC9441009.

74. D. He et al., "Evaluation of Effectiveness of Global COVID-19 Vaccination Campaign," *Emerging Infectious Diseases* 28, no. 9 (2022): 1873–1876; O. J. Watson et al., "Global Impact of the First Year of COVID-19 Vaccination: A Mathematical Modelling Study," *Lancet Infectious Diseases* 22, no. 9 (2022): 1293–1302.

75. V. Shankar Balakrishnan, "WHO Pandemic Treaty: The Good, the Bad, & the Ugly—an Interview with Larry Gostin," *Health Policy Watch*, September 14, 2023, https://healthpolicy-watch.news/who-pandemic-treaty-the-good -the-bad-the-ugly-an-interview-with-larry-gostin/.

76. A. Merlan, "Joe Rogan and Bret Weinstein Promote AIDS Denialism to an Audience of Millions," *Vice News*, February 15, 2024, www.vice.com/en/article /jg543y/joe-rogan-and-bret-weinstein-promote-aids-denialism-to-an-audience-of -millions.

77. Aldous J. Pennyfarthing, "Ron DeSantis Fed COVID Crow by Doctor He'd Ridiculed on Fox News," *Daily Kos*, July 15, 2021, www.dailykos.com

/stories/2021/7/15/2040161/-Ron-DeSantis-fed-COVID-crow-by-doctor-he
-d-ridiculed-on-Fox-News.

78. Zachary Pleat and Eric Kleefeld, "Wall Street Journal Editorial Board
Member Ignores the Newspaper's Reporting, Says Concern About Climate Change
Is a 'Real Mental Illness,'" *Media Matters*, August 1, 2023, www.mediamatters
.org/wall-street-journal/wall-street-journal-editorial-board-member-ignores-newspapers
-reporting-says.

79. Allysia Finley, "What Was Anthony Fauci's Top Aide Hiding?," *Wall Street
Journal*, May 26, 2024, www.wsj.com/articles/what-was-anthony-faucis-top-aide
-hiding-investigation-0d890911?mod=hp_opin_pos_4.

80. Mehdi Hasan (@mehdirhasan), "'Performative neutrality' is a great phrase
and so sadly apt for our current political and media moment," Twitter (now X), June
5, 2023, 12:35 p.m., https://twitter.com/mehdirhasan/status/1665759258166677507.

81. See Michael E. Mann, *The New Climate War: The Fight to Take Back Our
Planet* (PublicAffairs, 2021).

82. Bjorn Lomborg, "The Poor Need Cheap Fossil Fuels," *New York Times*,
December 3, 2013, www.nytimes.com/2013/12/04/opinion/the-poor-need-cheap
-fossil-fuels.html.

83. Susan Matthews, "Bret Stephens' First Column for the *New York Times*
Is Classic Climate Change Denialism," *Slate*, April 30, 2017, https://slate.com
/technology/2017/04/bret-stephens-first-new-york-times-column-is-classic
-climate-change-denialism.html.

84. Bret Stephens, "Yes, Greenland's Ice Is Melting, But...," *New York Times*, October 28, 2022, www.nytimes.com/interactive/2022/10/28/opinion/climate-change
-bret-stephens.html.

85. Molly Taft, "Bret Stephens' Bad Faith Climate Conversion," *Gizmodo*,
October 28, 2022, https://gizmodo.com/bret-stephens-new-york-times-greenland
-climate-essay-1849717244.

86. Amy Westervelt, Matthew Green, and Joey Grostern, "Reuters, New York
Times Top List of Fossil Fuel Industry's Media Enablers," *DeSmog*, December
5, 2023, www.desmog.com/2023/12/05/reuters-new-york-times-top-list-of-fossil
-fuel-industrys-media-enablers/.

87. David Wallace-Wells, "Time to Panic," *New York Times*, February 16,
2019, www.nytimes.com/2019/02/16/opinion/sunday/fear-panic-climate-change
-warming.html.

88. Michael Mann and Katharine Hayhoe, "The Antidote to Doom Is Doing,"
Sustainable Views, May 15, 2024, www.sustainableviews.com/the-antidote-to
-doom-is-doing-ac97ab67/.

89. "Meet Our Researchers: Alina Chan," Broad Institute, n.d., https://giving
.broadinstitute.org/broadignite/team/alina-chan.

90. "Matt 'King Coal' Ridley," *DeSmog*, n.d., www.desmog.com/matt-king
-coal-ridley/.

91. Jeremy Ashkenas, "Opinion Today" newsletter, *New York Times*, June 3, 2024.

92. Angela Rasmussen (angie_rasmussen), "Key points aren't actually very
key when they are factually incorrect. Good of the @nytopinion to help the mob
sharpen their pitchforks for Fauci's Select Subcommittee hearing by enshrining the
lies that he will be attacked with as truth in the paper of record," X, June 3, 2024,
9:17 a.m., https://x.com/angie_rasmussen/status/1797618535239405926.

93. David Wallace-Wells, "Why Are So Many People Sure Covid Leaked from a Lab?," *New York Times*, May 21, 2025, https://www.nytimes.com/2025/05/21/opinion/covid-lab-leak.html.

94. Aaron Rupar (@atrupar), "'We're gonna be paying for the Fauci effect for a generation. Perhaps more'—Laura Ingraham blames Fauci for parents dying, divorces, suicides, overdoses, and more. Pure insanity," X, June 3, 2024, 7:14 p.m., https://x.com/atrupar/status/1797768942288617655.

95. John Parkinson and Cheyenne Haslett, "Key Takeaways: Fauci Defends Against GOP Claims on COVID Origins, Response," ABC News, June 3, 2024, https://abcnews.go.com/Politics/republicans-poised-grill-anthony-fauci-covid-19-response/story?id=110677611.

96. Geoff Brumfiel, "As Republicans Probe COVID's Origins, Some See an Attack on Science; Others Say It's Long Overdue," NPR, June 2, 2024, www.npr.org/2024/05/30/g-s1-1788/republicans-probe-covid-origins-attack-on-science-or-long-overdue.

97. Antoinette Radford et al., "Fauci Testifies on the Origins of Covid-19," CNN, June 3, 2024, www.cnn.com/politics/live-news/anthony-fauci-covid-origins-hearing-06-03-24/index.html.

98. M. Worobey et al., "The Huanan Seafood Wholesale Market in Wuhan Was the Early Epicenter of the COVID-19 Pandemic," *Science* 377, no. 6609 (2022): 951–959, https://doi.org/10.1126/science.abp8715, PMID: 35881010, PMCID: PMC9348750.

99. J. E. Pekar et al., "The Molecular Epidemiology of Multiple Zoonotic Origins of SARS-CoV-2," *Science* 377, no. 6609 (2022): 960–966, https://doi.org/10.1126/science.abp8337, errata in *Science* 382, no. 6667 (2023): eadl0585, PMID: 35881005, PMCID: PMC9348752.

100. A. Crits-Christoph et al., "Genetic Tracing of Market Wildlife and Viruses at the Epicenter of the COVID-19 Pandemic," *Cell* 187, no. 19 (2024): 5468–5482.e11, https://doi.org/10.1016/j.cell.2024.08.010, PMID: 39303692, PMCID: PMC11427129.

101. Office of the Director of National Intelligence, "Unclassified Summary of Assessment on COVID-19 Origins," www.dni.gov/files/ODNI/documents/assessments/Unclassified-Summary-of-Assessment-on-COVID-19-Origins.pdf.

102. Office of the Director of National Intelligence, "Unclassified Summary."

103. J. E. Pekar et al., "The Recency and Geographical Origins of the Bat Viruses Ancestral to SARS-CoV and SARS-CoV-2," *bioRxiv* [preprint] (July 12, 2023), https://doi.org/10.1101/2023.07.12.548617, PMID: 37502985, PMCID: PMC10369958.

104. M. Hiltzik, "Column: U.S. Government Debunks COVID Lab-Leak Conspiracy Theory, Enraging Conspiracy Theorists," *Los Angeles Times*, June 26, 2023, www.latimes.com/business/story/2023-06-26/u-s-government-debunks-covid-lab-leak-conspiracy-theory-enraging-conspiracy-theorists.

105. Editorial Board, "China Pressured Experts Away from a Lab-Leak Investigation. What Is It Hiding?," *Washington Post*, August 14, 2021, www.washingtonpost.com/opinions/2021/08/14/china-pressured-experts-away-lab-leak-investigation-what-is-it-hiding/; Editorial Board, "Two Possible Theories of the Pandemic's Origins Remain Viable. The World Needs to Know," *Washington*

Post, May 17, 2021, www.washingtonpost.com/opinions/global-opinions/two
-possible-theories-of-the-pandemics-origins-remain-viable-the-world-needs-to
-know/2021/05/17/b87f0b0e-b737-11eb-96b9-e949d5397de9_story.html.

106. Editorial Board, "China Should Answer How Covid-19 Began. Propaganda Is No Substitute," *Washington Post*, March 2, 2023, www.washingtonpost.
com/opinions/2023/03/02/china-covid-origins-answer/.

107. Editorial Board, "There's New Light—and Lingering Questions—in the
Mystery of Wuhan," *Washington Post*, April 15, 2023, www.washingtonpost.com
/opinions/2023/04/15/covid-origins-china-mystery/.

108. S. G. Stolberg and B. Mueller, "Lab Leak or Not? How Politics Shaped
the Battle over Covid's Origin," *New York Times*, March 19, 2023, www.nytimes
.com/2023/03/19/us/politics/covid-origins-lab-leak-politics.html.

109. Michael Hiltzik, "House Republicans Give a Crash Course in How to Concoct a Conspiracy Theory About COVID's Origin," *Los Angeles Times*, July 11, 2023,
www.latimes.com/business/story/2023-07-11/column-house-republicans-give
-a-crash-course-in-how-to-concoct-a-conspiracy-theory-about-covids-origin.

110. Saul Elbein, "Catch-22: Scientific Communication Failures Linked to
Faster-Rising Seas," June 19, 2023, https://thehill.com/policy/energy-environment
/4057045-catch-22-scientific-communication-failures-linked-to-faster-rising-seas/.

111. Robert E. Copp et al., "Communicating Future Sea-Level Rise Uncertainty and Ambiguity to Assessment Users," *Nature Climate Change* 13 (2023):
648–660, www.nature.com/articles/s41558-023-01691-8.

112. S. Buxbaum and I. Flatow, "Science Journalism Is Shrinking—Along with
Public Trust in Science," *Science Friday* (January 5, 2024), www.sciencefriday.com
/segments/science-journalism-trust-in-science/.

113. S. Imbler, "Popular Science Ends and Science Journalism Keeps
Shrinking," *Defector*, November 30, 2023, https://defector.com/popular-science
-ends-and-science-journalism-keeps-shrinking.

114. Darrin Durant, "Are Honest Brokers Good for Democracy?," *Social
Epistemology* 37, no. 3 (2023): 276–289, www.tandfonline.com/doi/abs/10.1080
/02691728.2022.2139166.

115. Darrin Durant (@DarrinADurrant), "Pielke says issue advocacy is 'fundamental to a healthy democracy and is a noble calling', except when issue advocates don't follow his lead, then they're scurrilous dictators?," X, April 15, 2024,
8:15 p.m., https://x.com/DarrinADurant/status/1780027120519188669.

116. Michael E. Mann, "If You See Something, Say Something," *New York
Times*, January 17, 2014, www.nytimes.com/2014/01/19/opinion/sunday/if-you
-see-something-say-something.html.

117. Research!America, "Survey: Most Americans Cannot Name a Living Scientist or a Research Institution," Research!America, May 11, 2021,
www.researchamerica.org/blog/survey-most-americans-cannot-name-a-living
-scientist-or-a-research-institution/.

118. Delbert Tran, "The Fourth Estate as the Final Check," Yale Law School,
November 22, 2016, https://law.yale.edu/mfia/case-disclosed/fourth-estate-final
-check.

119. Sheldon Whitehouse and Jennifer Mueller, *The Scheme: How the Right
Wing Used Dark Money to Capture the Supreme Court* (New Press, 2022).

120. Charles Kaiser, "Paul Krugman on Leaving the *New York Times*," *Columbia Journalism Review*, January 24, 2025, www.cjr.org/analysis/paul-krugman-leaving-new-york-times-heavy-hand-editing-less-frequent-columns-newsletter.php.

121. Eric Boehlert, "After Forming Clinton Cash 'Exclusives,' *NY Times*, *Washington Post* Fail to Report on Book's Errors," *Media Matters*, May 1, 2015, www.mediamatters.org/new-york-times/after-forming-clinton-cash-exclusives-ny-times-washington-post-fail-report-books.

122. Elahe Izadi and Amy Argetsinger, "Sally Buzbee Steps Down as Executive Editor of the *Washington Post*," *Washington Post*, June 2, 2024, www.washingtonpost.com/style/media/2024/06/02/sally-buzbee-washington-post-steps-down/.

123. David Bauder, "*Washington Post* Columnist Quits After Her Opinion Piece Criticizing Owner Jeff Bezos Is Rejected," Associated Press, March 10, 2025, https://apnews.com/article/washington-post-resignation-marcus-bezos-8d6ce32b27f5c965fc972d73d0f95aac.

124. Will Bunch, "2024's Other Big Loser? The Mainstream Media. Is There Any Path Forward?," *Philadelphia Inquirer*, November 10, 2024, www.inquirer.com/opinion/commentary/trust-mainstream-media-2024-election-20241110.html?id=bpjtJ3uwA0BYL&utm_source=social&utm_campaign=gift_link&utm_medium=referral.

125. Camille Caldera, "Fact Check: Fairness Doctrine Only Applied to Broadcast Licenses, Not Cable TV Like Fox News," *USA Today*, November 28, 2020, www.usatoday.com/story/news/factcheck/2020/11/28/fact-check-fairness-doctrine-applied-broadcast-licenses-not-cable/6439197002/.

CHAPTER 7: THE PATH FORWARD

1. J. Alwine et al., "The Harms of Promoting the Lab Leak Hypothesis for SARS-CoV-2 Origins Without Evidence," *Journal of Virology* 98, no. 9 (2024), https://doi.org/10.1128/jvi.01240-24.

2. Julia Mueller, "Trump Says He'd Ban Government from Labeling Speech as Misinformation," *Hill*, December 15, 2022, https://thehill.com/homenews/campaign/3776629-trump-says-hed-ban-government-from-labeling-speech-as-misinformation/.

3. Damon Centola et al., "Experimental Evidence for Tipping Points in Social Convention," *Science* 360, no. 6393 (2018) 1116–1119, https://doi.org/10.1126/science.aas8827; https://science.sciencemag.org/content/360/6393/1116.editor-summary.

4. "Attitudes on Same-Sex Marriage," Pew Research Center, May 14, 2019, www.pewforum.org/fact-sheet/changing-attitudes-on-gay-marriage/.

5. Marie Snyder, "What Seems Impossible Can Become Inevitable," *Medium*, December 26, 2023, https://medium.com/through-the-fog/what-seems-impossible-can-become-inevitable-3069743db2a6.

6. "Quote Wrongly Attributed to Mahatma Gandhi," Associated Press, October 5, 2018, https://apnews.com/article/archive-fact-checking-2315880316.

7. "*The Ingraham Angle* on Jeff Zucker," Fox News, February 2, 2022, www.foxnews.com/transcript/ingraham-angle-zucker.

8. P. M. Sutter, "Opinion: Science Has a Communication Problem—and a

Connection Problem," *Undark*, October 20, 2022, https://undark.org/2022/10/20/science-has-a-communication-problem-and-a-connection-problem/.

9. E. Siegel, "Starts with a Bang: How America's Big Science Literacy Mistake Is Coming Back to Haunt Us," *Forbes*, September 9, 2021, www.forbes.com/sites/startswithabang/2021/09/09/how-americas-big-science-literacy-mistake-is-coming-back-to-haunt-us/?sh=289a3b86a16d.

10. Joshua Holland, "Six Things Michael Mann Wants You to Know About the Science of Global Warming," Billmoyers.com, June 12, 2014, https://billmoyers.com/2014/06/12/six-things-michael-mann-wants-you-to-know-about-the-science-of-global-warming/.

11. Sutter, "Opinion: Science Has a Communication Problem."

12. B. Kennedy and A. Tyson, "Americans' Trust in Scientists, Positive Views of Science Continue to Decline," Pew Research Center, November 14, 2023, www.pewresearch.org/science/2023/11/14/americans-trust-in-scientists-positive-views-of-science-continue-to-decline/.

13. M. Hiltzik, "House Committee Gives a Crash Course in How to Concoct a Conspiracy Theory About COVID Origin," *Los Angeles Times*, July 11, 2023, www.latimes.com/business/story/2023-07-11/column-house-republicans-give-a-crash-course-in-how-to-concoct-a-conspiracy-theory-about-covids-origin.

14. P. J. Hotez, "The Long Road to Repair: A U.S. National Post-doc for Science Education and Advocacy," *PLOS Speaking of Medicine and Health*, August 29, 2013, https://speakingofmedicine.plos.org/2013/08/29/the-long-road-to-repair-a-u-s-national-post-doc-for-science-education-and-advocacy/.

15. P. J. Hotez, "Combating Antiscience: Are We Preparing for the 2020s?," *PLOS Biology* 18, no. 3 (2020): e3000683, https://doi.org/10.1371/journal.pbio.3000683, PMID: 32218568, PMCID: PMC7141687.

16. "Mitigating Misinformation: Resources for Health Care Providers," Coalition for Trust in Health & Science, 2024, https://trustinhealthandscience.org/mitigating-misinformation-resources-for-health-care-providers/.

17. Jonathan Chait, "Donald Trump Has Finally Killed the GOP's Pro-Science Wing," *New York Magazine*, December 9, 2016, https://nymag.com/intelligencer/2016/12/donald-trump-has-finally-killed-the-gops-pro-science-wing.html.

18. "Mark Steyn, GB News, 4 October 2022, 20:00; 5 October 2022, 02:00: Summary," *OfCom Broadcast and On Demand Bulletin*," no. 473, May 9, 2023, www.ofcom.org.uk/siteassets/resources/documents/about-ofcom/bulletins/broadcast-bulletins/2023/issue-473/mark-steyn-gb-news-decision.pdf?v=329641.

19. Karen Antcliff, "GB News' Mark Steyne Programme Breached Broadcasting Rules 'Misleading' Public over Covid-19," *Nottinghamshire Live*, March 6, 2023, www.nottinghampost.com/news/celebs-tv/gb-news-mark-steyn-programme-8219635.

20. Haroon Siddique, "Former GB News Presenter Ordered to Pay £50,000 in Legal Costs," *Guardian*, October 29, 2024, www.theguardian.com/world/2024/oct/29/former-gb-news-presenter-ordered-to-pay-50000-in-legal-costs.

21. Elizabeth Grossman, "Legal Fund Helping Climate Scientists Draw Line in the Sand," *Inside Climate News*, March 13, 2012, https://insideclimatenews.org/news/13032012/climate-science-legal-defense-fund-skeptics-global-warming-michael-mann-hockey-stick-university-of-virginia/.

22. P. J. Hotez, *The Deadly Rise of Anti-science: A Scientist's Warning* (Johns Hopkins University Press, 2023).

23. L. Lafontaine, "CMA Recognizes One Year of Federal Law to Protect Health Workers, More Needs to Be Done," Canadian Medical Association, January 12, 2023, www.cma.ca/about-us/what-we-do/press-room/cma-recognizes-one -year-federal-law-protect-health-workers-more-needs-be-done.

24. Lafontaine, "CMA Recognizes One Year of Federal Law."

25. P. J. Hotez, "Science Tikkun: Science for Humanity in an Age of Aggression," *FASEB Journal* 35, no. 12 (2021): e22047, https://doi.org/10.1096/ fj.202101604, PMID: 34806227; Hotez, *Deadly Rise of Anti-science*.

26. "Promotion of Scientific Freedom and the Safety of Scientists: A New Programme and Call to Action," UNESCO, March 20, 2024, updated January 14, 2025, www.unesco.org/en/articles/promotion-scientific-freedom-and-safety -scientists-new-programme-and-call-action.

27. World Economic Forum, *The Global Risks Report, 2024: Insight Report*, 19th ed., https://www3.weforum.org/docs/WEF_The_Global_Risks_Report _2024.pdf.

28. Glenn Branch, "How to Support Climate Change Education in Your State's Schools," Yale Climate Connections, June 4, 2024, https://yaleclimateconnections .org/2024/06/how-to-support-climate-change-education-in-your-states-schools/.

29. Christopher Paul and Miriam Matthews, "The Russian 'Firehose of False-hood' Propaganda Model: Why It Might Work and Options to Counter It," RAND, July 11, 2016, www.rand.org/pubs/perspectives/PE198.html.

30. "Whitehouse, Cicilline Reintroduce DISCLOSE Act to End Corrupt-ing Influence of Dark Money in American Democracy," press release, Febru-ary 17, 2023, www.whitehouse.senate.gov/news/release/whitehouse-cicilline -reintroduce-disclose-act-to-end-corrupting-influence-of-dark-money-in-american -democracy/.

31. Sheldon Whitehouse and Jennifer Mueller, *The Scheme: How the Right Wing Used Dark Money to Capture the Supreme Court* (New Press, 2022).

32. Clarence Williams, "Activists Rally in D.C. on 2nd Anniversary of High Court's Abortion Ruling," *Washington Post*, June 24, 2024, www .washingtonpost.com/dc-md-va/2024/06/24/dobbs-anniversary-abortion-rights -protest-supreme-court/.

33. David Gelles and Manuela Andreoni, "A Seismic Supreme Court Deci-sion," *New York Times*, July 2, 2024, www.nytimes.com/2024/07/02/climate /supreme-court-climate-chevron.html.

34. Patricia Roberts-Miller, *Demagoguery and Democracy* (New York: Experi-ment, 2017).

35. B. Allyn, "Group Aiming to Defund Disinformation Tries to Drain Fox News of Online Advertising," NPR, June 9, 2022, www.npr.org /2022/06/09/1103690822/group-aiming-to-defund-disinformation-tries-to-drain -fox-news-of-online-advertis.

36. Associated Press, "Democratic Officials Criticize Meta Ad Policy, Saying It Amplifies Lies About 2020 Election," *Politico*, May 4, 2024, www .politico.com/news/2024/05/04/democratic-officials-criticize-meta-2020 -stolen-election-ad-policy-00156149.

37. Andrew Jaspan, "How Do We Get Serious About Fixing Australia's Lack of Media Diversity?," *Crikey*, May 28, 2024, www.crikey.com.au/2024/05/28 /newsmap-media-diversity-australia-advertising-universities/.

38. Taylor Noakes, "Pathways Alliance Website Scrubbed Ahead of New Greenwashing Law," *DeSmog*, June 20, 2024, www.desmog.com/2024/06/20 /pathways-alliance-website-scrubbed-ahead-of-new-greenwashing-law/.

39. Mark Sherman and Associated Press, "Supreme Court Rules in Biden's Favor, Tossing Out GOP Claims That Democrats Coerced Social Media Companies to Stamp Out Conservative Points of View," *Fortune*, June 26, 2024, https: //fortune.com/2024/06/26/supreme-court-rules-biden-favor-gop -claims-democrats-coerced-social-media-companies-censor-content/.

40. Mary Clare Jalonick, "Congress Eyes New Rules for Tech: What's Under Consideration," Associated Press, May 8, 2023, https://apnews.com/article /congress-technology-regulation-tiktok-155d696fd44a450a43ce8cb9802e40bb.

41. Vivek H. Murthy, "Surgeon General: Why I'm Calling for a Warning Label on Social Media Platforms," *New York Times*, June 17, 2024, www.nytimes .com/2024/06/17/opinion/social-media-health-warning.html.

42. Jalonick, "Congress Eyes New Rules for Tech."

43. Cat Zakrzewski and Elizabeth Dwoskin, "Facebook Quietly Bankrolled Small, Grass-roots Groups to Fight Its Battles in Washington," *Washington Post*, May 17, 2022, www.washingtonpost.com/technology/2022/05/17 /american-edge-facebook-regulation/.

44. "Facebook Funded American Edge Waged War Against www .washingtonpost.com/technology/2022/05/17/american-edge-facebook -regulation/.

45. Zakrzewski and Dwoskin, "Facebook Quietly Bankrolled Small, Grass-roots Groups."

46. Matt Egan, "AI Could Pose 'Extinction-Level' Threat to Humans and the US Must Intervene, State Dept.–Commissioned Report Warns," CNN, March 12, 2024, www.cnn.com/2024/03/12/business/artificial-intelligence-ai-report -extinction/index.html.

47. Mike Isaac, "Meta to Require Political Advertisers to Disclose Use of A.I.," *New York Times*, November 8, 2023, www.nytimes.com/2023/11/08/technology /meta-political-ads-artificial-intelligence.html.

48. See Eleanor Pringle, "Elon Musk Was Just Forced to Reveal Who Really Owns X. Here's the List," *Fortune*, August 22, 2022, https://fortune .com/2024/08/22/elon-musk-x-twitter-owner-list/ and https://eutoday.net/russian -oligarch-behind-musks-twitter-purchase/.

49. Kevin Roose, "Bluesky, Smiling at Me," *New York Times*, December 3, 2024, www.nytimes.com/2024/11/22/technology/bluesky-x-alternative.html.

50. John Naughton, "Closing the Standard Internet Observatory Will Edge the US Towards the End of Democracy," *Guardian*, June 29, 2024, www .theguardian.com/commentisfree/article/2024/jun/29/closing-the-stanford -internet-observatory-will-edge-the-us-towards-the-end-of-democracy.

51. Allison Neitzel, "Targeting 'Censors,' Far Right Plots Triumph of Online Disinformation," *Who What Why*, August 23, 2023, https: //whowhatwhy.org/science/health-medicine/targeting-censors-far-right-plots -triumph-of-online-disinformation/.

52. Naughton, "Closing the Standard Internet Observatory."

53. Naughton, "Closing the Standard Internet Observatory."

54. Michael Hiltzik, "Can Stanford Tell the Difference Between

Scientific Fact and Fiction? Its Pandemic Conference Raises Doubts," *Los Angeles Times*, October 15, 2024, www.latimes.com/business/story/2024-10-15/column -can-stanford-tell-the-difference-between-scientific-fact-and-fiction-its-pandemic -conference-raises-doubts.

55. Timothy Snyder, *On Tyranny: Twenty Lessons from the Twentieth Century* (Crown, 2017), https://timothysnyder.org/on-tyranny.

56. "Open Letter: Governments Should Act Now to Curb Climate Disinformation," November 13, 2024, https://caad.info/wp-content/uploads/2024 /11/Open-Letter_-Governments-Should-Act-Now-to-Curb-Climate -Disinformation.pdf.

57. www.whitehouse.gov/priorities/covid-19/ (no longer available).

58. Erin Schumaker, "Republicans Have a Post-pandemic Plan for the Scientific Establishment," *Politico*, September 30, 2024, www.politico.com/news /2024/09/30/republicans-covid-pandemic-nih-plan-00181512.

59. Lisa Mascaro, "Conservative Groups Draw Up Plan to Dismantle the US Government and Replace It with Trump's Vision," Associated Press, August 29, 2023, https://apnews.com/article/election-2024-conservatives-trump-heritage -857eb794e505f1c6710eb03fd5b58981.

60. *Loper Bright Enterprises et al. v. Raimondo, Secretary of Commerce, et al.*, www.supremecourt.gov/opinions/23pdf/22-451_7m58.pdf.

61. "To Protect our Rights and Democracy, We Must Expand the Court," Take Back the Court, n.d., www.takebackthecourt.today/court-expansion-overview.

62. Rachel Maddow, "Threat of Cyberattack on U.S. Infrastructure Takes More Serious Turn," MSNBC, April 14, 2022, www.msnbc.com/rachel -maddow/watch/threat-of-cyberattack-on-u-s-infrastructure-takes-more-serious -turn-137691717553.

63. Rachel Maddow, "US Piecing Together Russia Cyber Attack Goal," MSNBC, June 21, 2017, www.msnbc.com/rachel-maddow/watch/us-piecing -together-russia-cyber-attack-strategy-973300291540.

64. Sam Levin, "Did Russia Fake Black Activism on Facebook to Sow Division in the US?," *Guardian*, September 30, 2017, www.theguardian.com/technology /2017/sep/30/blacktivist-facebook-account-russia-us-election.

65. Danny Barefoot, "Here's What Interviewing Voters Taught Me About the Slogan 'Defund the Police,'" *Guardian*, November 20, 2020, www .theguardian.com/commentisfree/2020/nov/20/heres-what-interviewing-voters -taught-me-about-the-slogan-defund-the-police.

66. Alex Thompson, "First Look: RNC Slams Biden in AI-Generated Ad," *Axios*, April 25, 2023, www.axios.com/2023/04/25/rnc -slams-biden-re-election-bid-ai-generated-ad.

67. Paul Myers et al., "A Bugatti Car, a First Lady and the Fake Stories Aimed at Americans," BBC, July 2, 2024, www.bbc.com/news/articles/c72ver6172do.

68. Tim Reid and Sarah N. Lynch, "Hoax Bomb Threats Linked to Russia Target Polling Places in Battleground States, FBI Says," Reuters, November 5, 2024, www.reuters.com/world/us/fake-bomb-threats-linked-russia-briefly-close-georgia -polling-locations-2024-11-05/.

69. Michael E. Mann, "Welcome to the American Petrostate," *Bulletin of the Atomic Scientists*, November 7, 2024, https://thebulletin.org/2024/11 /welcome-to-the-american-petrostate/.

70. Matthew Dalton, "Trump Victory Leaves China Calling the Shots at COP29 Climate Negotiations," *Wall Street Journal*, November 11, 2024, www .wsj.com/world/trump-victory-leaves-china-calling-the-shots-at-cop29-climate -negotiations-f4161f7f.

71. Simon Tisdall, "How Should Europe Respond Now Its American Ally Has Turned Hostile?," *Guardian*, August 30, 20202, www.theguardian.com /commentisfree/2020/aug/30/how-should-europe-respond-now-its-american-ally -has-turned-hostile.

72. See https://standupforscience2025.org/dc-rally/. Mike's speech is available at www.youtube.com/live/KxeaXm7CIZE?t=9240s.

73. David Bauder, Randall Chase, and Geoff Mulvihill, "Fox, Dominion Reach $787M Settlement over Election Claims," Associated Press, April 18, 2023, https://apnews.com/article/fox-news-dominion-lawsuit-trial-trump-2020 -0ac71f75acfacc52ea80b3e747fb0afe.

74. The punitive-damage award against Steyn was recently reduced on March 4, 2025, to $5,000 by the judge. That decision is being appealed.

75. India Bourke, "Claudia Sheinbaum: What a Climate -Scientist Turned President Might Mean for Global Efforts to Tackle Climate Change," BBC, June 7, 2024, www.bbc.com/future/article/20240607-claudia -sheinbaum-mexicos-new-climate-minded-president.

76. Bill Cleverley, "B.C.'s Liquefied Natural Gas Plan a 'Pipe Dream,' Says Green MLA Andrew Weaver," *Times Colonist*, September 19, 2013, www .timescolonist.com/local-news/bcs-liquefied-natural-gas-plan-a-pipe-dream -says-green-mla-andrew-weaver-4599671.

77. Bourke, "Claudia Sheinbaum."

78. Lee Hedgepeth et al., "Climate Initiatives Fare Well Across the Country Despite National Political Climate," *Inside Climate News*, November 7, 2024, https://insideclimatenews.org/news/07112024/state-climate-initiatives-fare -well-across-country/.

79. Fiona Harvey et al., "Cop Summits 'No Longer Fit for Purpose,' Say Leading Climate Policy Experts," *Guardian*, November 15, 2024, www.theguardian .com/environment/2024/nov/15/cop-summits-no-longer-fit-for-purpose-say -leading-climate-policy-experts?CMP=Share_AndroidApp_Other.

80. Mike used this framing in a piece that he wrote for the *Bulletin of the Atomic Scientists* shortly after the result of the 2024 presidential election that brought Donald Trump back to power: "Welcome to the American Petrostate."

81. Mike wrote this paragraph for the book, but it first appeared in a commentary he wrote for the *Bulletin of the Atomic Scientists*, "Welcome to the American Petrostate."

INDEX

Dr. Michael E. Mann is presidential distinguished professor in the Department of Earth and Environmental Science and vice provost for climate at the University of Pennsylvania. He has a secondary appointment in the Annenberg School for Communication and is founding director of the Penn Center for Science, Sustainability, and the Media. Dr. Mann received his undergraduate degrees in physics and applied math from UC Berkeley, an MS degree in physics from Yale University, and a PhD in geology and geophysics from Yale University. His research interests include the study of Earth's climate system and the science, impacts, and policy implications of human-caused climate change. He is an elected member of the National Academy of Sciences and the Royal Society of the United Kingdom, has received numerous awards, and has authored more than three hundred scientific papers and seven books.

Dr. Peter J. Hotez is dean of the National School of Tropical Medicine and professor of pediatrics and molecular virology and microbiology at Baylor College of Medicine, where he is also the codirector of the Texas Children's Center for Vaccine Development and Texas Children's Hospital endowed chair of tropical pediatrics. He obtained his undergraduate degree in molecular biophysics and biochemistry from Yale University, followed by a PhD in biochemistry from Rockefeller University, MD from Weill Cornell Medical College, and pediatric residency training at Massachusetts General Hospital. Dr. Hotez is an internationally recognized physician-scientist in neglected tropical diseases and developing new vaccines for these conditions. He is an elected member of the National Academy of Medicine and American Academy of Arts and Sciences and has received multiple awards from scientific and medical societies. He is the author of seven hundred scientific papers and five books.